MW01506267

"Are you convinced by the latest claims about brain power, hypnosis, memory decline or the dangerous effects of video games? Are you tempted to waste money on a fancy new therapy and convinced that at least it can do no harm? Then read this book first and you may change your mind."
—Professor Susan Blackmore

"This book is well worth reading by all those who appreciate the skeptical attitude and want to develop critical thinking skills – so important in today's world flooded with informational noise, bombarding our brains from all sides. The book describes detailed analysis of abuses, dishonesties and frauds in some areas of psychology and related areas. This is not a book for psychologists only, it is a book for everybody who wants to progress their scientific thinking. There can never be enough skepticism and criticism."
—Professor Czesław S. Nosal, psychologist

"The author had a simple goal that required enormous knowledge and determination: to reveal the areas of psychology and psychotherapy that promise results that can never be delivered. Witkowski wants to dismantle false claims and show that despite the abundance of pseudoscientific theories, increasing numbers of tinhorns and constant challenges to science – psychology remains a sound knowledge on human's behavior. (...) Such books are rare – publications that help to distinguish fraudsters from genuine therapists. It's reassuring that Tomasz Witkowski does not stop in his crusade against superstitions and stands for those who consider psychotherapy and psychology as science."
—Jakub Winiarski, *Wiedza i życie*

"You might choose to agree or disagree with Tomasz Witkowski, but you cannot choose to ignore him. This book is definitely worth reading carefully! Witkowski, an enfant terrible of Polish social sciences, devoted this entire volume to contemplate the miserable condition of psychology."
—Professor Dariusz Doliński, social psychologist

"*Psychology Gone Wrong* sketches a dark picture of intellectual abuse, scientific misconduct that harms—sometimes irreversibly—of those who are seeking help. Civilizational changes in recent years brought an increase of scientific swindles, not only in social sciences. We celebrate the anniversary of the cold fusion scandal, while renowned journals (like *Science* and *Nature*) recently were forced to retract several publications on miraculous microelectronic findings. All those scientific sins, hidden in the shadows, must be brought to daylight."
—Professor Łukasz Turski, physicist

ABOUT THE AUTHORS

Dr. Tomasz Witkowski is a psychologist and science writer. He is the founder of the Polish Skeptics Club and specializes in debunking pseudoscience, particularly in the fields of psychology, psychotherapy, and diagnosis. He is the author of numerous books, dozens of scientific papers, and over 100 popular articles (some of them are published in *Skeptical Inquirer*). As an expert in science-pseudoscience issues, he is frequently called by the media to comment on frauds and abuses witnessed in psychology, psychotherapy, and other areas of scientific activity. In 2010, he was awarded Rationalist of the Year in Poland.

Dr. Maciej Zatonski is a surgeon and researcher, known for debunking unscientific therapies and claims in clinical medicine. He is actively involved in encouraging people to trust in scientifically proven therapies rather than trusting fairies, myths, and tricksters. He is the founder of the Polish Skeptics Club and a leader in public understanding of science in Poland. His struggle to clean up medical curricula from obsolete or bogus therapies was recently noted by the Polish Academy of Sciences. Zatonski is also known for promoting evolution and evolutionary sciences. He is an author of over 30 scientific papers, and over 40 popular science articles.

PSYCHOLOGY GONE WRONG

PSYCHOLOGY GONE WRONG

THE DARK SIDES OF
SCIENCE AND THERAPY

TOMASZ WITKOWSKI
MACIEJ ZATONSKI

BrownWalker Press
Boca Raton

Psychology Gone Wrong: The Dark Sides of Science and Therapy

BrownWalker Press
Boca Raton, Florida • USA
2015

ISBN-10: 1-62734-528-0
ISBN-13: 978-1-62734-528-6

www.brownwalker.com

Cover art by Jacob Peter Gowy [Public domain], via Wikimedia Commons

Publisher's Cataloging-in-Publication Data

Witkowski, Tomasz.
Psychology gone wrong : the dark sides of science and therapy / Tomasz Witkowski & Maciej Zatonski.
pages cm
Includes bibliographical references.
ISBN: 978-1-62734-528-6 (pbk.)
1. Psychotherapy. 2. Psychology. 3. Mental health. 4. Pseudoscience. I. Witkowski, Maciej. II. Title.
RC480.5 .W495 2015
616.89`14—dc23

2015930186

In the temple of science are many mansions, and various indeed are they that dwell therein and the motives that have led them thither.
—Albert Einstein

TABLE OF CONTENTS

Introduction.. 11

Part I: Fraudsters in the Temple of Science:
Some Sins of Academic Psychology... 17
 Chapter 1: In Search of the Super Human: Wanderings of Eugenics........ 19
 Chapter 2: The Fraud as a Nobleman: From Lies to Academic Glory...... 27
 Chapter 3: Epigones of Sir Cyril Burt: Other Scientific Fraudsters 37
 Chapter 4: The Hidden Part of the Iceberg:
 The Problem of Raw Data and the Scope of Scientific Misconduct.......... 51
 Chapter 5: Lying is the First Step to the Gallows:
 Plagiarism and Other Misconduct in Scientific Publications....................... 61
 Chapter 6: Scientific Conspiracy? The Myth of Replication...................... 75
 Chapter 7: A Big Jar of Cookies:
 The Reasons for the Current State of Academic Psychology..................... 87

Part II: Conquering Patients' Souls: Sins of Psychotherapists 107
 Chapter 8: Psychoanalysis: Castle Built on Sand 109
 Chapter 9: The Myth of Childhood:
 Foundation of Therapies Exploring the Past... 123
 Chapter 10: The Nightmare of Recovered Memories 141
 Chapter 11: Of the Need for... Prostitution:
 A Different Perspective on Psychotherapy 159
 Chapter 12: Light at the End of the Tunnel:
 The Effectiveness of Psychotherapy Today.. 173
 Chapter 13: First, Do No Harm.. 187
 Chapter 14: With What Can We Replace Psychotherapy?..................... 199
 Chapter 15: Ship of Fools:
 The Reasons for the Present State of Psychotherapy 211

Part III: Beyond Control: Psychobusiness.. 225
 Chapter 16: Fair of Illusions, Miracles and Hopes............................... 227
 Chapter 17: The Art of Hammering Nails or the Theory and Practice of
 NLP in the Eyes of a Social Psychologist... 237
 Chapter 18: Hows and Whys of Inventing a New Therapy:
 A Psychological Sokal Hoax. 259
 Chapter 19: From Keeping Silent to Arrogant Hostility:
 Strategies Employed by Scientists with Regard to Pseudoscience............. 277

To be continued.. 303

INTRODUCTION

In his report on the state of the church from 1987, Joseph Cardinal Ratzinger, future pope, expressed his concerns that psychology posed a real threat to religion, that it was responsible for empty monasteries and superseded theology:

> On the other hand, active orders and congregations are in grave crisis: the discovery of professionalism, the concept of 'social welfare' which has replaced that of 'love of neighbor', the often uncritical and yet enthusiastic adaptation to the new and hitherto unknown values of modern secular society, the entrance into the convents, at times wholly unexamined, of psychologies and psychoanalyses of different tendencies: all this led to the burning problems of identity and, with many women, to the collapse of motivations sufficient to justify religious life. Visiting a Catholic bookshop in Latin America, I noticed that there (and not only there) the spiritual treatises of the past had been replaced by the widespread manuals of psychoanalysis. Theology had made way for psychology.[1]

The future pope was not the only person to raise concerns on substituting religion with psychology. In 1977, Paul Vitz published a book entitled *Psychology as Religion*, where he argued that "selfism" (as he called the dominant trend in psychology) had replaced religion. Many similar books were published, including the book by William Epstain *Psychotherapy as Religion* where the author shows that psychotherapists and their offices had replaced priests, preachers, churches and chapels in the United States.

Can psychology really have such a strong influence that a pope's fears could be justified? Can psychology influence people's lives as strong as religion and can it shape human lives according to its own assumptions? Are we as connected to psychology with such strong emotions as we are to religion? Do we devote so much attention to it?

In order to answer those questions, let's try to look at our everyday lives through the eyes of visitors from another galaxy who just recently landed on our planet. What would they see? They would be stunned by the enormous popularity and spread of psychology. They would notice that most illustrated magazines have columns edited by a psychologist giving advice to readers. They would notice that many humans have scheduled appointments with their psychologists or psychotherapists. Some humans even spend a few hours every week in such offices! Our guests would notice that TV programs often feature a local expert (a psychologist) who explains why somebody had killed, raped, defrauded money or committed suicide. They would notice – with great interest – that psychologists always appear AFTER some event took place in order to "interpret" what has happened, but that none of them actually accurately predict any upcoming events. In most bookstores, many shelves are overloaded with psychology support and advice. Celebrities usually discuss their psychological problems publicly and openly talk about therapeutic programs that they participated in. During social events, people exchange recommendations of various psychotherapists. Psychologists often show up at crash sites or in epicenters of natural disasters. They work at schools, hospitals, hospices, social support sites and human resource departments of most corporations. They can be found in the police and in the army, but also in churches, parishes and prisons.

Undoubtedly, these strangers from another planet must notice the omnipresence of psychology. They would most likely quickly conclude that its popularity must come from the enormous powers of this discipline. How could they justify such great fascination and trust that vast majority of representatives of our civilization put in it? How else could they explain the sudden (lasting barely a century and escalating in the recent decades) surge of psychological knowledge?

Those conclusions might be premature. Humankind has followed bizarre paths all too often in its history where representatives of our species showed a complete lack of rationality and common sense. Crowds of people were lost in the abyss of wars and revolutions driven by ridiculous ideas. Generations devoted their lives to defend values that turned out to be nothing but a mirage. Unfortunately, the popularity of a particular phenomenon does not necessarily confirm its practical value. Perhaps it would be worthwhile to take a closer look at psychology, look at the hands of academic psychologists conducting psychological research, take a peek into therapists' offices and see what hides under the marketing slogans? Perhaps it is time – just as in a poker game – to say "call"?

We, the authors of this book have already done it. Many times we were simply astonished, sometimes terrified. If you are not afraid to shake your own beliefs about psychology and psychotherapy, we encourage you to follow in our footsteps.

In part one, titled "Fraudsters in the Temple of Science: Some Sins of Academic Psychology," we present examples of one of the most dangerous scientific abuses: the creation of *false theories* with wide impact. Why do we consider them the most dangerous? Because science has a cumulative character. Scholars share their discoveries and publish findings of their research. Each scholar then has the right to use the accomplishments of others to state further hypotheses, to build new theories or to create tools used in practice – such as therapy. If a particular concept becomes highly recognized and widely accepted, other researchers use this concept as a sort of figure of authority on the subject – just as rats do when following the leader of the pack. We act in the same way when we trust researchers, scientists and scholars. Such trust in and blind following of authority, as we will show in the Chapter 1 dedicated to eugenics, can even lead to crime.

But creating false theories, though dangerous, is not the only abuse. Later in the part one, we analyze other common sins committed by scientists – from forgeries and frauds, to plagiarism, to a common lack of attention to detail in research or in citing literature, to insufficient replications of research results, etc. At the end of the part one we will analyze the causes behind such conditions in academic psychology.

The products expected from science include all kinds of inventions, technologies, medicines, therapies, etc., that are made available to average citizens. This concept will be addressed in the second part, titled "Conquering Patients' Souls: Sins of Psychotherapists." It is dedicated to what psychology most often offers to their potential customers (patients) and what for many is the synonym of psychology – psychotherapy. We begin the second part with a description of psychoanalysis, created behind the walls of academic institutions and which deserves the title of oldest and most widely accepted (still!) psychological pseudoscience. Nowadays, the recognition and /or acceptance of pseudoscientific concepts occurs most often outside of the official academic system, or sometimes, during attempts to cheat that system. Chapters dedicated to psychoanalysis, the childhood myth and recovered-memory therapy show how, by imitating the academic system, these concepts gained the status of universally accepted and trendy concepts, despite being completely pseudoscientific.

In later chapters, we consider the essence of psychotherapy and discuss its efficacy; we will also show its negative aspects and the threats it carries. At the end of the part two, we will show a rather different perspective by considering what can be proposed as a replacement to therapy, and if therapy is at all necessary. Part two ends with an analysis of the causes of the current poor condition of psychotherapy.

Therapy and other forms of practical applications of psychology – despite their progressively more frivolous relation to dependable, fair science, have surprisingly significant market value. The activity of all sorts of therapists, personal development trainers, educators and coaches is a serious source of income and represents a significant but hard to estimate segment of the economy. As is every market-related activity, certain practices are important: market access, customer acquisition, competition, marketing, public relations, advertising, etc. In spite of the almost complete lack of regulations in the therapy industry, a global phenomenon known as psycho-business is unfolding rapidly. We dedicated the third part of the book, entitled "Beyond Control: Psycho-business" to this particular topic. This part also contains a description of a hoax that one of us conducted and published in the monthly periodical, *Charaktery* in 2007, revealing how easily pseudo-therapy can be injected into minds of the public. This very hoax allowed us to reveal the relationships between academia and the world of psycho-business, to which we have dedicated the last chapter of the part three.

Many people and some institutions would prefer that the contents of this book never get to see the light of day, at least, not during their lifetime. Readers will find an excellent example supporting this claim in Chapter 8 of this book where we mention a time capsule on psychoanalysis – a collection of letters and other documents in the United States Library of Congress that is inaccessible until the twenty-second century. Similarly, specific documents that could reveal the history of institutions and people working on eugenics are hard to access. Further, access to many research results conducted at universities around the world is usually denied. This "no access policy" is not written anywhere, and yet certain research data is virtually impossible to review. During our work on this book, we stumbled upon various people and institutions that were annoyed by us digging into certain subjects, therefore some fragments of this book are literally written by life. Wherever it was possible, we have documented those events to our readers.

The first edition of this book was published in 2009 in Poland. Since that time, much has happened and changed in science. We have worked hard to recognize these changes in this edition. We removed many references relevant to Polish audiences that we believed would be of little interest to an English-speaking audience. In this edition, we tried to emphasize the problems typical for modern psychology on a global scale.

A global system of social control introduced in science via psychology is progressively less effective and allows more and more nonsense and frauds into the field. Sometimes it even encourages them, rather than prevent them. The goal of this book was not to seek sensations – we leave this to tabloid journalists. By writing and publishing this book, we were looking to achieve our personal goal: to safeguard the years of wisdom dedicated to exploring our specialties, research and educational work. We began our studies with the belief that science is able to deepen humanity's understanding of the world around us. This fundamental belief that we share was tested numerous times by imposters wearing different masks. One of us is a doctor, the other a psychologist – often embarrassed by that fact, embarrassed because the representatives of this discipline around the world greatly abuse their positions and the trust put in them by society. Accusations against psychological studies or therapies are often valid and we must be able to hang our heads in shame. Being representatives of science, we do not want to spend our lives as con artists or members of a scam group; yet, we often feel this way as professionals in our field. It is also not our goal to personally rationalize years of our research and studies. We know that even an exceptionally critical analysis of our disciplines will leave a significant number of valid, concrete and extremely useful knowledge. Such a handful of knowledge is far more beautiful than the magnificent walls of a university building made of Styrofoam and painted attractively.

Additionally, it is not our goal to undermine the essence of academic research and the images of reality derived from them. We would not want anyone to use the facts gathered in this book to ridicule certain academic perspectives or to use our examples to generalize to across the entire spectrum of science. These examples are meant to refer only and explicitly to the limitations of the societal system of science control. We ourselves continue to nurture our faith in one of the most beautiful ideas ever created in human minds: science. We do admit though, that sometimes keeping that faith is tremendously difficult.

The fundamental goal of this book is to describe and discuss the most spectacular misconducts and abuses, to point out the larger gaps in our control system, to make our audience aware of the negative occurrences in science, and of the new dangers and pitfalls that result from interactions between science, market and media. It is possible that this book will help to increase academic standards. Perhaps a few more people will ask therapeutic shamans a few uncomfortable questions before they use their services.

"In the temple of science are many mansions"... so let's peek through the keyhole to the temple of science – specifically into the mansion with the big inscription over the front gate: *psychology – the study of the soul.*

---

[1] J.C. Ratzinger and V. Messori, *The Ratzinger report.* (San Francisco: Ignatius Press, 1987), 99-100.

PART I

FRAUDSTERS IN THE TEMPLE OF SCIENCE:
SOME SINS OF ACADEMIC PSYCHOLOGY

CHAPTER I

IN SEARCH OF THE SUPER HUMAN:
WANDERINGS OF EUGENICS

Why do makers of commercial film images pertaining to science rarely use their imagination? Why are they so often ignorant of history, the history of science in particular? Why, whenever a thriller movie is made, a "mad scientist," usually a medic, chemist or experimental physicist, has to star in it? Some mutated organism escapes from mad the scientist's laboratory, reproduces at an incredible pace, and threatens the entire human race. Or it could be an alien substance that allows humans to cross beyond their ordinary capabilities. Or a device that provides some sort of significant advantage that falls into the wrong hands.... The conventionality of those scenarios is unbearably boring to anyone who has ever had anything to do with science or to anyone who understands where some intriguing discoveries can actually be made. Why hasn't a single director ever dared to put a truly intelligent character in the main role? What is wrong with a hero who has access to nothing but a pen and a piece of paper?

Psychologists are scientists who, until very recently, did not need anything more than a sheet of paper and a pen or a pencil to do all their work. With those simple tools, they have managed to create disasters that many directors and screenwriters wouldn't even imagine. Some of those scientists armed with writing tools could even be charged with genocide. All this happened in the name of scientism, using simple tools such as the afore mentioned piece of paper, pencil and few columns of figures only.

It all started pretty harmlessly with a man who wasn't even a scientist, although the methodology he adopted affected numerous scientists. Obsessed with quantifying anything and everything that he could see, and blessed with a bright, insightful mind, Francis Galton was looking for regularities and patterns everywhere he could. He used to count which people in the audience fidget in their seats trying to measure the spectators' interest in the show. He counted waves during his baths to check if they formed any patterns. Most of his findings sound ridiculous now, but among them are some that proved to be actual correlations that survived up to today. Not many remember that it was Galton himself who established the basis for weather forecasting; he also discovered the fact that human fingerprints are unique and could be used to identify an individual.

Some of Galton's observations significantly affected other scientists, and in consequence, influenced the fates of hundreds of thousands of people worldwide. By analyzing biographical dictionaries, encyclopedias and the genealogies of 415 outstanding poets, artists, authors, scientists, explorers, judges and military officials, he concluded that the majority of them came from the same genealogical lines. Based on this observation, he postulated a hypothesis that we inherit more than only physical features (such as a color of our hair, height, etc.), also our emotional, intellectual and creativity-related traits. He calculated that 48% of the sons of those outstanding individuals, 7% of their grandsons and only 1% of their great-grandsons were also exceptional people. He published his observations in a work entitled *Hereditary Genius.*[1]

Galton did not stop there. From these observations, he drew the conclusion that human development, guided by a marriage control system that would pair together talented individuals, would yield talented people, just as it is done when breeding animals where farmers select the best individuals for reproduction at their own discretion. He decided to give a new name to this area of knowledge. The archives of University College London still hold a sheet the size of a palm on which Galton inscribed the Greek letters combining two words *good* and *born*. This is how the term *eugenics* originated – a word that inspired Galton's contemporaries and also the one word that became an obsession of his numerous followers. This one word was written into the credo of numerous institutions and broke the continuity of generations in the name of law, deprived thousands of basic human rights and brought death and misery to millions.

We do not want to put the entire blame on Galton. There were many historical reasons that contributed to spread of eugenics, but it was Galton's fault to draw far-reaching conclusions from insufficient and lacking research materials. It was his unacceptable impatience and desire to test his mere assumptions on a large scale. The imitators, adopters, successors and followers of eugenics are guilty of not conducting reliable investigation and research on heredity. They did not bother to verify or question Galton's hypotheses, did not try to replicate his results. Instead, they started a giant social experiment, an experiment in which psychologists played a significant role. After all, the enhancement of intellectual capabilities of the human race was at stake. In order to achieve this goal, a measurement of these capabilities was required. From that moment on, until today, the development of eugenics is inseparably connected to the history of measurements of human intelligence. The catastrophic campaigns that followed depended on the reliability of intelligence measurements provided by psychologists. All their methodological mistakes had significant and direct impact on thousands of people. Let's have a closer look at this marriage.

The first known attempts to assess intelligence were performed by Galton himself. During the International Health Fairs in the South Kensington

IN SEARCH OF THE SUPER HUMAN

Museum, he had set up a booth where all visitors could test their intellectual capabilities for the moderate price of three pence. Approximately nine thousand men and women participated in this pioneering project. The tests measured the time it took for the subjects to respond to auditory, visual, or touch stimuli, and also to other aspects of sensory-motor functioning that could have been easily and precisely measured. Galton believed that intelligence is a general trait. He expected outstanding individuals to score better in all measurements. Unfortunately, specific measurements did not correlate with each other; there was also no correlation with independent indicators of intelligence. The project failed to produce the expected results and researchers were forced to develop intelligence measuring tests in a more direct way.

The first useful test measuring intelligence was developed in France in 1905 by Alfred Binet, a psychologist, and Theodore Simon, a physician. It was the first known public contract addressed to psychologists. As commissioned by the Ministry of Education, they prepared a test that consisted of 30 increasingly difficult questions, which allowed assessment of the intellectual capability of the person tested. Binet emphasized several times that the test results were not permanently conclusive, as a child's measured intelligence levels may change. He developed a unique set of exercises and teaching methods for children to improve their intellectual performance. Binet openly opposed the hypothesis of a hereditary nature of intelligence.[2] Unfortunately, his concepts soon got distorted and used for the purpose of eugenic "improvement" of the human race.

The German psychologist William Stern introduced the concept of the intelligence quotient (IQ) in 1912. He noticed that the intellectual age of a tested person accurately represents the intelligence of an individual only when it is related to the current, chronological age of a given person. Let's consider three hypothetical people: Peter, Mark and Mary. Assessment of their intellectual capability test indicated they were 10 years old. Let's assume their biological age is 20, 10, and 5 years, respectively. In such scenario Peter displays severe mental retardation, Mark has an average intelligence, and Mary is a genius. Therefore, Stern suggested that the ratio between the level of intellectual development and biological age would be much more accurate. He called this ratio an intelligence quotient. The popular abbreviation IQ was introduced later by the American psychologist, Lewis Terman, who further suggested multiplying the calculated ratio by 100 to express the quotient as a percentage and avoid inconvenient fractions. According to this formula that soon became broadly accepted, Peter's IQ was $(10/20) \times 100 = 50$, Mark's IQ $(10/10) \times 100 = 100$ and Mary's – $(10/5) \times 100 = 200$. According to such representation, the intelligence quotient close to or equal to 100 represents average value of the current population at this age. In the case of an IQ lower than 100 we are dealing with mental underdevelopment and retardation, while results above 100 indicate progressively more capable and outstanding individuals.[3]

Binet's method of measuring intelligence and Stern's concept of calculating the intelligence quotient were welcomed in the US with great enthusiasm. Lewis Terman from Stanford University adapted the Binet-Simon scale, known today as Stanford-Binet, and it soon became a prototype for all IQ tests developed later. Intelligence tests soon fell into the hands of scientists blinded by various ideologies. Two psychologists are worth mentioning here: Henry Goddard from New Jersey and Robert Yerkes from Harvard University, (then the acting chairman of the American Psychological Association). They were deeply convinced that those tests measure intelligence as a permanent and congenital predisposition. It is also worth mentioning that they both were active members of eugenic associations. They spread a concept that was greatly opposed by Alfred Binet. They claimed that genetically inferior people represent a threat to the social, economic and moral condition of the country. Tests measuring intelligence became a tool for identifying, and consequently, for eliminating "inferior" individuals. Goddard and Yerkes were joined by Terman in 1916, who stated that the ultimate result of the common use of intelligence tests would be to free society from tens of thousands of mentally retarded people – African Americans and representatives of other racial minorities in particular.

Very quickly, the fact that measuring intelligence became possible provided the basis for the introduction of laws that enabled one of the most rigorously concealed events in modern history: the mass sterilization of thousands of people. A test result provided an "objective" basis for making a judgment on who had the right to reproduce. In over 30 states, laws were passed that allowed forced sterilization as a means of preventing the reproduction of people with low intelligence in order to support the "enhancement" of the human race. The first legal act that allowed sterilization was passed in Pennsylvania in 1905, but it was vetoed by the governor and never came into force. The first law became effective in Indiana in 1907. New Jersey followed track in 1911☐ and later the same year, the state of Iowa implemented a law depriving criminals, mentally retarded, and so forth of the possibility to reproduce. The sterilization bill passed in the state of Washington was appealed in the Supreme Court in 1912, but the court unanimously upheld its lawfulness, making reference to research on eugenics available at that time. The law passed in California in 1918 allowing compulsory sterilization, provided it was approved by a team of experts. The teams had to include a psychologist with a PhD, (at that time, Terman was one of a very few people who actually had such qualifications in California). A similar law passed in Virginia led to over 7☐ 500 forced sterilizations in the years 1924 – 1972. A scalpel was in constant use, but the people subjected to these procedures were rarely aware of what was going on. Doris Buck Figgins, a young woman sterilized in 1928 against her will in accordance with effective legislation, was informed that she actually underwent an appendectomy. She finally discovered that she was sterilized in 1980: "I never knew anything about it.... I am not mad, just broken hearted is all. I just wanted babies bad....I don't know why they done [sic] it to me. I tried to live a good life."[4]

Since the first law was passed, by 1940, a total of 35,878 people in the United States had been sterilized against their will or even without their knowledge. At least, that is the number of cases that had been accurately documented. It was not the end, though. Today, the number of all sterilized people amounts to approximately 60,000. Not so long ago, between 1972 and 1976, hospitals in only four cities in the USA sterilized 3,406 women and 142 men, mainly Native Americans.[5] The numbers may be quoted endlessly, but they will never reflect the suffering of the victims of "science" built on insufficient or misinterpreted evidence. Compulsory sterilization of the mentally ill is rare nowadays (if practiced at all), but there are still some states where sterilization laws are in effect. In North Carolina for example, the laws were even updated in 1973, and then again in 1981.[6]

Give me your tired, your poor, your huddled masses yearning to breathe free; The wretched refuse of your teeming shore. This is the inscription with which the Statue of Liberty welcomes travelers disembarking from ships in New York. In 1912, a scary shadow appeared between the harbor and the Statue: the shadow of a psychologist obsessed with the idea of the improvement of the human race – Henry Goddard. Before they left the harbor, immigrants had to solve intelligence tests. Goddard's conclusions were that 79% of the Italians, 87% of the Russians and 83% of the Jews are mentally retarded. Publication of these findings not only fuelled the existing prejudices, but also led to tightening of immigration legislation. The Immigration Act of 1917 provided an option to reject people considered mentally inferior, barring them from entering into the US. In the same year, Goddard reported with pride that huge numbers of foreigners were deported because of their mental retardation. It happened thanks to the efforts of physicians who were convinced that intelligence tests may be used for detecting mentally impaired foreigners. Finally, the Immigration Restriction Act that passed in 1924 was a direct result of the lobbying of various groups and individuals opposing immigration from Southern, Central and Eastern Europe. The Act specified the annual number of immigrants from any country not to exceed 2% of the total population of immigrants from that country who were already living in the US in 1880, i.e., from the year prior to the wave of immigration from the Southern, Central and Eastern Europe.[7]

In the 1930s, hundreds of thousands of Jews, refugees escaping the extermination by Nazis, were seeking asylum in the United States. Many of them were deported back to Germany where they faced the nightmares orchestrated by the Nazis – faithful disciples and followers of American eugenicists. A significant number of those who were killed in the Nazi concentration camps could have survived if the 1924 Act had not been introduced; Terman, Goddard, Yerkes and other American psychologists who followed are intellectually responsible for "scientific" foundations of the Immigration Restriction Act.

Uncritical application of tests prepared by psychologists blinded by eugenic ideas led to further distortions of reality. It was very soon "discovered" that black citizens were getting on average 60% worse scores than the white population. The application of tests developed by Yerkes, Goddard and Terman for the US Army demonstrated that 47% of white and 89% of black recruits demonstrated intelligence levels lower than average 13-year-old teenagers. Expelling them all from the Army would make it impossible for the US Army to participate in any military conflict!

The work of Yerkes, Goddard and Terman was continued by another eugenic movement activist, Carl Brigham, a psychologist from Princeton. His studies provided "unquestionable evidence" of the superiority of the Nordic race. The German eugenicists, initially lagging far behind the achievements of their American colleagues, turned out to be avid and enthusiastic students. They soon overtook the Americans by making the eugenic scenario of improvement of the human race come true not only by sterilization, but also by mass extermination of millions of people, wiping out entire nationalities and races, all based on solid "scientific" evidence from a handful of psychologists.

When we look at those events from a historical perspective, it seems obvious that Galton was too quick to introduce postulates based on weak hypotheses into real-life practice. We can see clearly, too clearly, the lack of sufficient research and evidence for the hypothesis of a hereditary nature of intelligence. The disputes in this regard continue today, and despite thousands of research studies in this area, we are still far away from any definite answer. Even very superficial knowledge of the structure and application of intelligence tests is enough to conclude that all assumptions of these eugenic psychologists were simply untrue because their research failed to take into account the cultural background of their test subjects. This horror, presented here in an extremely abridged way, should, however, teach us – psychologists – and teach us once and for all, the following rules:

1. One must not, under any circumstances, formulate any laws or rules when the evidence is hardly sufficient to form a hypothesis.
2. One must not present research results that may affect the lives of other people if we are uncertain that they were acquired in a reliable, repeatable, ethical and verifiable manner and with the best methodological standards currently available.
3. Any methodological ignorance that prevents the detection of errors in other scientists' research and in conclusions formulated from such research is a burden solely on one's soul.

It is not enough, we know, but it is better than nothing. Have psychologists learned the above lessons? We doubt it, as we continue to see great impatience in publishing findings that often only present preliminary results. Later in this book, readers will see that there are still millions of research studies performed carelessly, without any attention to methodology, or by ignoring methodological standards completely. Methodological ignorance is common and tolerated in the scientific community. The majority of graduates in psychology are incapable of correctly interpreting results of simple surveys; only a few are able to pinpoint methodological errors. If this is the case, the horror may repeat itself at any time. Why? The majority of psychology textbooks do not even include the word "eugenics" in their indexes. Most of the aforementioned names are referred in textbooks, not in the context of a warning, but in the context of important milestones in the development of modern psychology. Last but not least, the popularity and spread of numerous pop-psychology concepts without any support or evidence derived from reliable research is more proof of the common acceptance and tolerance of methodological ignorance.

---

[1] F. Galton, *Hereditary Genius* (London: Macmillan and Co., 1892), 317.

[2] R. E. Fancher, *The Intelligence Men. Makers of the IQ Controversy.* (New York, London: Norton, 1985), 100-104.

[3] A.M. Colman, *Facts, Fallacies and Frauds in Psychology.* (London: Hutchinson, 1987), 19-20.

[4] S. Noll, *Feeble-Minded in Our Midst: Institutions for the Mentally Retarded in the South, 1900-1940.* (Chapel Hill, NC: UNC Press, 1996), 73.

[5] See: E. Black, *War Against the Weak: Eugenics and American Campaign to Create a Master Race.* (New York: Dialog, 2012).

[6] Ibid., 385-411.

[7] Ibid., 63-87.

CHAPTER 2

THE FRAUD AS A NOBLEMAN: FROM LIES TO ACADEMIC GLORY

It naturally could have looked like this… On that day in fall 1946 in London, the air was damp and the weather rather gloomy. Shy, rare rays of sunshine peeked from behind the clouds as if they wanted to brighten this day of well-deserved praise and glory. Cyril Lodovic Burt arrived at Buckingham Palace knowing that when he left castle's gates later that day, he would be a different person: not only bigger and stronger than he had ever been, but also some-what different – finally joining the ranks of nobility and cutting away, once and for all, his ties to the lower classes. He would now be able to forget his childhood memories of growing up in the working-class neighborhoods, just as you forget about a film you just saw when you leave a movie theater. The Main Ballroom, in which the ceremony was scheduled, did not impress Cyril Burt with its splendor. It seemed no more than a proper environment for the recognition of him on this day. Despite this, when the King entered the hall with two Gurkha officers at his side and in the presence of Lord Chamber-lain, he felt a thrill of excitement. A thought that his father would have been extraordinarily proud—if he could have seen him at that moment—flashed in Cyril's head. Five soldiers from the Royal Guard were already at the podium when the national anthem sounded. When it stopped, Lord Chamberlain took his place at the King's right side and began reading the names of candi-dates along with their honorable accomplishments. Finally it was his turn. As Lord Chamberlain read Burt's accomplishments, his worries and anxieties, his fears and struggles for respect slowly faded away. With every solemn word spoken by Lord Chamberlain, a new, different Cyril Burt was being born – Sir Cyril Burt, the first knighted psychologist in the world.

Besides the knighthood granted by the King, Burt was recognized for his accomplishments by being appointed as the Head of Psychology Department at University College London, where he worked before finally retiring in 1950. Shortly before his death, the American Psychological Association rec-ognized him with the prestigious Thorndike Award.

Cyril Lodovic Burt was born on March 4, 1883 in one of the working-class districts of London. According to Burt's biographers, his father, having sufficient medical background, ran a prosperous pharmacy. Young Cyril went to school and spent many hours in the streets with friends from poor, working-class families. He even learned to speak cockney. His street experiences often helped him in his future work, especially when he was dealing with teenage and juvenile criminals. He remembered how his colleagues experienced similar harsh lessons in their youths. Despite such an environment, there was an emphasis placed on intellectual values in Cyril's home. He was often reminded of his middle-class status. The Burt's family situation changed significantly when Cyril was nine years old. His father took up the post of a countryside doctor. Even Francis Galton's siblings were among his patients! Burt later recalled in his autobiography that his father always inspired him with stories of famous patients. He talked a lot about Galton as the successor of Milton and Darwin.[1]

He later had the opportunity to meet Galton in person. When Burt borrowed Galton's publication on *Inquiries into Human Faculty* from the school library, he discovered with excitement that it was published in 1883, the year Cyril was born. From that moment on, Burt started to strongly identify himself with Galton and his views. This, combined with the strong pressure from his father, lead Cyril into the walls of Oxford University, where he graduated with honors in 1907. Even during his studies, Burt encountered advocates of eugenics who reinforced his ideas. He successfully reached for the prestigious position of Lecturer in Experimental Psychology at the University of Liverpool.[2]

His first academic work titled, *Experimental Tests of General Intelligence* was published in 1909.[3] Paired with Burt's childhood experiences, it set a course for his entire future work. His greatest recognition came from research conducted on twins. The results suggested a hereditary nature of intelligence and perfectly matched the expectations of advocates of eugenics. In his work, Burt repeatedly achieved results suggesting up to 80% influence of hereditary factors in the development of intelligence.

At this point, certain readers who are less familiar with the methodology of psychological studies deserve some clarifications in regard to research on the heritance of intelligence. Researching the heritance of intelligence means that we aim to answer how strongly the development of intelligence depends on genetic factors and how strongly it depends on the environment that we live in. Researchers were blessed with an excellent, purely natural tool – identical twins. Identical twins always have 100% identical genetic makeup.[4] Identical twins who are raised separately present a unique research opportunity. If their intelligence is similar after being raised in different environments, we would have the basis to assume that intelligence is influenced by genetic factors. If it is significantly different, we will be more likely to accept that the environment influences it.[5]

The result of most of Burt's research and theories clearly pointed to heredity as the main intelligence-influencing factor and was met with widespread acceptance. As a renowned scholar and expert in the 1930's and 40's, Burt became a government adviser and committed himself to introducing a system of education known as the 11 Plus plan. This system intended to administer intelligence-measuring tests to 11-year-olds, organizing them into one of three learning levels. Thus, the small upper percentile of 11-year-olds, who achieved the highest grades, almost automatically gained access to elite schools. The remaining majority was pushed to inferior schools and had practically no chance of progressing to more elite or more advanced institutions in the future. Today we know that the developmental differences in children at the age of 11 cannot accurately predict future potential. It is common for children who develop at a slower pace at that age to rapidly accelerate their learning and intellectual capabilities at a slightly older age. Yet such children, in the system co-founded by Burt never stood a chance. This system was rigorously implemented in England and Wales and lasted more than an entire generation.

Burt's thoughts were influential to such an extent that a renowned intelligence researcher, Arthur Jensen from the University of California, suggested in an article published in the *Harvard Educational Review* in 1969 that the hypothesis of intelligence depending on racial heredity could explain the failures of programs struggling to address the educational gap among minority groups (at that time described as racial minorities) in the USA. This claim was met with widespread approval from other scholars and researchers.[6]

The need for adoration and Cyril Burt's relentless attempts to climb to the top of the mountain of fame became obvious during a particular event hardly ever mentioned in official records. His behavior during discussions was such that Burt tried to win at any cost, not always adhering to the common rules of honest debate; this resulted in his younger friends giving him a mocking nickname of 'the old delinquent'. In the late 1930s, a small misunderstanding with Charles Spearman took place. (Spearman is regarded nowadays as the most influential intelligence researcher of that era.) Burt wrote in a footnote of his article in 1937 that he was the first to suggest using a specific equation for factor analysis[7] during his work in 1909. Meanwhile, Spearman, who was already retired at that time, was still vigilant and wrote a personal letter directly to Burt stating that he was the one to have discovered the equation, and reminded that he shared it with Burt a few months before the work was published. Burt apologized to Spearman, but that event was the beginning of Burt's increasing attempts to re-write the history of factor analysis, highlighting his role in its creation and minimalizing Spearman's contribution in development of this statistical method.[8]

Later, Burt asked Hans J. Eysenck, at that time one of his top students, for help on his article dedicated to factor analysis. Burt wrote the text while Eysenck was supposed to take care of the calculations. "Burt ... showed me the paper he had written under our joint names, and I thought it was very good. I was rather surprised when it finally appeared in the *British Journal of Educational Psychology* in 1939 with only my name at the top, and with many changes in the text praising Cyril Burt."[9]

After Spearman's death in 1945, Burt's self-promoting campaign grew in strength. He exploited his position as the editor of the *British Journal of Statistical Psychology*. He published many of his own articles there, in which he emphasized his own importance in developing the foundations of statistical psychology and minimalized Spearman's involvement. Furthermore, he did not hesitate to publish articles in his journal that portrayed him in a very positive light. Some might argue that this is acceptable for a journal's editor, but in reality, the articles were written by Burt himself, but published under someone else's name. A great example is a praising acclamation dedicated to Cyril Burt, which appeared in 1954 and was signed by Jaques Lafitte, a French psychologist who was very familiar with Burt's accomplishments, and who, as it turned out many years later, never actually existed.

In spite of those fabrications, few psychologists became somewhat suspicious of Burt, yet not one of them publically stated their doubts. Burt's contributions to mathematical psychology, to the development of statistical methods and to theories of statistical psychology were so widely known that such minute details may have seemed petty and of too little importance to be publically discussed. Furthermore, his reputation was additionally reinforced by the knighthood granted to him in 1946. Yet dark clouds were slowly building on the horizon of his career. The first attack on the 11 plus system and his methods of researching intelligence appeared soon after his retirement in 1950. Burt's answered in 1955 to those allegations by publishing his findings from research on twins in the field of hereditary intelligence – one of the most disgraceful studies in the entire history of psychology.[10]

The publication of those findings was a response specifically aimed to answer accusations of Brian Simon, a lecturer from the University of Leicester, and of Alice Heim, a psychologist from Cambridge University. Simon, a former teacher, claimed that the "11 plus" system discriminated against gifted children who developed later than their peers for their entire lives. He also claimed that this discrimination was done using questionable methods based on poor quality tests. Heim criticized Spearman's model of intelligence and questioned Burt's assumptions of the inheritance of intelligence, arguing for Binet's approach, who believed that development of intelligence can be externally influenced.

Burt's views, to which he was deeply devoted, were attacked again in the early 1960s. Psychologist John McLeish came with the most crushing criticism. In his book *The Science of Behavior*,[11] he questioned Spearman's entire tradition of measuring intelligence. The attacks intensified over time. Through various publications, Burt fiercely defended his opinions; the most important one was another study on twins in 1966.[12]

Up until this point, Burt's studies conducted on twins did not raise significant suspicions, even despite the fact that his correlations were higher than any other found in similar studies. They did not differ enough from other research to provoke suspicion. The article from 1966 demonstrated the relationships between social position and intelligence based on Burt's research on monozygotic twins raised separately, but pursuing very different career paths and living in different social environments. No one was able to do such research before simply due to logistics of such enterprise: it is very hard to find any large enough sample of twins raised apart, let alone a large enough sample of twins separated, and raised apart, and doing different jobs, and living on different levels of the socio-economic ladder.

The first doubts began to appear. Sandra Scarr-Salapatek from the University of California brought to attention the fact that Burt's data sets seemed a bit odd. She asked him to explain his procedures, but never received a satisfactory response. Burt never presented a single detailed case study of the twins he analyzed, even though researchers who were working in the same areas, such as Newman, Freeman and Holzinger, did. When other researchers wrote to Burt with requests for more detailed data, they were politely, yet firmly referred to little known documents from the 1910's and 20's, or they received excuses about the inaccessibility of the data and the inability to decode it. When sociologist Christopher Jencks directly requested the list of fifty-three pairs of twins and their IQ results with evaluation and descriptions of their job circumstances, he received (after several weeks) a list with all data erased, except for those that Jencks specifically requested. To this date, this is the most detailed document available of Burt's research work.

These doubts were not enough to change Burt's recognition, prestige or adoration. We write about them here because they provide a background for the events that took place after his death. In 1972, Burt's articles fell into hands of Leon Kamin, a psychologist from the University of Princeton. Kamin immediately noticed internal discrepancies in the published works; he pointed out the lack of basic systematic data, such as the sex of children studied, type of tests used, etc. Kamin also noticed that in Burt's publication from 1939, he had claimed that the methods and procedures he used "were described in more detail in the dissertations of researchers involved in the study" or that the methodology could be found in some other inaccessible dissertation. In another paper from 1943, Burt claimed that a comprehensive collection of sources and calculations, together with detailed charts and tables could be found in the works of J. Maver, conducted in the psychology lab at University College London (UCL). In reality, Kamin discovered that such works or dissertations or any other publications on that topic were never sent, presented or archived at UCL. It turned out there were many more similar missing links in Burt's works.[13]

The most important fact that Kamin brought to public attention was regarding the correlation coefficients that Burt allegedly obtained and described in three different publications dedicated to twins. Despite the fact that he compared groups of twins of different sizes, the resultant coefficients were accurate up to three decimal places! They turned out to be 0.771 for separated twins and 0.994 for twins raised together. Anyone who has ever done any statistical calculations knows that getting an identical result twice is already extremely rare. Three identical results seem to be virtually impossible. But three sets of identical results from research conducted on different populations, over many years? This must have been a miracle! When we now take into account the inaccuracies in documentation, non-existent works and unwillingness to allow access to raw data, we might be led to believe (as researchers across the world did) that it was not a miracle; it was an ordinary fraud.

Even Burt's worshippers who rushed to defend their idol, such as Arthur Jensen, were soon forced to face harsh facts and accept the available evidence against Burt. Jensen traveled to Great Britain specifically to collect an anthology of Burt's research in order to compile and publish an overview of his work. To his surprise, he found 20 cases of identical sets of values of correlation coefficients in Burt's articles...twenty identical sets of numbers in research conducted on different populations. This was impossible to explain. Despite this fact, Jensen and Eysenck concluded that fabricated data was only a background meant to illustrate the more important theoretical foundations of qualitative genetics. They assumed that "mistakes" in Burt's works were the result of lack of attention of the 72-year-old scholar, rather than a conscious and planned deception.

Kamin however stubbornly supported his theory that Burt intentionally deceived the scientific community. He analyzed every piece of Burt's available work and concluded that he had already started forging and fabricating all of his data in 1909! The controversies surrounding Burt's monumental figure grew and eventually led to the decision to study all of his original records and notes. It turned out that shortly after his death, some of Burt's close friends took a number of his books and other documents from his home. They left however six large chests, previously used to store tea, filled with documents, notes and calculations. The owner of the house, not knowing what to do with the remains, asked Burt's friends for advice. They suggested to burn the evidence, which was precisely what the landlord did. It would seem that's how history puts future disputes of the researcher's work to final rest, surrounding Burt with a mist of uncertainty, or better yet – with smoke from burning documents. This would have happened, if not for an especially inquisitive journalist...

In October of 1976 an article appeared in *The Sunday Times* written by Oliver Gillie, who, after thoroughly investigating Burt's situation, concluded that women such as M. Howard and J. Conway never existed! Who were these women to whom Gillie dedicated so much attention? They were the co-authors of Burt's most important works. They were supposedly both leading the research in his name. When Burt was the editor of the *Journal of Statistical Psychology*, both women wrote reviews and articles praising Burt's accomplishments, frequently acknowledging his ground-breaking ingenuity and fiercely criticizing his opponents. After Burt stepped down from the editor's position at the journal, the reviews and articles from both women ceased to appear. These facts cemented the case. Sir Cyril Burt became recognized by the academic community as a fraud who actually faked most of his data, although Jensen and Eysenck stayed faithful to their opinion of the old researcher's "lack of attentiveness."

In 1978, a serious academic analysis of Burt's works appeared, conducted by Donald D. Dorfman from the University of Iowa. He thoroughly analyzed Burt's publication from 1961 on the links between intelligence and advancement up the socio-economic ladder and concluded that Burt's data was… faked. Dorfman showed that Burt, instead of obtaining new data, simply copied numbers from old charts that he published in 1926 in an article devoted to career counseling. And how did Burt get the numbers for this career-counseling article? Well, he copied them from a population census published in 1921.[14]

The best epilogue to Burt's history was presented by Thomas F. Gieryn and Anne Figert in their article: "Scientists Protect their Cognitive Authority: The Status Degradation Ceremony of Sir Cyril Burt."[15] Their analysis showed the ceremony of Burt's degradation, the ceremony that did not occur in the splendor of Buckingham Palace's ballroom, the ceremony that was not accompanied by the sounds of the national anthem, nor ever attended by the Queen or the Lord Chamberlain. It proceeded slowly, and its focus was not so much Cyril Burt but rather on the entire authority of the world of science. As Leslie Hearnshaw, Burt's biographer, noticed,

> The most serious harm done by Burt is the discredit he has brought on the profession of psychology. He has undermined the public's faith, slowly and laboriously built up, and still far from achieved in the work of psychologists. It is a setback that may take some time to repair. Psychologists will find it harder to gain credence for their findings. Burt has generated an aura of distrust, which may affect public support, financial and moral, for some time to come. Distrust of this sort is hard to dispel.[16]

Gieryn and Figert showed that the academic community had done much to minimalize the impact of Burt's actions on their societal and professional image. At the time of his death in 1971, Burt was placed in the elite group of top psychology researchers. Ten years later, his colleagues excluded him from their ranks and stripped him of his academic status and scientific recognition. According to the authors' opinions, Burt's degradation had eight distinct stages. The first stage was called *feigned ignorance*, when researchers had to confront their image of Burt with the devastating facts for the first time. The next stage was the *denial of charges* – an attempt to save an already historical figure and the search for explanations of the continuous stream of uncovered and inconvenient facts. As the charges accumulated, most psychologists had to choose a side: guilty or not guilty? This stage was called *stacking the jury*. The stage of *plea bargaining: guilty of lesser charges* was the time when Eysenck,

Jensen and a few other defenders of the accused tried to diminish the weight of the charges in ways already known to us. This stage rather quickly moved to the phase of *blaming the accusers*. Its appearance was inevitable, as the attack on Burt was an attack on all of the supporters of the concept of hereditary intelligence. In this battle between supporters of hereditary and environmental origins of intelligence, good old reason prevailed. Both sides agreed that empirical, hard evidence obtained from reliable studies was needed to resolve the difference in opinions.

However, before the final judgment was passed on Burt's case, psychology had to deal with one more stage called *perpetrator as victim*. This was Burt's defenders' last attempt. It appeared when it was obvious to everyone that Burt was guilty of the charges of fraud brought against him. His defenders pointed to a long life full of crises that prevented Burt from conducting proper research. They claimed that events such as his marriage falling apart in 1932, the loss of many research materials during London bombing in 1941, a deteriorating sickness that included hearing loss and in fact many other similar critical misfortunes had led Burt to believe that he was a sick and tormented person. His defenders tried to convince the public that Burt's actions were not those of a rational person but rather of a desperate victim. They did not however, try to rehabilitate his status as a researcher. In a certain way, that stage finally convinced everyone that Sir Cyril Burt stopped being treated as a worthy representative of the scientific community – the judgment was finally passed. The seventh stage was thus called, *the sentence: banishing Burt*. Gieryn and Figert end their description of the degradation ceremony with *recovering the authority of science: 'the truth will out'*. In their view, the ceremony had two goals: excluding Burt from the realms of science and rebuilding public trust in psychologists as scholars. In their opinion, both goals were met.

Two issues are especially intriguing in Burt's history. One of them is tied directly to Burt himself, the second is linked to conclusions from Burt's history. Jensen was the first to focus on the former issue. Why did Burt, despite his undeniable intelligence, knowledge in the field of statistics, and incredible scrutiny (reflected in the reviews he wrote for authors of research papers in statistical and mathematical psychology), fake his results so incompetently? After all, he must have known that obtaining 20 identical correlation coefficients was so unlikely that, sooner or later, it would catch someone's attention. Jensen and Eysenck attributed this to Burt's lack of attention, old age and to the way he treated research data as of secondary importance in relation to his theories. However, when we read the first publications about Burt, we were convinced that he had gained a lot of satisfaction from fraud and deception. There are some criminals for whom the results of the crimes are not as important as the emotions experienced while committing them. This type of motivation is characteristic for many thieves, like pickpockets who often steal completely useless or worthless items just for the thrill associated with it. Burglars can also be found among this group; they accept the risks of navigating their way through various security systems only to steal an item the value of which is disproportionate to the risks necessary to obtain it. Many computer hackers also work this way, entertained solely by the fact of breaking into a computer system and then not making any use of it.[17] Now imagine Burt with a roguish grin, inserting the same sets of data into yet another article and thinking: "When will this group of losers realize that I'm making fools out of them? What will their faces look like when they finally find out about this?" What if he was intentionally provocative to demonstrate what this book actually focuses on: to point out the ineffectiveness of science's quality control systems?

The more facts we learned while studying his full biography and autobiographical recollections, we saw an entirely different psychological profile. Cyril Burt was a dominating individual with a very strong need to succeed and attain fame, of most likely, a fairly narcissistic personality. He despised rather than respected people, although he was intelligent enough to not show this to them outright. He was certainly aware that his methods of deception were rather simplistic. The fact that he did not make a concerted effort to cheat by using more sophisticated methods came most likely from his hidden contempt for others. Today we imagine Burt with a bitter expression on his face, furious at yet another attack on him. He who persistently inserted the same sets of coefficients "calculated" from data which never even existed, derived from studies that were never conducted, and his words, which resonate in his head, but were never said out loud: "This band of idiots is too stupid to discover even such a simple thing as identical correlation coefficients in my every article. They are not worth the effort."

The second dilemma lies with the fact that we don't know if we should be satisfied by the end of Burt's history or should we rather sound the alarm because of it. On the one hand, the fact that all of Burt's work was x-rayed and his frauds were unmasked should be comforting, as it demonstrates the efficiency of the self-cleansing mechanisms within the scientific community. However, we are terrified of the conclusions drawn from this tale of a fraud who became a nobleman. Let us not forget that during Burt's entire life, only a few scholars were brave enough to directly confront him. Any doubts or questions rarely appeared. If Burt had made even a tiny effort when falsifying his results, if he had fabricated a few more documents or if he hadn't exaggerated with the fictional people who upheld his narcissistic convictions, today he would most likely be mentioned in every psychology textbook, right next to Spearman, Binet or Yerkes. His concept of hereditary intelligence would continue to be taught at all universities, at least as an important and meaningful concept in the history of the evolution of differences between individuals.

Is it possible that there were more figures such as Burt in psychology? Perhaps some of them were more thoroughly deceptive? Or maybe they are still working and are quite well off? Is there any chance of identifying them, especially when they have a much lower position than Burt had and perhaps the focus of their research is not as controversial or as sensitive as intelligence? The following chapter will help us find the answers to those questions.

---

[1] C. Burt, "Autobiography," in *A History of Psychology in Autobiography, Vol. 4*, eds. E. G. Boring and H. S. Langfeld, (Worcester, MA: Clark University Press, 1979), 54-73.

[2] Fancher, *The Intelligence Men*. 169-172.

[3] C. Burt, "Experimental Tests of General Intelligence." *Journal of Psychology, 3*, (1909): 94-177.

[4] Today we know that so called epigenetic effect is responsible for inhibition of transcription of DNA and can lead to permanent cessation of gene expression. Genetic sequence remains the same, but parts of it may be inactivated. This is the main argument against identicality of monozygotic twins. This effect was not known at the time Burt conducted his research.

[5] This description of twin research is simplified by necessity. Readers who are interested in this issue should read Fancher, *The Intelligence Men*.

[6] A. Kohn, *False Prophets: Fraud and Error in Science and Medicine*, (New York, NY: Basil Blackwell, 1986), 52-57.

[7] Statistical method used to describe variability among observed, correlated variables. Charles Spearman pioneered the use of factor analysis in the field of psychology and is sometimes credited with the invention of factor analysis.

[8] Factor analysis has significant meaning in psychology during the analysis of phenomena studied using questionnaire tools. It is a valuable tool, but has its limitations. Factor analysis can be only as good as the data allows. In psychology, where researchers often have to rely on less valid and reliable measures such as self-reports, this can be problematic. More than one interpretation can be made of the same data factored the same way, and factor analysis cannot identify causality.

[9] L.S. Hearnshaw, *Cyril Burt: Psychologist.* (Ithaca, NY: Cornell University Press, 1979), 285.

[10] C. Burt, "The Evidence for the Concept of Intelligence." *British Journal of Educational Psychology, 25*, (1955): 158-177.

[11] J. McLeish, *The Science of Behavior.* (London: Barrie and Rockliffe, 1963).

[12] C. Burt, "The Genetic Determination of Differences in Intelligence: A Study of Monozygotic Twins Reared Together and Apart," *British Journal of Psychology, 57*, (1966): 137-153.

[13] Kohn, *False Prophets*.

[14] D. D. Dorfman, "The Cyril Burt Questions: New findings," *Science, 201*, (1978): 1177.

[15] T.F. Gieryn and A. Figert, "Scientists Protect their Cognitive Authority: The Status Degradation Ceremony of Sir Cyril Burt," in *The Knowledge Society, Sociology of the Sciences Yearbook. 10*, eds. G. Bohme and N. Stehr, (1986): 67-86.

[16] L. Hearnshaw, "Balance Sheet on Burt." in *Supplement to the Bulletin of the British Psychological Society, 33*, (1980), 7. As cited in Gieryn and Figert, Scientists Protect.

[17] See K. Mitnik and W. Simon, *The Art of Deception: Controlling the Human Element of Security* (Indianapolis, IN: Wiley, 2003).

CHAPTER 3

EPIGONES OF SIR CYRIL BURT: OTHER SCIENTIFIC FRAUDSTERS

An everyday client entering an electronics store in Rochester back in November 1988 could not possibly suspect that just few years ago, the owner of the shop was a well-known, brilliant scientist. With perfect knowledge of Hi-Fi equipment and his top-notch, professional service, nobody could possibly think that they were dealing with a specialist in… mental retardation in children. Perhaps the business owner even had some impact on his clients' family members? Indeed, various degrees of cognitive delay are not that rare. Who would have imagined that the person taking the payment for the just purchased audio system was in fact the first scientist in the world to be prosecuted for gross scientific misconduct?

On September 19th, in Federal Court in Baltimore, psychologist Stephen Breuning – accused of falsifying research results and defrauding funds from federal research grants – testifies on two cases of scientific misconduct. The third accusation of the prosecution is the obstruction of investigations conducted by National Institute of Mental Health in his workplace. The final court session is scheduled for 10th of November. Prosecutors believe that this is the first case ever where a criminal court has been involved in a case of scientific misconduct. Thirty-six year old Breuning had faced the court before on April 15th where he defended himself against charges of forging his own research results – twice. In 1983, NIMH awarded a total of over $200,000 to fund Breuning's research grants. The NIMH Audit Committee concluded that in the research report submitted by Breuning in May 1987, many research results financed by the institute were falsified. The research was supposed to be focused on the use of stimulants in treatment of children with mental retardation. As a result of the audit, the committee [has] prohibited any further funding of Breuning's research for the next 10 years and referred the case further to the government attorney in Baltimore for investigation. According to the prosecution, Breuning was facing 10 years behind bars and a fine of $20,000. Alleviating circumstances could reduce the sentence to 5 years in prison and $250,000 fine. Assistant US Attorney Thomas Roberts said that he will also press for [a] complete employment and research ban in the field of experimental psychology for Breuning for 10 years, including [an] injunction of $20,000 from Breuning's remuneration that he received when working at the University of Pittsburgh, where he conducted most of [the] alleged research. The university returned to NIMH over $163,000 of federal funds "used" by Breuning.[1]

Stephen Breuning made a rapid career rise in his field. It was mostly based on research that showed promising results: using stimulating drugs to modify self-injuring behavior in children with mental retardation. At the time of his trial, he was already a well-known, respected scientist in his field. His fraud was uncovered by psychologist Robert L. Sprague from the University of Illinois. Many years before the fraud was brought to daylight, Sprague, a former colleague of Breuning, had written a multi-page letter with his suspicions related to Breuning's work. It took five years from the moment Sprague reported Breuning's misconduct in research until the fraudster finally faced justice. Sprague wrote about the difficulty of the entire investigation process in a bitter article describing his unmasking operation. In his opinion, regarding the myth that science is beautifully self-correcting, he says, "The science mechanism either refuses to adjudicate fraud or is very, very sluggish. I've been at Illinois for 25 years. I'm a full professor, a research professor at a major university, and I almost didn't get the job done."[2]

The fact that the case involved a renowned and famous psychologist caused significant turmoil, drew a lot of attention and eventually resulted in public condemnation of Breuning's actions. The US Attorney Breckenridge L. Wilcox stated that the court's decision in this case must issue a clear warning to all scientists that research fraud and misconduct will not be tolerated. Sprague officially announced that he expected a heavy sentence and that he could not imagine a probation for Breuning, as there should be no empathy for those who think of "cutting corners in science."

The severity of his criticism was additionally emphasized by the fact that Breuning's research had a significant impact on mentally retarded children – an extremely vulnerable, defenseless group of patients. Furthermore, some of the people engaged in this case pointed out that Breuning destroyed the careers of at least few of his assistants. They were engaged in projects that would never allow them to progress their careers, and when the case was over, they were left without any significant scientific achievements. Similarly, as in Cyril Burt's case, an unanswered question remains: Why did he do it? Robert Sprague assumed that Breuning was guided by "ordinary human lust for power, prestige, fame and money. For the same reason, people embezzle money from banks, lie as witnesses in courts or cheat on the stock market."[3]

Stephen E. Breuning was sentenced to 60 days in a halfway house and five years of probation period. The court also ordered him to pay back $11,352, serve 250 hours of community service and abstain from psychological research for at least the period of his probation. Breuning agreed not to undertake any work as a psychologist for at least ten years, just as the prosecution wanted. He became the world's first scientist accused and sentenced for scientific misconduct and for falsifying research data. The scientific community received this verdict as being far too soft. Some even argued that it would encourage similar transgressions in the future. After serving his time, Breuning opened the aforementioned electronics store near his home in Rochester. Today he runs a private counseling practice, or at least this is what information found on the internet suggests.[4]

It would be difficult to prove the thesis that mild sentencing in Breuning's case encouraged other attempts of scientific fraud. However, the following examples show that it clearly did not serve as a warning to other scientists as Attorney Wilcox wanted:

> A psychology professor recruited by the University of Texas in Austin last year has resigned amid charges that she falsified research data when she worked for her former employer – Harvard University. Karen M. Ruggiero also asked two journals, which published her research on discrimination, to withdraw her previously published articles. Ms. Ruggiero was recruited to Austin in September with a $100,000 start-up package as an assistant professor primarily for her work on the attitudes of women (and other groups) towards discrimination. She resigned in June, after Texas officials were notified that Harvard University was investigating fraud charges against her.[5]

Ruggiero's research results were questioned during attempts to replicate her studies. A number of institutions were involved in the investigation, including the US Public Health Service, Office of Research Integrity, Harvard University and the National Institute of Mental Health. The investigators concluded that Ruggiero fabricated data in three experiments, including data allegedly obtained from 240 subjects participating in her trials. The fake results were published in one of the world's most prestigious and recognized journals, *Journal of Personality and Social Psychology*.[6] She later used the same faked results in her application for a research grant submitted to National Institute of Mental Health in October 1999. Ruggiero admitted to fabricating her results. However this was not the end of her sins. In a similar fashion, she simply made-up all the results of a trial conducted on 360 "participants" that supposedly contributed to two experiments published in her next article. The "results" were later included in the application form for another research grant. She also confessed to misconduct in this case. She later used fake research results three more times in her efforts to obtain financing for her work.

Her penalty was not severe. Ruggiero agreed that for five years she would not sign any contract with any institution funded by the US government, neither as a contractor nor as a subcontractor. She also agreed not to participate in any research financed by the Public Health Service. The period mentioned in her sentence ended on November 26, 2006. However, already in February 2005 she appeared on the *15th Anniversary Conference On State Mental Health Agency Services Research* under the aegis of the Texas Department of State Health Services. Early in 2006, she had already worked as an Editor-in-

Chief of *Behavioral Health* journal, published by the same institution. The journals where Ruggiero published her fake results issued appropriate notifications (errata) informing of the retraction of her articles. Unfortunately, full-text versions of her fraudulent work are still publically available, not only in printed versions of the journals, but also in electronic databases. It seems that her fraud did not prevent her from continuing her career in the field. Another scientist accused of fraud was slightly less fortunate:

> Christopher Gillberg, the psychiatrist at the center of a dispute about granting access to research data in Sweden, has been given a conditional sentence and fined 37,500 kronor (£2,800; $4,800; €4,000) plus legal costs of 80,000 kronor for "misuse of office" by the criminal court in Gothenburg. The conditional sentence obliged him not to commit any similar offences for a period of two years. The court ruled that Gillberg, [a] professor of child and adolescent psychiatry at Gothenburg University and world-class expert in attention deficit hyperactivity disorder (ADHD), had acted "willfully" when he refused to follow his employer's directives to make his research data available to the university management.
>
> Two other university employees, vice chancellor Gunnar Svedberg and board chairman Arne Wittlöf, were charged together with Professor Gillberg for "misuse of office" over their alleged role in withholding access to the research data. The case against Arne Wittlöf was dismissed. He successfully argued that it was the vice chancellor's responsibility to hand over requested documents, not the board's. Professor Svedberg, by contrast, was found guilty, fined 32,000 kronor and ordered to pay legal costs of 32,500 kronor.[7]

Swedish psychiatrist and scientist Christopher Gillberg is a world-renowned researcher on autism in children and a leading, world-class expert on Asperger's syndrome. He was the Editor-in-Chief of *European Child & Adolescent Psychiatry* journal. He is mentioned as an author or editor in numerous scientific publications and books. He has received multiple scientific awards, including Philips Nordic Prize in 2004. He has built his international reputation largely on discoveries in the area of the genetic aspects of autism. The court declared Gillberg guilty, mostly because of certain contradictory laws in Swedish legal system. On one hand, all the data and information collected and acquired with public funds must be made publically available to all citizens. On the other hand, the medical data included in the study could not be made completely anonymous. Researchers collecting the data also assured the patients and their families about the confidential nature of gathered information. Gillberg's decision not to disclose sensitive medical data of his patients was supported by many of his colleagues and by the families of people involved in his studies. The court's judgment raised a nationwide debate and consequently led to changes in the laws regulating the disclosure of data, particularly those of a medical nature. We need to ask however, if such ethical issues can serve as sufficient explanation for ordering the destruction of research material gathered for at least 12 (and in some cases even 27) years.

A neuroscientist from Denmark, Milena Penkowa, is the hero of another particularly adventurous story. She made a brilliant and rapid academic career, starting with a doctorate in 2006 and becoming a full professor in 2009 the same year Crown Princess Mary awarded her with the prestigious Elite Researcher prize. Apart from the honors, the prize included 1.1 million Danish Krone in cash. A private foundation, the IMK General Fund, awarded her with 5.6 million Danish Krone for research. At the same time, Penkowa earned a visible public profile. She frequently appeared on radio, on television and in popular magazines. Unfortunately, at the peak of her career in 2010, problems started to appear:

> A high-profile neuroscientist in Denmark has resigned after facing allegations that she committed research misconduct and misspent grant money. Meanwhile, the administration of the university where she worked has been accused of ignoring her alleged misdeeds for the better part of a decade. Milena Penkowa, a 37-year-old researcher who was lauded in 2009 by the Danish science ministry, denies all the accusations against her and stands by her work, but left her post as a full professor at the University of Copenhagen in December.[8]

Questions over Penkowa's research began at least as early as in 2002 with some suspicions over the first version of her doctoral thesis. Since that time, she has published nearly 100 articles focused mainly on brain-repair mechanisms. How many of them contained false data? Well, till now, four of her articles have been withdrawn, and two other expressions of concern were published, but the administration of the University of Copenhagen suspects that she embezzled grant funds and possibly committed other forms of scientific misconduct in at least 15 papers.[9] The case is even more complicated. The university had to return two million Danish Krone (about $380,000) to the IMK General Fund, which is convinced that Penkowa used the money to pay lawyers, to buy designer clothes, to eat at expensive restaurants and to fund her international journeys. In 2010 she was sentenced for three months of suspended jail-time for embezzlement, and in 2012 the Danish Committees on Scientific Dishonesty found her guilty of gross scientific misconduct.

Moreover, two different researchers filed complaints against four of her co-authors. One of them was Bente Klarlund Pedersen, who has been found by the Danish Committees on Scientific Dishonesty to act in a "scientifically dishonest" and "grossly negligent" manner.[10] What is grotesque is that Pedersen immediately reported James Timmons (one of the two complaining researchers) to the Danish police for harassment; the police neither took the complaint seriously nor investigated it any further.

In 2013, Police dropped charges against Penkowa for document forgery and fraud. At one point she reported on her web site that, "After a massive media frenzy over the last several years, I am happy to finally have my name cleared and look forward to putting the matter behind me. I am now considering seeking compensation for damages, as the case proceeding against me—lasting for nearly 2 1/2 years—has had a great impact on both my personal and professional life."[11]

But this is not the end of her story. Milena Pankova reappeared in the news, this time speaking at an exhibition organized by a Scientology-founded group. She opened the exhibition entitled *Psychiatry, Industry of Death* with a keynote speech. She was presented in the Church of Scientology's press release as "physician and professor of neurology" although she was actually an ex-professor.[12] The exhibition took place at the museum of the Commission on Human Rights, which, according to the press release: "...was established in 1969 in the United States by the Church of Scientology and psychiatry professor Thomas Szasz, the world's most famous psychiatry critics, and in 1972 in Sweden, to investigate and expose abuses of human rights in mental health care and to clean up the field of mental healing."[13] Well, this is a quite different playing field than the scientific one, but she is still in the game.

It would seem that such things could only happen at universities that are very low on the list of highest-ranking universities in the world, but nothing could be further from the truth. Do you remember the case of Karen M. Ruggiero? She was employed by Harvard University when she fabricated data. Again, in August 2010 in an e-mail sent to Harvard University faculty members, Michael Smith, dean of the Faculty of Arts and Sciences, had to confirm that cognitive scientist Marc Hauser was guilty of scientific misconduct.

> No dean wants to see a member of the faculty found responsible for scientific misconduct, for such misconduct strikes at the core of our academic values. Thus, it is with great sadness that I confirm that Professor Marc Hauser was found solely responsible, after a thorough investigation by a faculty investigating committee, of eight instances of scientific misconduct under FAS (Faculty of Arts and Sciences) standards. The investigation was governed by our long-standing policies on professional conduct and shaped by the regulations of federal funding agencies. After careful review of the investigating committee's confidential report and opportunities for Professor Hauser to respond, I accepted the committee's findings and immediately moved to fulfill our obligations to the funding agencies and scientific community and to impose appropriate sanctions.[14]

Houser's case is not clear. He has admitted to making "mistakes" in his research that led to the findings of research misconduct, but never admitted to fraud. "'I let important details get away from my control, and as head of the lab, I take responsibility for all errors made within the lab, whether or not I was directly involved,' says Hauser in a statement sent to Nature by e-mail. The ORI says that he neither admits nor denies committing research misconduct."[15]

Hauser's paper, published in the journal *Cognition* was retracted. Harvard refused to release the investigation report, and because the nature of his transgression was unclear, numerous Hauser supporters stepped to his defense. Most of his supporters claimed that sloppy record keeping should not be considered as fraud. The Office of Research Integrity (ORI) findings concluded that Hauser fabricated data published in *Cognition* and falsified results of other (unpublished) experiment on tamarin monkeys. The conclusion lacked a statement on whether his fabrication and falsification were intentional. According to the National Institutes of Health website, the US government spent $223,860 in one year on just 1 of Hauser's 4 grants that were used in part, to fund research affected by the misconduct.

Hauser was banned from teaching at Harvard University. His articles were retracted. Yet he remained very active in his field. He writes his own blog,[16] and in October 2013☐ his new book titled *Evilicious: Cruelty = Desire + Denial* was released. Viking/Penguin originally scheduled the book for publishing, but then withdrew it from their schedule. *Evilicious* is now available at Amazon as a digital download and also as a paperback (print-on-demand) at Createspace. What is quite intriguing is that his book is praised by prominent representatives of scientific world. MIT's Noam Chomsky, Hauser's former mentor, highly praised the book. So did Randy Cohen, who used to write "The Ethicist" for the *New York Times Magazine*.[17] Michael Shermer, founding publisher of *Skeptic* magazine, the Executive Director of the Skeptics Society, and a monthly columnist for *Scientific American* wrote an enthusiastic blurb. A sentence from his praise even appeared on the cover of Hauser's book: "Every Congressman, Senator, and journalist voting or writing on what to do about violence should read this book first." Nicholas Wade, former science editor of the *New York Times*, who covered the Hauser's case and co-authored 1983's *Betrayers of the Truth: Fraud and Deceit in the Halls of Science*, also wrote a blurb: "What Steven Pinker has done for violence, Marc Hauser has achieved with evil – this book brings the light of science to illuminate the heart of darkness."[18]

Which is Marc Hauser—a genius who simply tripped, or a well-connected fraudster? This case does seem to raise a concern that the authority of Harvard University has quite an impact on other members of the world of science. This impact seems so overwhelming that most independent minds engage in justifications of all the actions of its representatives. We might be right or wrong. Time will tell. Meanwhile let's just take a look at another case.

It is difficult to make a fast and spectacular career in the world of science. This truism does not apply to Dutch social psychologist Diederick Stapel. "Wunderkind," "bright, thrusting young star of Dutch social psychology," "charismatic, friendly and incredibly talented," "a leader with great dedication to his students and colleagues." These are just a few of many praises he received. Nobody spared him honors: in 2009 he received the Career Trajectory Award from the Society of Experimental Social Psychology. During his years of power and glory, he was appreciated by other scientists and students. He also made regular tv appearances and was one of the most well-known psychologists in the Netherlands. Too good to be true? Unfortunately—yes—just like his research results published in dozens of scientific articles.

Diederick Stapel's career was ruined in one day in September 2011 when Tilburg University's press release was published. It uncovered Stapel's frauds: "The Executive Board of Tilburg University has suspended Prof. D.A. Stapel from his duties with immediate effect. Dr. Stapel, who is a Professor of Cognitive Social Psychology and Dean of the Tilburg School of Social and Behavioral Sciences, has committed a serious breach of scientific integrity by using fictitious data in his publications."[19] The press release also contained information about a special commission appointed by the rector of Tilburg University to investigate which of Stapel's works were based on fabricated data. The first report was published after a month, on October 31, 2011. The report described how Stapel falsified his data and estimated the scale of his fraud. As Professor Pim Levelt, chair of the committee that investigated Stapel's work said, "We have some 30 papers in peer-reviewed journals, where we are actually sure that they are fake, and there are more to come."[20]

The disclosure of the scale of Stapel's falsifications forced Philip Eijlander, the Rector of Tilburg University, to reveal his will (at a press conference) to pursue criminal prosecution for Stapel. It seemed that Stapel could be a new world record holder in the number of falsified research papers in psychology. The reality, however, exceeded all expectations. *Retraction Watch* reported that the up-to-date number of retracted articles is now 54, plus an additional 10 PhD dissertations written by his students.[21] We still cannot be sure that those are all of his falsified articles. Yes, Diederik Stapel is unquestionably one of the record holders in scientific fraud, beaten perhaps only by Japanese anesthesiologist Yoshitaka Fujii, (who fabricated at least 170 retracted scientific papers)[22] and an anthropologist from Netherlands Mart Bax, (who published at least 60 fake research papers).[23] Certainly he is the number one in psychology. But how is it possible that he could deceive so many for so long?

As we have written earlier, Stapel was known as a charismatic leader, greatly dedicated to his students and colleagues. He used his power and prestige in a very sophisticated way. He was very helpful during the early phases of his students' research; he told them that data would be collected at local secondary schools where he was supposed to have some connections. Coming back with the data, Stapel would then advise his young researcher to "be careful, as you have gold in your hands."[24] He made them feel as if he were doing them a favor. Stapel controlled the whole process of collecting and analyzing data in his lab, and when sometimes students asked to see completed questionnaires or the raw data, they were often given excuses. For example, they were told that neither the schools nor he had enough room to store them, but what helped him most was a simple lack of scientific scrutiny and skepticism.

Stapel produced articles that are sometimes called 'sexy studies' because of their supposed potential to attract media attention. For example, one of his faked studies "showed" that meat eaters are more selfish and less social than vegetarians. This was widely popularized by Dutch media.[25] In another article he "proved" that a disorderly physical environment promotes stereotyping and discrimination.[26] He knew what the media expected and he skillfully produced appropriate results.

Apart from the severe damage he caused to science and to public confidence in the scientific method, the individuals most directly affected were his students (from both masters and doctoral courses). Significant numbers of their dissertations were believed to be based on fabricated data. Similarly, as in the Breuning case, he had caused serious harm to young scientists at the beginning of their careers. How did Stapel end up? Well, he initially admitted his guilt, repented and apologized by pointing out some of his motives:

I failed as a scientist. I adapted research data and fabricated research. Not once, but several times, not for a short period, but over a longer period of time. I realize that I shocked and angered my colleagues, because of my behavior. I put my field, social psychology in a bad light. I am ashamed of it and I deeply regret it. ... I think it is important to emphasize that I never informed my colleagues of my inappropriate behavior. I offer my colleagues, my PhD students, and the complete academic community my sincere apologies. I am aware of the suffering and sorrow that I caused to [sic] them. ... I did not withstand the pressure to score, to publish, the pressure to get better in time. I wanted too much, too fast. In a system where there are few checks and balances, where people work alone, I took the wrong turn. I want to emphasize that the mistakes that I made were not born out of selfish ends.[27]

In a settlement with the prosecution in June 2013, Stapel agreed to perform 120 hours of community service and to lose the right to some benefits associated with his former job (equivalent to 1.5 years of salary). This allowed him to avoid further criminal prosecution. What helped his case greatly was the fact that he voluntarily returned his PhD to the University of Amsterdam, stating in a letter that his conduct "does not fit with the duties associated with a doctorate."[28]

You might think that such a clever avoidance of responsibility would be enough to close the case, or would be enough to end the career of scientific fraudster, but you'd be wrong. As in Pankova's case, our hero resurfaced again – this time in a shameless attempt to use his situation to continue his career, not as a scientist, but as a con man. In 2013 he resurfaced in TEDx Maastricht Brain Train program where he gave a lecture entitled, *What I Lost: The Importance of Being Connected*.[29] But there is more. In 2012, he published his memoirs! No, that is not a mistake; a book called *Ontsporing* (English: derailment) is a detailed description of the events that led to one of the largest scientific frauds to ever be uncovered. We are far from recommending purchasing his book (we believe that financial support for fraudster is, well, immoral). We do however encourage you to read it if otherwise possible. You might find it interesting for at least a couple of reasons. First of all, just like network security officers can learn a lot from computer hackers, everyone who wants to know how to improve the control mechanisms in science should devote a lot of attention to this book. In his description of the academic realities of social psychology, Stapel points out the virtually complete absence of any control structures. "Nobody ever checked my work. They trusted me… I did everything myself, and next to me was a big jar of cookies. No mother, no lock, not even a lid… Every day, I would be working and there would be this big jar of cookies, filled with sweets, within reach, right next to me — with nobody even near. All I had to do was take it."[30]

Another good reason to read (not buy) the book is to learn about the motives of the fraudster, well, at least those motives he reveals. The third reason is the book is an excellent example of how our brains rationalize, self-clean and direct our thoughts in order to protect our good self-image and self-integrity. In the Dutch media, Stapel's motivations for publishing *Ontsporing* were widely questioned and ridiculed. The book, however, is still a fascinating illustration of a narcissistic personality craving for attention or partial rehabilitation.

You might already be tired, but it is still not the end of Stapel's giant fraud. The biggest surprise appears in the last, unexpectedly beautiful, poetic chapter of his book. As many literary critics noticed, it is composed of sentences that Stapel *copied* from the fiction writers Raymond Carver and James Joyce! Furthermore, he presented them without quotation marks and only acknowledged the sources separately in the appendices. Mockery? Another ordinary fraud? A hidden message, 'be careful, deceivers cannot stop deceiving'? Or nothing but a wretched theft? There is another possibility, identified by Sylvain Bernès, who commented one of the notes about Stapel published in *Retraction Watch*: "Most of the RW posts about Stapel's retractions are illustrated with the same photo. I've now seen 21 times this guy [sic] with this mocking smile. Maybe I'm turning paranoid, but I imagine he says: 'yeeeeesss, I'm back! I've soooooo many papers to retract, you will see me for ever and ever!'"[31]

Probably nobody will be able to assess the amount the damage that fraudsters have done to science over the years. It is hard to imagine how it is even possible to clearly communicate the fact that a significant amount of articles published in reputable journals is invalid. The sheer amount of published scientific information makes it even harder. What's worse is that scientific information systems do not gather all retracted papers in one place.

An attempt in this matter is a previously quoted blog, founded by two journalists and doctors, Adam Marcus and Ivan Oransky. It is called *Retraction*

Watch. Their motto is, "Tracking retractions as a window into the scientific process."[32] Although the blog was only started in 2010, it already contains descriptions of over 1000 cases of retracted papers, including 71 papers in psychology. The authors of the blog claim that it has been a struggle for them to keep up with retractions as they happen. How large a complete database of retracted, withdrawn and invalidated articles would be? How many more would we have to add if we added all the research papers based on the retracted articles? Are we even able to identify all textbooks and popular science books infested with false conclusions drawn from fraudulent papers?

When seeking answers to those questions, a metaphor of falling domino blocks pops into mind; when they fall, they knock tens or hundreds of other tiles, even those put up many years before. Perhaps a more appropriate metaphor would be the Butterfly Effect, derived from the chaos theory, where the flap of a butterfly's wings in Brazil sets off a tornado in Texas?

In 2001, Margaret Gibelman and Sheldon R. Gelman published an article entitled "Learning from the Mistakes of Others: A Look at Scientific Misconduct in Research"[33] which contains a list of frauds in psychology, psychiatry and nursing. We can add to their list many more examples of scientific misconduct, especially from the field of psychology and psychiatry. Before the scientific world even stopped discussing Diederik Stapel's case, in 2012 another scandal erupted in the psychology community. Professor Dirk Smeesters, a Belgian-born social psychologist who specialized in consumer behavior, resigned from his post at the Rotterdam School of Management, part of Erasmus University, when serious questions were raised regarding his work. Shortly after his resignation and the retraction of four articles, Lawrence Sanna mysteriously resigned from his professorship at University of Michigan in Ann Arbor. Odd statistical patterns in Smeesters's and Sanna's work were found.[34] It is hard to tell how many more cases will emerge after this book goes to press. Taking into account the scale of presented misconduct, we should ask ourselves another question: who learned more? Those who should guard and protect the system of scientific control, or the fraudsters and scammers who abused the system's vulnerability? Did we learn anything from Cyril Burt's case? Or perhaps it is only fraudsters, like Diederik Stapel and others whose names we don't yet know, who completed their homework?

---

[1] J. Bales, "Breuning Pleads Guilty in Scientific Fraud Case," *Science 242* (1988): 27-28.

[2] R. L. Sprague, "Whistleblowing: A Very Unpleasant Avocation," *Ethics & Behavior, 3* (1993): 103-133.

[3] Bales, "Breuning Pleads Guilty."

[4] Healthgrades.com, (n.d.): http://www.healthgrades.com/provider/stephen-breuning-x8kkv#tab=background-check&scrollTo=BackgroundBackgroundCheck_anchor

[5] J. Jacobson, "A Psychology Professor Resigns amid Accusations of Research Fraud at Harvard," *Chronicle of Higher Education 47* (2001): A10.

[6] K. M. and D. M. Marx, "Less Pain and More to Gain: Why High-Status Group Members Blame their Failure on Discrimination," *Journal of Personality and Social Psychology 77* (1999): 774-784.

[7] C. White. "Swedish Court Rules Against Doctor at Centre of Row Over Destroyed Research Data," *BMJ, 331,* 180.7. (2005): from http://www.bmj.com//content/331/7510/180.7

[8] E. Callaway, "Fraud Investigation Rocks Danish University," *Nature* ((January 2011): http://www.nature.com/news/2011/110107/full/news.2011.703.html

[9] M. Young, "University: Penkova Could Lose PhD, Doctorate," *University Post.* (August 2013): http://universitypost.dk/article/university-penkowa-could-lose-phd-doctorate

[10] Danish Committees on Scientific Dishonesty, "Draft Ruling for Consultation," (July 2013): http://retractionwatch.files.wordpress.com/2013/07/timmons-uvvu.pdf

[11] V. Yakimov, "Media Reports: Case Against Penkova Dropped," *University Post* (May 2013): http://universitypost.dk/article/media-reports-case-against-penkowa-dropped

[12] Pressmadelande, *"Professor i neurologi talade vid invigning av KMR-utställning.* Kommitten for Manskliga Rattigheter," (October 2013): http://www.mynewsdesk.com/se/kommitten_for_manskliga_rattigheter/pressreleases/professor-i-neurologi-talade-vid-invigning-av-kmr-utstaellning-916444?utm_source=rss&utm_medium=rss&utm_campaign=Alert&utm_content=pressrelease

[13] Ibid.

[14] G. Miller, "Harvard Dean Confirms Misconduct in Hauser investigation," *Science* (August 2010): http://news.sciencemag.org/2010/08/harvard-dean-confirms-misconduct-hauser-investigation

[15] E. S. Reich, "Marc Hauser Admits to Errors as US Government Finds Misconduct," *Nature* (September 2012): http://blogs.nature.com/news/2012/09/marc-hauser-admits-to-errors-as-us-government-finds-misconduct.html

[16] See: http://mdhauser.blog.com/

[17] I. Oransky, "Marc Hauser's Second Chance: Leading Science Writers Endorse His Upcoming Book," *Retraction Watch* (September 2013): http://retractionwatch.com/2013/09/26/marc-hausers-second-chance-leading-science-writers-endorse-his-upcoming-book/#more-15800

[18] Ibid.

[19] Press Release, "Prof. Diederik Stapel Suspended," Tilburg University (September 2011): http://uvtapp.uvt.nl/fsw/spits.npc.ShowPressReleaseCM?v_id=4082238588785510

[20] R. T. Gonzalez, "Psychologist Admits to Faking Dozens of Scientific Studies," *io9* (February 2011): http://io9.com/5855733/psychologist-admits-to-faking-dozens-of-scientific-studies

[21] I. Oransky, "Measure by Measure: Diederik Stapel Count Rises Again, to 54," *Retraction Watch* (August 2013): http://retractionwatch.wordpress.com/2013/08/02/measure-by-measure-diederik-stapel-count-rises-again-to-54/#more-15141

[22] D. Normile, "A New Record for Retractions?" *Science Insider* (July 2012), American Association for the Advancement of Science.

[23] "Professor Faked 61 Pieces of Research: Volkskrant," *DutchNews.nl.* (September 2013):
http://www.dutchnews.nl/news/archives/2013/09/professor_faked_61_pieces_of_
r.php#sthash.nmcumyZM.dpuf

[24] Levelt Committee, *Interim Report Regarding the Breach of Scientific Integrity by Prof. D. A. Stapel,* Tilburg University, (October 2011): http://en.wikipedia.org/wiki/Died
erik_Stapel

[25] Stapel et al., "Meat Eaters are Selfish and Less Social," *Dutch Daily News* (August 2011): http://www.dutchdailynews.com/meat-eaters-selfish-less-social/

[26] D. Stapel and S. Lindenberg, "Coping with Chaos: How Disordered Contexts Promote Stereotyping and Discrimination," *Science, 332* (2011): 251–253.

[27] "Stapel betuigt openlijk 'diepe spijt'," (2011, October 31). *Brabants Dagblad.* Retrieved December 1, 2013, from: http://en.wikipedia.org/wiki/Diederik_Stapel
#cite_note-26

[28] M. Verfaellie and J. McGwin, "The Case of Diederik Stapel," American Psychological Association (December 2011): https://www.apa.org/science/about/
psa/2011/12/diederik-stapel.aspx

[29] TEDx Maastricht, (n.d.): http://tedxmaastricht.nl/braintrain/speakers2013/

[30] D. Stapel, *Ontsporing.* (Uitgeverij; Prometheus, 2013) 164.

[31] Oransky, *Measure by Measure.*

[32] I. Oransky, "The Retraction Watch Transparency Index," *Retraction Watch* (August 2012): http://retractionwatch.wordpress.com/transparencyindex/

[33] M. Gibelman and S. R. Gelman, "Learning from the Mistakes of Others: A Look at Scientific Misconduct in Research," *Journal of Social Work Education, 37* (2001): 241-254.

[34] E. Yong, "Uncertainty Shrouds Psychologist's Resignation," *Nature* (July 2012): http://www.nature.com/news/uncertainty-shrouds-psychologist-s-resignation-
1.10968

THE HIDDEN PART OF THE ICEBERG: THE PROBLEM OF RAW DATA AND THE SCOPE OF SCIENTIFIC MISCONDUCT

Despite the extensive scale of fraud documented in the previous chapters, they seem to represent only the tip of an iceberg. It immediately begs the question: how many frauds and abuses remain undetected? What about honest scientific mistakes, negligence and omissions? We can get some insight by taking a closer look at an event that began very innocently...

It will soon be nearly half a century since Leoy Wolins, a psychologist from Iowa State University, authorized one of his students to write a letter to 37 authors of original research articles, asking them to submit the raw data that their studies were based on. His student intended to make practical use of the data in a study of their own. 32 out of 37 authors replied to the request. However, 21 of those 32 researchers who replied informed "with tremendous regret" that their data was accidentally destroyed, lost or archived in such a way that it was impossible to retrieve them. Only nine researchers, (i.e. 24% of the initial group) appeared to be willing to make their data available. Doctor Wolins, (an expert in statistics), took a closer look at the data received and concluded that only seven of them met the requirement of something that we might call a reliable statistical analysis. Three datasets contained critical mistakes that completely undermined all the conclusions drawn from them.[1] You should note that Wolins' discovery was not a result of a planned investigation; it was a side-effect of another project. His student was really trying to get access to the requested data as it was necessary for his own research. It sounds scary, doesn't it? What is the actual extent of this problem? Is the data really missing? Or maybe the evidence was never there? If one-third of voluntarily released data sets contained critical mistakes, what shall we expect from those that never reach daylight? Can those results be extrapolated on all research conducted in the field of psychology? If the answer is yes, does it mean that most of the researchers have something to hide?

Wolins, surprised by the scientists' responses, described the experience in the American Psychologist, thus triggering a heated scientific debate over the availability of raw data. This debate has been dragging on for nearly half a century now. Eleven years later in 1973, three scientists from Western Kentucky University, James R. Craig and Sandra C. Reese, decided to replicate Wolins' inadvertent research. Their study was thoroughly planned and was designed to check whether there had been any improvement in the availability of raw data since Wolins' attempt nearly a decade earlier. They each chose one of the most influential psychological journals. Their choice assumed different levels of scientific rigor. The list below was arranged by the authors in order from the toughest to the weakest in terms of methodological rigor and requirements:

1. *Journal of Comparative and Physiological Psychology*
2. *Journal of Verbal Learning and Verbal Behaviour*
3. *Journal of Personality and Social Psychology*
4. *Journal of Educational Psychology*

Craig and Reese asked 53 researchers to provide them with raw data that they used in their research. The requests were made for studies published in the last month only. The results were more optimistic that those presented by Wolins. Nine researchers refused to reveal the data, claiming it was lost, destroyed or otherwise unavailable. One author even said, "I follow my own rule to never disclose my research data." Eight did not bother to reply at all. Some even asked for money. In one case, instead of requested data, the author received a job offer. Only about a half of researchers declared a will to cooperate. Twenty researchers sent analyzed or summarized data. Seven declared a will to cooperate, but under certain conditions. Craig and Reese concluded their data ironically, but optimistically: "In summary, it appears that psychologists are currently striving to improve the quality of life for mankind by saving and sharing potentially important research data and by creating mounds of recyclable data."[2]

Warren Eaton, a scientist from the University of Manitoba, obtained even more optimistic results in 1984. The success rate of 73.5% (25 of 34) was much better than Wolins' and Craig's & Reese's.[3] However, in 2006, a group of Dutch psychologists from Amsterdam obtained exactly opposite results, virtually identical to those of Wolins from 1962. In an attempt to reanalyze data sets to assess the robustness of research findings, they contacted 141 authors of 249 studies. They received 38 positive answers and managed to obtain actual data sets from 64 studies. This figure represents 25.7% of the total number of 249 data sets; 73% of the authors did not share their data.[4]

Inspired by examples of Wolins and other researchers, we decided to replicate their studies on our own.[64] In order to replicate the research, we randomly selected 50 empirical studies out of the PsychINFO database, all of which had been published in the last 12 months. We contacted the authors, requesting them to share the raw data from their studies. We explained that we were conducting similar research and that our intention was to compare the outcomes. The random sample included articles written by American and European authors, as well as researchers from Israel and Australia. They represented a vast diversity of psychological fields of study and a variety of scientific journals in which their articles had been published. The only inclusion as criterion was that a particular work had to be an empirical study based on quantitative data. We excluded case studies, literature reviews and meta-analyses. Here is what we found out.

Out of fifty requests, we have received 27 replies, which accounted for 54% of the entire sample. We have only received seven raw data sets (sorted and described in a manner consistent with the respective articles). A further seven responses could be labeled as "willing to cooperate." Some of these researchers asked for more information about the variables we wished to reanalyze, and others pointed out that the results were written in, for instance, Hebrew or Flemish and that it would take quite some time for the results to be translated into English. Others inquired as to the nature of our alleged research, our job position, etc. We found such replies to be a valid response to our request for information, as these authors clearly had the right to inquire about the reason and nature of our study. We assumed that those researchers would be quite willing to share their data upon providing the requested information in detail. Therefore, our success rate was 30% of the total sample. It is remarkably close to results presented by Wolins in 1962, and subsequently of Wicherts et al. in 2006. The remaining four authors indicated that some third party was the actual data owner or holder. We have resent our requests to these "third parties" with the following outcome: the total number of data sets that we have received rose to a total of 9; those categorized as "willing to cooperate" increased by 1.

There were only four explicitly negative responses. Two authors explained that the data was still being analyzed. We felt that the odds of receiving the requested data any time in the near future was very low. The third response invoked prohibitive regulations of the institution acting as a sponsor—virtually preventing disclosure of any data. The fourth negative response cited ethical principles that apparently prohibit this type of information sharing.

We labeled five other responses as "ambiguous or evasive." These authors insisted on the necessity of contacting fellow researchers involved and obtaining their permissions, which they suggested would be extremely difficult or impossible. Among those, we received replies claiming that the format of the data sets excluded any further analysis. We have no clue what that means.

The table below shows the steps taken in the course of our study and their respective outcomes.

Table 1

Action	n	%
Requests for data sent	50	100
Responses received	27	54
Unanswered requests	23	46
Data sets received	7	14
Willing to cooperate on certain conditions	8	16
Responses requiring further contact	4	8
Ambiguous or evasive responses	5	10
Negative responses	3	6
Data received as a result of further contact	2	4
Willing to cooperate following further contact	1	2
Negative reply following further contact	1	2
Total data sets received	9	18
Total positive responses	18	36
Total unanswered, ambiguous or negative responses	32	64

To complete this procedure, we disclosed our true intention by sending debriefing information to all the authors contacted. We also explained that there was absolutely no need for them to process the data in preparation for reanalysis. Furthermore, we assured them of our willingness to share our results after the study is completed. Most of the respondents were genuinely interested in the results. One author was quite surprised: "Wow! Who could have thought? I bet, most people do not check your publications after replying to your requests…"

In addition to the findings reported by Wollins in 1962, Craig & Reese in 1973, Eaton in 1984, and Wicherts and his coworkers in 2006, our own observations indicate a mood of cautious optimism. On one hand, of all the previous attempts, ours had the lowest general number of responses, merely half of those contacted. On the other hand, the study came out ahead of all the previous attempts in terms of the percentage of those willing to cooperate: more than half of those who replied agreed to provide the requested data. We were impressed by how quickly some of the authors we had contacted offered their data; the first data sets arrived within days after sending our request. We were also impressed by their eagerness in offering practically unconditional assistance with our research.

Today's mailboxes are filled with hundreds of messages; some of the contacted scientists may have simply sent my request straight to trash simply in an attempt to avoid unnecessary communication. It is also possible that they may have regarded our request as a spam, despite all our efforts for it not to be perceived as such. Lastly, we believe that scientists might be more inclined to respond to a request from researchers from their own country, rather than from a little known scientists in Eastern Europe.

After we collected and summarized all data for this study, we wrote a commentary article and submitted it to *American Psychologist*, where a discussion on this particular topic has been ongoing for over half of a century. To our disbelief, we were told that this particular subject is actually now completely explored and further discussions from now on will be closed: "As you know, every member of APA receives AP. The limited space we have available constrains us, and we can accept only articles that are of interest to a broad range of psychologists and that have broad consequences for the science and practice of psychology. ... the AP has recently published a comment on this topic, and the present proposed comment does not really add any new additional information to the discussion."[5] Is it possible that for thousands of psychologists who receive the *American Psychologist*, the problem of accessing raw data has been finally solved? Or perhaps this primary source of possible distortion of knowledge about human beings is of no interest to the editors?

Problems related to accessing raw data are most likely the cause of at least some voluntary retractions. Analysis of such cases can shed light on this problem. This is an example where the author of a paper retracted it after claiming to be unable to repeat his own analysis, and therefore being unable to fix an issue he noticed during his own study. This is what he said to editors of the *Retraction Watch* blog, "The main problem is that we used the mean estimate of coin sizes (the measure of desire for money) to conduct analysis. However, participants were asked to estimate four different coin sizes. Unfortunately, my research assistant (who left my lab last year), did not keep the original questionnaires. Hence, we cannot have correct data to rerun the data.[6]

Such explanations hardly dispel any doubts, and we cannot simply abstain from asking further questions. Why does only one of the corresponding authors want to retract the paper? Were the authors asked for repeated analysis on their data and then filed for a retraction of the article because they feared being exposed? Or perhaps they did this on their own guided by their scientific integrity and honesty? This is not an isolated case.

A group of researchers at the University of Leuven in Belgium decided to retract a paper after realizing there were some serious problems with one aspect of their work:

The authors retract this publication. Because of human errors by the first/corresponding author, the fMRI data reported in this retracted paper were not analyzed properly. The errors were detected when other lab members reanalyzed the data for another purpose. At that point, it turned out that the original data analyses by the first author included several operations, which are hard to replicate, and which do not fit fully with the methods as agreed upon with the co-authors and as described in the paper. Because of this, we no longer consider these results trustworthy.[7]

Once again questions arise. Why didn't the other authors notice the problem before? Did one member of the team trick the others? Or was it simply, as they describe, human error? Such questions always pop into our minds in similar cases. They will probably remain unanswered forever. The fact that such questions can be raised is definitely not due to adequate transparency of gathering, storing and sharing of raw data.

Yet still, in the Wollins' sample, among the respondents contacted by Craig and Reese, as well as in mentioned research by the author, there were those who deliberately chose not to disclose their data. How many felt the data would not stand up to scientific scrutiny due to internal flaws? This question will probably remain unanswered, despite the attempts undertaken in the past. In 1976 *New Scientist* asked their readers if they knew about scientific fraud or if they suspected someone of intentional manipulation.[8] The journal received 204 returned questionnaires, and among those who answered, 92% admitted that they know directly or indirectly about a scientific fraud.[9] Is it possible that people who decided to send their answers were motivated to do so because they witnessed unethical behavior? Perhaps, but it might be also worth taking a look at another study conducted with more attention to scientific methodology on scientists at various stages of their academic career in the USA. The results are shocking. About 33% among thousands of anonymous respondents admitted to personally breaching at least one out of ten listed examples of scientific misconduct.[10]

The *Project on Professional Values and Ethical Issues in Graduate Education of Scientists and Engineers* surveyed 2,000 doctoral candidates and 2,000 professors who were awarded federal grants. Among them, 6% of doctoral students and 9% of faculty members reported that "they have direct knowledge of faculty members involved in data falsification or plagiarizing."[11] Additionally, 7.4% of principal investigators funded by the American national agency, the National Institutes of Health, also reported that they witnessed at least one case of scientific misconduct during last three years. Moreover, they reported that 36% of misconducts they observed were never reported.[12]

The data obtained in this way is still far from being methodologically unquestionable, but they paint a terrifying picture. When searching for higher quality research, we came across on what appears to be the only published meta-analysis, conducted by Daniele Fanelli. The author analyzed 69 research papers devoted purely to investigating the honesty of other scientists. He included 21 of those articles for further systematic review. Inclusion criteria for meta-analysis were met for 18 papers. The final results complete the picture we have seen sketched by previously analyzed works. Two percent of all examined scientists voluntarily admitted to committing the most serious of all crimes against science, the fabrication of data. Stunningly, 14% admitted that they have personally witnessed such practice among other scientists. Almost 34% plead guilty to other, less serious offences and 72% (!) have said that they had personal evidence of such behavior in their colleagues, suggesting that it is underreported.[13]

Even more credible data comes from a study conducted by Lesli K. John and his team. They questioned 2155 research psychologists who published in *Psychological Science*, a journal of the Association for Psychological Science, using a method known as the Bayesian-truth-serum.[14] It incorporates a scoring algorithm to provide incentives for truth telling. This algorithm uses respondents' answers about their own behavior and their estimates of the sample distribution of answers as inputs in a truth-rewarding scoring formula. Similarly, as in the previously described meta-analysis, 7% of researchers admitted fabricating research results and 66.5% confessed to other serious wrongdoings.[15] When World Health Organization announced a pandemic of swine flu in 2009 (H1N1), the reported CFR (case/fatality ratio) turned out to be 0.03%. What results do we need to sustain in order to announce an epidemic in our discipline? Aren't the results presented sufficient to talk about a pandemic problem?

In our opinion, the results are so alarming that we desperately need to seek solutions that can effectively deal with the uncertainty of published results and speculation about these results. A number of recommendations have already been made. A psychologist from University of North Dakota, Richard W. Johnson, proposed possible solutions in 1964:

> Second, develop a filing system that enables you to readily locate previously collected data. This can be easily done with a little extra effort. All data should be carefully labeled, especially data that have been collected on IBM cards. All IBM cards or raw data collected on one research project should be kept together as a complete set. Third, seek to deposit significant original data with the American Documentation Institute (ADI), a service of the Library of Congress, if the research findings have been published as an article for general circulation. Upon the recommendation of the journal editor, ADI accepts raw-data tables *in extenso*, analyzed data, extensive contingency tables, data summaries, detailed drawings of apparatus, large-sized charts, and photographic illustrations. This method is most helpful in providing easy access to one's data by other investigators.[16]

Although it presented a serious technical problem when Johnson wrote his propositions, technological developments today allows psychologists to almost effortlessly put the mentioned ideas into everyday practice. Wicherts and his colleagues discussed them in a previously mentioned article as well:

> Fortunately, there is a rather simple solution to this problem. That solution lies at the editorial end of the publishing process. It seems to us that in the electronic age, it would be easy to modify the manuscript submission process to avoid this problem. Only two modifications are necessary, namely: (a) that upon the acceptance of a manuscript for publication, authors be required to submit an ASCII file with anonymized data, as well as a codebook in standardized format

and (b) that the journal publish the data and codebook on the World Wide Web as an electronic appendix to the article.[17]

Without such solutions being put into practice, progress in the field of scientific research integration may be seriously hindered, affecting, for example, the replication of major scientific research. If such publications are collected and put into meta-analyses, we get a severely distorted picture of reality backed up with "evidence" of very poor quality. Proposed solutions would, beyond any doubt, contribute to strengthening psychology's most significant weaknesses: the fragmentation of knowledge and the failure to synthesize the research output of generations of scientists.

Fortunately, nothing less than some 65,000 people, including researchers, librarians and advocates of information sharing understand this problem. In May 2012, they signed a petition urging the Obama administration to adopt open access policies that would make the results of taxpayer-funded scientific research freely available to the public. In response, the White House issued a memorandum in February to almost two dozen federal funding agencies, instructing them to create individual plans for ensuring that research papers will be available within 12 months of publication. It also required agencies to make the data in those papers "stored and publicly accessible to search, retrieve, and analyze."[18]

Since the original Polish version of this book was published, new initiatives that might change the attitude of researchers towards the disclosure of data appeared. In 2011, the APA Publications and Communications Board approved the recommendations of a data-sharing task force. It called for journals to reveal all the data on which published research reports are based. One such journal is APA's *Archives of Scientific Psychology*, which debuted in 2013. It is the first open-access, open-methods, open-data journal. The *Archives* requires authors to submit their full data sets and make them available for any researcher who requests them.[19] Still, many scientists resist these requests to make data available. "It is threatening to scientists to think that their data will be that available," says Heather Piwowar, who describes herself as a "scientist who studies scientists." She is a co-founder of ImpactStory, a non-profit organization that attempts to ameliorate some of the issues raised by open data. "Multiple studies show that scientists are not sharing their data when their peers request it, or even when it is required by a journal.[20]

---

[1] L. Wolins, "Responsibility for Row Data," *American Psychologist 17*, (1962): 657.

[2] J. R. Craig and S. C. Reese, "Retention of Raw Data: A Problem Revisited," *American Psychologist, 28*, (1973): 28.

[3] W. O. Eaton, "On Obtaining Unpublished Data for Research Integrations," *American Psychologist, 39*, (1984): 1325–1326.

[4] J. M. Wicherts, D. Borsboom, J. Kats and D. Molenaar, "The Poor Availability of Psychological Research Data for Reanalysis," *American Psychologist, 61*, (2006): 726–728.

[5] Private correspondence from the editor of the *American Psychologist* to Tomasz Witkowski.

[6] I. Oransky, "A Real Shame: Psychology Paper Retracted when Data Behind Problematic Findings Disappear," *Retraction Watch* (August 2013): http://retraction watch.com/2013/08/15/a-real-shame-psychology-paper-retracted-when-data-behind-problematic-findings-disappear/

[7] I. Oransky, "Doing the Right Thing: Psychology Researchers Retract after Realizing Data were Not Analyzed Properly," *Retraction Watch* (May 2013): http://retraction watch.com/2013/05/22/doing-the-right-thing-psychology-researchers-retract-after-realizing-data-were-not-analyzed-properly/#more-14270

[8] I. James-Roberts, "Are Researchers Trustworthy? *New Scientist, 71,* (1976): 481-483.

[9] I. James-Roberts, "Cheating in science," *New Scientist, 72,* (1976): 466-469.

[10] B. C. Martinson, M. S. Anderson and R. de Vries "Scientists Behaving Badly," *Nature,* (June 2005): 435.

[11] E. Altman, "Scientific and Research Misconduct," in *Research Misconduct: Issues, Implications, and Strategies,* eds. E. Altman and P. Hernon, (Greenwich, CT: Ablex, 1997), 1-32.

[12] J. A. Wells, *Final Report: Observing and Reporting Suspected Misconduct in Biomedical Research,* (Rockville, MD: The Office of Research Integrity, 2008).

[13] W. Fanelli, "How Many Scientists Fabricate and Falsify Research? A Systematic Review and Meta-Analysis of Survey Data. *PLoS One, 4,* (2009), 1-11.

[14] D. Prelec, "A Bayesian Truth Serum for Subjective Data," *Nature, 470,* (2004), 437.

[15] L. K. John, G. Loewestein, and D. Prelec, "Measuring the Prevalence of Questionable Research Practices with Incentives for Truth Telling," *Psychological Science, 25,* (2011), 524-532.

[16] R. W. Johnson, "Retain the Original Data," *American Psychologist, 19,* (1964), 350–351.

[17] Wicherts et al., "The Poor Availability."

[18] V. Schlesinger, "Scientists Threatened by Demands to Share Data," *Aljazeera America,* (October 2013): http://america.aljazeera.com/articles/2013/10/10/scientists-threatenedbydemandstosharedata.html

[19] American Psychological Association, "*APA Launches 'Archives of Scientific Psychology,' its First Open Methodology, Open Data, Open Access Journal,*" (August, 2012): http://www.apa.org/pubs/newsletters/access/2012/08-21/first-journal.aspx

[20] Schlesinger, "Scientists Threatened."

CHAPTER 5

LYING IS THE FIRST STEP TO THE GALLOWS: PLAGIARISM AND OTHER MISCONDUCT IN SCIENTIFIC PUBLICATIONS

The main character of the story *Lucky Jim* written by Kingsley Amis, is a medieval history lecturer at a redbrick university in England, Jim Dixon. The problem is, Jim does not have significant academic credentials, and for that reason he cannot be tenured and might lose his position at the end of his probationary period. In order to establish his credentials, he is advised to publish his first scholarly article. Any article. Any journal. If he does, the befriended head of his department would treat his publication as sufficient proof of academic credibility. Jim tries to publish his article entitled "The Economic Influence of the Developments in Shipbuilding Techniques, 1450–1485."

The author was convinced that his work "was worth the amount of frenzied fact-grubbing and fanatical boredom" and that the title "crystallized the article's niggling mindlessness." Jim, who considers himself a hypocrite, describes his creative work as a "funereal parade of yawn-enforcing facts" shining "the pseudo-light ... upon non-problems."[1] His manuscript gets repeatedly rejected, and he finally tries his luck in a newly established and advertised journal. As weeks go by without any response from the journal, Dixon becomes increasingly worried. In the meanwhile, he hears very disturbing gossip and rumors about the journal's editor-in-chief. He finally calls the editor and hears a plethora of typical (for a scientific journal) excuses, but he finally manages to get a vague promise of publication. Dixon's anxieties turn into helpless rage when he discovers an article on shipbuilding techniques in an Italian journal. Even without any knowledge of Italian, he is able to recognize an exact copy of his work published under the name of... the editor of the newly established and advertised journal!

Unfortunately, this brilliant parody of the world of science, as presented by Amis, is not that far removed from reality. We are convinced that the real publishing process can get even more grotesque. The case of Paraskevi Theofilou, a psychologist and editor of *Health Psychology Research* can serve as an excellent illustration. Theofilou published seven papers which have been retracted for either plagiarism or duplication.[2] It did not seem to discourage him however, from holding his position as editor-in-chief of the abovementioned journal.[82] But before we share more about the abuses related to the publishing process, let us take a more systemic look at the crimes, frauds, misconducts, and neglectfulness perpetrated by scientists.

Previously described frauds, such as forging and fabricating data, are the heaviest of all scientific crimes. They contradict the essence of what science is all about: the search of the truth. Scientific fraudsters, scammers and crooks are nothing more than traitors to the ideas that science stands for. In addition to most serious frauds, such as those described in the previous chapters, various scientists commit other offences against science. Some commit plagiarism, that is, theft of intellectual property; others are involved in corruption that usually appears where science and business, or science and politics interact. Sometimes scientists deliberately hide research results or intentionally misinterpret them. It is difficult to assess the impact of those common sins against reliability and integrity on accumulated knowledge. We believe that the difficulty in detecting those 'little crimes' and a sense of impunity lead to these problems becoming prevalent, while their importance is largely underestimated.

An acknowledgment of the fact that this problem exists and is real has led to the establishment of specialized institutions that take initiatives to counter scientific fraud. The Office of Research Integrity is the most engaged and widely recognized institution among the life sciences, such as psychology or medicine. Established by the US government, its origins go back to the year 1980, when some of the largest and most outrageous scandals involving scientists surfaced and appalled the American public. This is when Albert Gore, Jr., Chairman of the Investigations and Oversight Subcommittee of the House Science and Technology Committee, conducted his first investigation into the scientific misconduct of four well-known research centers. All four cases were related to the biomedical sciences. The first case concerned Eliasa Alsabti, a scientist from Iraq who fabricated medical documentation as well as medical research results. He was accused of plagiarizing about 60 scientific articles on immune response in patients with cancers. As a result of the investigation, his license to practice medicine was revoked in 1988; so was his PhD. The second case is Marc Straus, an oncologist from Boston University charged with falsifying records that allowed patients to apply for medical treatment. He denied the accusations and few years later applied for another research grant. Vijay R. Soman, an endocrinologist from Yale Medical School, was accused of fabricating data and plagiarizing his rival's article. Following the disclosure of the investigation, he was forced to resign from his post and later returned to his home country, India, abandoning all contacts with the United States. The fourth and last of the accused scientists was John Long, a researcher investigating cancer treatments at Massachusetts General Hospital. He was accused of fabricating and falsifying medical data and of numerous errors in research procedures. As the result of the investigation, he was only forced to abandon a research grant of $305,000.[3]

These events started an avalanche of reports of scientific misconduct that flooded congress until 1985☐ when they passed a Health Research Extension Act. It required the Secretary of Health and Human Services to present reports on all scientific fraud. Finally, in March 1989, the Office of Scientific Integrity, and later Office of Scientific Integrity Review were established. In May 1992, both of those institutions were consolidated under the name of Office of Research Integrity (ORI).[4]

This institution operates very actively, and not only in US territory. ORI conducts investigations into all dubious cases, presents annual reports and organizes conferences. It is quite interesting to take a look at how the ORI classifies scientific fraud:

- Fabrication – simple act of "inventing" data, making up results and recording or reporting them in any way.
- Falsification – manipulating research material, data, equipment or processes and/or omitting data or results in such way that the actual results of the study do not adequately or accurately represent the research records (not mentioning the reality).
- Plagiarism – appropriation of someone else's ideas, processes, results, data or words without pointing their origin or without giving appropriate credit to the original author.

According to ORI, unethical behaviors do not include honest mistakes or differences in opinions. This classification, very useful for our future discussion, was created in the United States of America during a two-day conference related to the promotion of good scientific practice in the behavioral science. Scientists, representatives of the government and other participants of the conference at Vanderbilt University made an effort to classify scientific abuse in order to reflect the severity of the offenses. Here are the results of their work, (starting with the most egregious misconduct):

- Data fabrication
- Data falsification ("cooking" or altering data)
- Plagiarism
- Unethical treatment of animal or human subjects
- Undisclosed conflicts of interest
- Violation of privileged material
- Irresponsible authorship credit (honorary authorship or exclusion of major contributor)
- Failure to retain primary data
- Inadequate supervision of the research products
- Sloppy recording of data

- Data-dredging
- Undisclosed repetition of unsatisfactory experiments
- Selective reporting of findings
- Failure to publish
- Unwillingness to share data and research materials
- Inappropriate statistical tests and procedures
- Insufficient or misleading reporting
- Redundant publication
- Fragmentary publication ("salami-science")
- Inappropriate citation
- Intentional submission of "sloppy manuscripts"[5]

The first two misconducts (fabrication and falsification) were covered in previous chapters. The delegates agreed that plagiarism is the third most important offense against science. We have a slightly different view on this matter. This particular form of scientific dishonesty does not have a significant nor direct impact on the description of reality formulated by science. It does not distort the image of reality; it does not betray the truth. It is rather a sign of personal pettiness, greed and the moral poverty of people who would steal someone else's ideas. Nevertheless, if one scientist copies (without mistakes) the results of his reliable colleague's research data, the scientific truth will not be compromised. Sometimes inaccurate recording of research results can harm the truth more than insolent plagiarism. Stealing intellectual property however allows the fraudsters and imposters to enter the temple of science. If a student obtains his Master's degree by copying someone else's work, he will probably not hesitate to plagiarize his doctorate and future publications as well. Such disguised priests in the temple of science certainly would not despise fabricating or falsifying research data in order to quickly climb the steep ladder of a scientific career. This is the reason why plagiarism earned a high place in the official ranking of scientific misconduct.

Numerous other sins can be derived directly from plagiarism. They all have a common denominator: they can be made during the publication of research results. This is an area where covering the truth may have far more serious consequences than simply signing one's name under someone else's work. We would like to take a closer look at those phenomena.

Misconduct related to the process of disseminating research results, which can have a huge impact on scientific progress, yet is rarely mentioned, are those committed by the reviewers and editors of scientific journals. Unfortunately, they often use their privileged position for the development of their careers. The story of the manuscript about shipbuilding, mentioned at the introduction of this chapter, perfectly describes one of the ways of this happens.

Amis' story, apart from being a simple mockery, points out another area prone to abuse. We have mentioned a couple of times already that modern science is extremely specialized. There is also enormous competition and a huge need for new discoveries in certain areas of science. Quite often, there are only a few experts worldwide in a specific field, and they usually know each other. When one of the authors of this book worked at the University of Bielefeld in Germany, his mentor, who was also a reviewer for at least a few scientific journals, said that when he receives an article to review, even when the authors' names were censored out, he knew exactly who wrote it. Years of experience allowed him to know the views and writing styles of all the experts currently working within his area of expertise.

All those factorss mean that reviewers and editors face numerous temptations every day. Imagine a reviewer who receives an article that describes research results surprisingly consistent with his own current research. How bad could it be to stall this other review slightly? Perhaps he could stall it just enough to publish his own results first? He could also write a negative review of his rival's work. This would make it impossible for the author to publish his work, at least in that particular journal. If he is an editor, he could reject valuable work after stalling it for long enough (obviously for very "important" and "objective" reasons), and use the time to simply steal someone else's work instead, just as the editor did in Amise's fictional story. Our previously mentioned "hero" Cyril Burt, has also taught us a lesson on how to benefit from being a journal editor. Remember the excellent reviews of his own work and the violent attacks on his rivals—all signed with names of non-existent, made-up characters?

On the other hand, authors can also slow down scientific progress by delaying the publication of their own discoveries. In some areas, the path to a significant scientific breakthrough resembles a chain of events, where particular links in the chain represent the results of scientific research. It is impossible to complete the chain without completing all the links first. Usually, many research teams work on each link on their own. Publishing partial results makes all other teams move one step ahead towards the major breakthrough. Withholding results gives an advantage to the authors who made the discovery, as they can move one step ahead while their rivals are still researching this particular current link. Such practices are most common in areas such as cancer research, the development of vaccines and so forth.

Psychologists usually commit another sin: unnecessary publications. Articles with virtually identical content but with different titles are published in numerous journals. This creates an artificial impression of an author's achievements and boosts his academic credentials. It also creates stacks of unnecessary information that his followers will have to dig through. Particularly, psychologists seem unable to refrain from publishing completely worthless, insignificant findings, often focused on trivial problems. The annual Ig Nobel Prize Awards irreverently shows how absurd those research problems can be. A good example is a study published by the Ig Nobel Prize winner - Harold Hillman, "The Possible Pain Experienced during Execution by Different Methods" published in *Perception*. Another example is a study done by Juan M. Toro, Josep B. Trobalon and Núria Sebastián-Gallés, entitled, "Effects of Backward Speech and Speaker Variability in Language Discrimination by Rats." And one more example, Anita Eerland, Rolf Zwaan, Tulio Guadalupe, "Leaning to the Left Makes the Eiffel Tower Seem Smaller." This study was realized as an international project. By the way, Anita Eerland and Rolf Zwaan married after finishing this study. Probably it was the greatest, and the only, significant impact of this research.

In the next chapter devoted to the problem of replication of research results, we present accurate data on the frequency of citations of scientific articles. They are quite overwhelming: over half of all scientific research in respected journals is never cited or quoted. This means either that nobody ever reads them or that they discard them as worthless after reading. These articles only accumulate on the pile of scientific junk and contribute to the increased workload of librarians, people processing scientific information, and other researchers, especially those just starting their scientific careers, as they have to read it all in order to separate the wheat from the chaff.

The lack of integrity and neglect of the reviewer must be mentioned here as well. Adam Eyre-Walker and his colleague Nina Stoletzki put in a lot of effort into analyzing two post-publication peer review databases containing 5,811 and 716 papers respectively. They have subjectively scored papers in each of those databases on the basis of merit. The expectation was that papers of similar merit would get similar scores. Surprisingly, they found that reviewers assigned the same scores to papers half the time, just slightly more often than expected by chance. Eyre-Walker and Stoletzki have also found a strong correlation between the impact factor of the journal in which papers were published and the merit scores that reviewers assigned to the particular paper.[6] This is what Eyre-Walker said about the problem of poor reviews, "Scientists are probably the best judges of science, but they are pretty bad at it."[7]

Unfortunately, his point of view was confirmed by John Bohannon's (*Science* magazine journalist) provocation. He wrote an article where he described the discovery of a molecule capable of inhibiting the growth of cancer cells. This molecule was supposed to be present in a certain species of lichen. He fabricated the research, made up methods and faked all the conclusions. Since he intended to send his "revelations" to 304 journals, he decided to mix-up all of his data. In numerous versions of his articles, the "discovered" molecule had different names, took effect on different cells, or came from different species of lichens, all this in order to avoid exposing his hoax. Each journal's editor received an original and unique article. While plotting his intrigue, he consulted his work with a group of scientists from Harvard University to make sure that he did not accidentally create anything that would even remotely resemble a real scientific work. Every version of Bohannon's article was grossly erroneous.

Of course he didn't publish under his own name. He made up names of other authors as well. For example, he signed his works as Ocorrafoo M. L. Cobange from Wassee Institute of Medicine in Asmara. Most of his made-up authors had African names and affiliations, (or other typical names from the developing world). Names were randomly taken from a database and a computer software randomly generated initials.

Consultants from Harvard University even pointed out that his English was grammatically "too correct." To add more authenticity, Bohannon translated his articles from English to French, and then from French to English using Google translator. After correcting only gross mistakes, he ended up with understandable text, but full of typical phrases and mistakes that could be expected from a non-native English speaker.

His prepared articles were sent to the editors of open access journals. They were assumed to be critically evaluated. As John Bohannon described it in *Science*, the mistakes in the articles were so obvious that even a cursory reading would be more than enough to reject them...if someone was reading them of course, because it turned out that almost nobody had. The articles were accepted for publication by 157 journals, 49 journals never replied, and 98 rejected Bohannon's revelations. Some editors requested upfront payment for processing the manuscript, others promised to send an invoice after accepting it for publication. He even received some feedback: somebody didn't like the format, or the size of images were inappropriate, and other similar unimportant details.

Who fell for the scam? According to *Science*, even journals published by Elsevier, Wolters Kluwer or Sage accepted fabricated articles. However, the majority of editors who agreed to publish his papers came from Turkey, India, China and Pakistan. Also, the IP addresses found in e-mails to the editors and some journals' bank account numbers also traced back to those countries. Price? Depending on the journal's "dignity," anywhere from $200 to over $3,000.[8]

Open access journals were supposed to be the remedy for many problems in the world of science, such as long waiting times for publication in traditional journals, ability to publish inconclusive results, unlimited and free access to the journals and research results previously limited by various subscriptions and fees, etc. Meanwhile, the growth of open access journals took the pathological path of a cancerous tissue: instead of solving problems, it created new ones, causing irreparable damage to science, and above all, further diminishing trust and confidence in the publication process. We will be watching closely the development of this new phenomenon, including all opportunities and threats which it creates.

There are many temptations to scientists who have just left their labs and now sit behind a desk, and many researchers do not remain indifferent. Reliable and comprehensive description of those temptations, together with detailed cases of those who fell for them, was presented by Marcel C. LaFollette in her book, *Stealing Into Print: Fraud, Plagiarism and Misconduct in Scientific Publishing.*[9]

If the temple of science could be cleared of all crooks, thieves and villains, some might expect that only honest and diligent researchers would remain in their fields. We imagine that scholars are those for whom every little detail is important, and their labs are immune to the destructive power of incidents and fortuity. Unfortunately, such an image is far removed from reality. There are tons of negligent bunglers constantly botching their work. Carelessness, incorrect reporting or improper citations are the common and everyday sins of many authors. It has been proven that a vast amount of citations are in fact secondary and tertiary citations.[10] Many authors choose to copy entire blocks from someone else's work, but cite the primary source without ever referring to the original and omit the author that they have copied their passages from. If you add in some mistakes and mishaps in the translations from various languages (sometimes multiple), then you will end up with far more interesting curiosities than those described above.

Does this ring a bell? Of course! Children across the world play "Chinese whispers," where a message is passed through a line of people and errors typically accumulate in the re-telling, so that the statement announced by the last person often significantly or amusingly differs from the one uttered by the first person. Scientists however do not do it for fun, and the following story may shed some light on the problem.

It happened at the beginning of twentieth century, where classic experimental psychology, with consciousness as the central subject of study and introspection as a method of researching it, was in crisis. John B. Watson, a thirty-five year old researcher of animal behavior, entered the realm of psychology on the February 24, 1913 by presenting his lecture during the meeting of the American Psychological Association in New York. His lecture at the University of Columbia was later published in his article "Psychology as the Behaviorist Views It" and became known as the "behaviorist manifesto."[11] Watson viewed psychology as an objective branch of the natural sciences. He advocated focusing psychological research on human behavior and to replace introspection with the experimental methods used in analyzing behavior in animals. This event started the successful battle for behaviorism and the behavioral approach to psychology that dominated scientific psychology for many years.

From this turbulent period, the most interesting from our point of view, is the a case of Albert, a 9-month old boy, also known in the literature as *Little Albert*. This case is worth mentioning not only because Albert was the first human ever to be experimentally conditioned. Little Albert is still alive in psychological literature, yet his fate would probably surprise Watson and Rosalie Rayner, with whom he conducted the experiment. Their work was supposed to answer three questions:

1. Can a newborn child be conditioned in such a way to become afraid of an animal presented simultaneously with the loud clanging of an iron rod?
2. Can the fear be transferred to other animals or objects?
3. How long would this effect last?[12]

In order to answer those questions Watson and Rayner picked a child – Albert B., whom they described as a healthy, passionless and unemotional subject. At the age of 9 months, Albert was subjected to tests that showed that he displayed no signs of fear or anxiety when he was presented with live animals (such as a rat, a rabbit, a dog or a monkey,) nor when presented with other objects (such as cotton, facemasks or burning newspaper). He displayed however significant anxiety and fear every time someone suddenly and unexpectedly clanged an iron rod with a hammer.

Two months after the tests, Watson and Rayner attempted to condition Albert with fear of a white rat by presenting it to the 11-month old boy, but every time Albert tried to touch the rat, the researchers immediately struck an iron rod with a hammer. After seven such events, (touching of the rat plus loud noise), which took place during two sessions separated with seven days of rest, Albert was crying and tried to avoid the rat every time he saw it, even when no loud sounds were presented.

In order to check how generalized Albert's reaction has become, 5 days later, the boy was presented with a rabbit, a set of wooden blocks (well-known to him), a rat, a short-furred dog, a furry coat, a pack of white cotton, Watson's head and the heads of two of his assistants (turned in such way, that Albert could touch their hair) and a mask of Santa Claus. Albert showed strong fear when presented with the rat, the rabbit, the dog and the furry coat. He also presented a "negative" reaction towards Santa's mask and Watson's hair and a moderate response to cotton. Albert had no problems playing with the wooden blocks and showed no anxiety when touching the heads of Watson's assistants.

After five days, Albert underwent another, single attempt of conditioning (rat plus sound). Also, researchers tried to condition a reaction to the previously presented rabbit (one attempt) and the dog (one attempt). When the effects of those procedures were later verified in another, larger room, only weak anxiety was observed after showing Albert a rat, a dog and a rabbit. Watson later tried to "refresh" boy's reaction to rats by another exposure to both stimuli (rat plus sound). During this attempt, the dog started to bark at Albert and wounded him and the researchers. This event ultimately complicated entire experimental procedure.

To answer their third question, after 31 days of break, during which they did nothing, Watson and Rayner repeated the last series of observations. This time Albert reacted with fear when touching the Santa's mask, the furry coat, the rat, the rabbit and the dog. At this time however, he also voluntarily touched the coat and the rabbit, showing a mixed reaction between a desire to retreat and a willingness to manipulate the objects. After this set of observations, Albert's mother took him away from the hospital where the experiment was conducted and subsequent contact with the child was lost. The authors admitted that they had known since the beginning when Albert would be leaving the hospital.

The Little Albert Experiment was unique because it empirically verified a theory of behavior and development that Watson had crafted over many years. Today Watson's work raises many doubts and objections, but we would like to analyze how Little Albert was treated by other fellow psychologists. In 1979, Ben Harris, an American psychologist, thoroughly analyzed literature that quoted the Little Albert experiment.[13] Among other publications, he also analyzed all textbooks that served as a teaching foundation in introduction to general, developmental and clinical psychology at American colleges. Only four textbooks accurately described the experiment. A few textbooks mentioned the experiment without mentioning its authors. Inaccuracies of least importance, because they had no or little impact on the results of the experiment, were related to Albert's age (2 textbooks), his name (1), Rosalie Rayner's name (4) and whether Albert was initially conditioned with a rat or rabbit (2). Mistakes in description of the generalization of Albert's fears and what happened to the boy after the experiment ended are far more important. The list of made-up stimuli that the boy allegedly reacted to is quite impressive. American psychology textbooks mentioned animal's fur (1), male beard (1), a cat, a puppy, a furry muff (1), white furry glove (1), Albert's aunt, who unexpectedly wore a furry coat (1), Albert's mother's furry collar (3), a teddy bear (1). Three textbooks added a happy end to the story, and few textbooks described in detail how Watson unconditioned Albert.

The author of above analysis stressed that such errors can occur not only from blind reliance on secondary sources without ever reaching for the original work, but probably other authors tried to make the history more believable to their students, or more in line with their current vision of research. Some "mistakes" may have resulted from their personal idea of Watson's work and of the process of conditioning emotional reactions. None of the textbooks described the rather brutal fact that Watson and Rayner knew exactly when Albert was leaving the hospital and that it would have been impossible to continue the research, and that they therefore consciously chose not to un-condition Little Albert. Another untold fact is that Little Albert died at the age of 6 from congenital hydrocephalus; therefore it cannot be concluded what (if any) effects the experiment had on his life.[14]

Some of the distortions can be attributed to Watson himself. In various subsequent descriptions of his own experiment, he omitted or changed some details of the experimental procedures. The most important and strongest factor influencing the creation of similar stories is the deeply rooted need for the experiment's compliance with the particular textbook author's theory. Textbook authors adjusted the description of Watson's experiment to match and confirm theories they presented. Strange interpretations could be nothing more than funny anecdotes if they ended up in a textbook on introduction to psychology, but it becomes scary when we read how some people understood, interpreted and used Little Albert's experiment for constructing a new therapeutic system.

A good example would be a model of phobia as a disorder of fear conditioning or a system of treatment of neurotic disorders based on this model. For its creator Joseph Wolpe, Little Albert's case is proof that a single event can induce a phobia![15] Eysenck also assumed that Albert developed a phobia not only to white rats, but actually to all fur-coated animals.[16] Similar interpretations of Watson's experiment can be found in at least four other works of behavioral therapists. The problem is that critical review of the experiment does not even remotely allow one to draw any conclusions about any sort of phobia. Don't forget that such generalizations and misinterpretations were the basis of an entire therapeutic system. This problem becomes more serious when we realize that many researchers after Watson tried to replicate the conditioning of emotional reactions, and most of them failed. According to Harris, creators of therapeutic concepts didn't even mention (let alone include) all those failed attempts during the construction of their new therapeutic system. Shouldn't those failed attempts be included in their research? Would that change their conclusions? Or are those just little mistakes of no importance, and we are just being picky?

In his article, Ben Harris mentions many more "findings" related to Little Albert and describes the strange conclusions drawn from those first, very awkward and methodologically flawed attempts of conditioning emotional reactions. The catalogue of scientific sins expanded when Watson's original research was published just a year later in *American Psychologist*.[17] Both articles show a distortion of scientific evidence that resembles the game of Chinese whispers rather than serious scientific concepts.

Another good example of a gossip that spread in the field of business training is the reinterpretation of Albert Mehrabian's research conducted over 40 years ago. At that time, a newly employed doctor of psychology at the University of California conducted a few experiments that became famous as the "7%–38%–55% rule."[18] It is commonly known among people engaged in various forms of coaching and psychological training. The rule tells us something like this: of all the information we receive, 7% comes from actually spoken words, 38% comes from the tone of voice and as much as 55% is derived from the body language. While the rule was based on numbers from Mehrabian's research, their meaning and interpretation are completely misunderstood. In order to actually understand it, we need to take a closer look at one of the experiments.

It involved students of the University of California. The researchers set up the following materials for their experiment: three photographs of a face of the same person: (1) smiling, (2) neutral, (3) sad; three sets of words: (1)

positive: *dear, great, honey, love, thanks*, (2) neutral: *maybe, oh, really, so, what* and (3) negative: *brute, don't, no, scram, terrible.* The words were recorded by a lector in such way that every word was pronounced in a positive, neutral and negative way. Next, the researchers constructed sets of stimuli to create incongruent expressions. An example of such incongruence would be asking someone "How was your day, darling?" and then replying "Fine!" in a way that means trouble rather than expresses genuine well-being.

The participants of the study were presented with sets of stimuli. A sample set included a photograph of a smiling face and a neutral word like 'so,' said in a negative tone. The participants were asked to rate how they find (or like) the person represented by the set of stimuli. The aim of the study was therefore to determine the significance of particular elements of incongruent (inconsistent) messages (mimics, words, vocalization) that influence our perception of sympathy. The results of the study showed that the perceived level of sympathy is determined 55% by facial expression, 38% by the tone of voice and only 7% by actual words. Therefore the "7%–38%–55% rule" applies only to ambivalent situations, where inconsistent messages are received by different channels.

Multiple repetitions, improper citations and a lack of scientific scrutiny resulted in the popular belief that this rule applies to all human communications. This is obviously absurd. If this rule was true, people could hardly communicate using phones, or radio. All those distortions differ in their severity and impact. When they are used to create new therapeutic procedures, the consequences are of far greater importance than in situations where people simply waste their time and money trying to improve non-verbal communications. Regardless of the direct consequences, myths that arise in this way replace scientific knowledge and cause havoc in rigorously built and gathered scientific evidence. Education based on distorted reality does not bring anyone a single step closer to the truth.

Before we close this chapter, we would like to focus your attention on a research process that helps in tracking dishonesties related to the improper quoting of scientific literature in a far more systematic way than described accidental findings or Harris' analysis of Little Albert's case. Researchers devoted to bibliometrics track random mistakes in citations, such as simple mistyping, misspellings or misprints. Particularly important for them are errors in important and often cited articles.

A prime example in this regard is the story of the "Ortega hypothesis." Heidi Hoerman and Carole Nowicke described in their article the fate of quotes about the hypothesis by the Spanish philosopher José Ortega y Gasset. Jonathan R. Cole and Stephen Cole quoted Ortega's hypothesis in four of their works.[19] Tracing the fate of his quote was possible because of mistakes made by Cole and Cole in the original citation. We will not analyze here this funny but also disturbing story of misrepresenting quoted words. It is worth mentioning some numbers however. In over 30 analyzed articles on the Ortega hypothesis, 14 mentioned and cited the original article in their references, but all but one repeated all three of Coles' mistakes. Those publications are over 20 years old! For all this time, scientists were not able to correct their mistakes and they mindlessly continued to copy them in following publications.

Purely quantitative research of random mistakes in citations reveals puzzling, large numbers. In a sample of 2933 citations, 1888 mistakes were found, which is more than 64%.[20] Later analyses of 1189 citations published in four different scientific journals showed 519 mistakes, which amounts to 43.6%.[21] Of course, it is difficult to conclude how many errors are a result of simple, honest mistakes and how many of them appeared during copying citations from other papers. However, the scale of this seemingly innocent dishonesty is so great that it can easily severely distort the scientific mosaic, not so carefully laid (as it turns out) by scientists.

---

[1] K. Amis, *Lucky Jim,* (London: Penguin Books, 1961), 5.

[2] I. Oransky, "Psychology Journal Editor has Seven Articles Retracted for Duplication or Plagiarism," *Retraction* (November 2013): http://retractionwatch.com/2013/11/20/psychology-journal-editor-has-seven-articles-retracted-for-duplication-or-plagiarism/#more-16604

[3] See H. F. Judson, *The Great Betrayal. Fraud in Science.* (Orlando: Harcourt, 2004).

[4] See http://ori.dhhs.gov/.

[5] T. Adler, "Outright Fraud Rare, But Not Poor Science," *The Monitor, (American Psychological Association),* 22, (1991): 11.

[6] A. Eyre-Walker and N. Stoletzki, "The Assessment of Science: The Relative Merits of Post-Publication Review, the Impact Factor and the Number of Citations," *PLOS Biology* (October 2013).

[7] M. Clune, "Scientists 'Bad at Judging Peers' Published Work,' Says New Study," University of Sussex (October 2013): http://www.sussex.ac.uk/newsandevents/?id=21204

[8] J. Bohannon, "Who's Afraid of Peer Review?" *Science, 342,* (October 2013): 60-65.

[9] M.C. LaFollette, *Stealing into Print. Fraud, Plagiarism and Misconduct in Scientific Publishing.* (Berkeley: University of California Press 1992).

[10] H. L. Hoerman and C. E. Nowicke, "Secondary and Tertiary Citing: A Study of Referencing Behavior in the Literature of Citation Analysis Driving from the Ortega Hypothesis of Cole and Cole," *Library Quarterly, 65,* (1995): 415-434.

[11] J. B. Watson, "Psychology as the Behaviorist Views It," *Psychological Review, 20,* J. B. (1913): 158-177.

[12] J. B Watson and R. Rayner, Conditioned emotional reactions. *Journal of Experimental Psychology, 3,* 1-14, 1920. As in B. Harris, (1979). Whatever happened to Little Albert? *American Psychologist, 34,* 151-160.

[13] Harris, *Whatever Happened to.*

[14] H. P. Beck, S. Levinson, and G. Irons, "Finding Little Albert: A Journey to John B. Watson's Infant Laboratory," *American Psychologist, 64,* (2009): 605–614.

[15] J. Wolpe and S. Rachman, "Psyhoanalytic 'Evidence': A Critique Based on Freud's Case of Little Hans," *Journal of Nervous and Mental Disease, 130,* (1960): 135-148.

[16] H. Eysenck, "Learning Theory and Behaviour Therapy," in *Behaviour Therapy and the Neuroses: Readings in Modern Methods of Treatment Derived from Learning Theory*, ed. H. J. Eysenck (Oxford: Pergamon Press 1960), 4-21.

[17] F. Samelson, "J. B. Watson Little Albert, Cyril Burt's Twins, and Need for a Critical Science," *American Psychologist, 35,* (1980): 619-625.

[18] A. Mehrabian and M. Wiener, "Decoding of Inconsistent Communications," *Journal of Personality and Social Psychology, 6,* (1967): 109-114; A. Mehrabian and S. R. Ferris, "Inference of Attitudes from Nonverbal Communication in Two Channels," *Journal of Consulting Psychology, 31,* (1967): 248-252.

[19] Hoerman and Nowicke, "Secondary and Tertiary."

[20] A. White, "Reference List Inaccuracies: A Four-Decade Comparison," *Journal of Counseling and Development, 66,* (1966): 195-196. As cited in Hoerman and Nowicke, "Secondary and Tertiary," 415-434.

[21] H. N. Rae and A. White, "Pass it On: Errors in Direct Quotes in a Sample of Scholary Journals," *Journal of Counseling and Development, 67,* (1989): 509-512. As cited in Hoerman and Nowicke, "Secondary and Tertiary," 415-434.

CHAPTER 6

SCIENTIFIC CONSPIRACY?
THE MYTH OF REPLICATION

We would like our readers to imagine a group of villain scientists who cooperate in order to falsify and distort all the knowledge about humans. They are completely devoted to their conspiracy. It is not an easy task. After all, the world's most brilliant minds will immediately mobilize to counter their devious plot. The world's most penetrating gazes will be upon them and the most advanced technologies will be used against the conspirators. Such an evil plot would seem impossible to succeed. It would have to remain hidden from watchful eyes; the plot's mechanisms would have to work flawlessly, even without supervision. You have probably already spotted that our science fiction, even after just three sentences, is nothing but a potboiler. But you'd be surprised to learn that such a mechanism actually exists! People created it, although they were not in a conspiracy. It is not conspicuous, though nobody even tries to hide it. It works without any supervision and it works flawlessly, perhaps with minor exceptions. Furthermore, the majority seems to actually appreciate the fact of its existence. How is this possible? What is it? Who created it?

Every scientist who attempts to test any hypothesis should start his work with a detailed literature review in order to check whether someone has already conducted similar studies. It is possible that his hypothesis was verified long time ago. Following this simple rule saves tremendous amounts of time and work, thus protecting scientists from wasting their scientific resources. This is not always the case, as sometimes, if someone failed to prove a given hypothesis, the researcher has almost no chance of finding out about it! Even if the same hypothesis was analyzed and studied by many scientists, and none of them reached interesting results, the chance to learn about their failures are still close to none. Why? Psychology journals do not publish the results of negative studies, especially those considered as inconclusive. The majority of the journals mention it even in the instructions for authors. If the study did not verify assumed hypotheses, most journals will reject them, devoting precious journal space for "publishable" results.

Why do they do it? This might be justified to a certain degree. Imagine a group of researchers making some completely irrational assumptions about human nature and conducting intensive studies that will, obviously, yield negative results. They would soon flood all journals' pages with their complete and utter nonsense. Similarly, authors unable to properly plan, conduct and conclude a project would demand to be published as well. Those reasons we understand. But the unwillingness (it became an unwritten law) to publish negative results applies and extends to all research, *even if they exactly replicate previously conducted studies*. It appears that another custom in science somehow developed spontaneously. Most editors and reviewers believe that if there are reasons not to publish negative results, it is best not to publish them at all. Let's take a look at the consequences of this common approach, as there is a possibility that such behavior is actually the proverbial throwing the baby out with the bathwater.

Lack of access to negative research results causes an enormous waste of time, energy and scientific resources, simply because researchers cannot know if someone has ever worked on the issues that interest them. Therefore, numerous scientists all around the world waste their efforts, time and public money from research grants trying to investigate the same problems. For this reason alone, a couple of decades ago, attempts were undertaken to actually create a journal that would publish negative research results only. Marvin Goldfried and Gary Walters even proposed a name for such journal: *Journal of Negative Research* printed in a similar way to *Psychological Abstracts*.[1] After many years, the situation did not change much. Actually, it is getting worse. "Negative results' now account for only 14% of published papers, down from 30% in 1990."[2]

For readers who managed to read our book up to this point, the gigantic waste of time and energy executed by psychology researchers isn't really a major surprise. However, the consequences of non-publication of negative results undermine the very objectives of science: the pursuit of truth. In the following parts of this book, we will discuss, among other things, the efficacy of various psychotherapies. Nevertheless, let's now try to imagine a researcher who checked several therapies for their efficacy and did not achieve any statistically significant results. For most editors of psychological journals, such work is not worth publishing, as it is inconclusive. However, if the research was methodologically correct, it does present a tremendously important discovery, one that has no chance to reach wider public.

Furthermore, we could witness a plethora of articles with so-called meta-analyses. A researcher who conducts a meta-analysis does not research the topic on his own. Instead, he uses articles previously published by other scientists to reach his conclusions. Meta-analysis can therefore be extremely useful, as it gathers experiences and results of many scientists in one, single analysis. If a researcher was interested, for example in the efficacy of Gestalt therapy, he could collect tens or hundreds of articles that include similar parameters and he could draw a common conclusion from all previous research. You can probably see the problem already; this powerful tool becomes useless if researchers can only collect articles that were considered publishable by journal editors. Without the insight into all conducted research on a given topic, including all negative trials, a meta-analysis will only further distort our perception of reality.

This phenomenon of severe distortion of knowledge has gained a few original labels. "File drawer effect" was first proposed by R. Rosenthal.[3] It is also sometimes called a publication bias.[4] An unimaginable amount of knowledge is lost in desk drawers, archive shelves and document shredders. After all, storing useless, un-publishable research results is pointless.

Can you now see the mysterious intrigue that successfully attempts to distort all knowledge about humans? It would be hard to think of anything more original, less conspicuous and as effective as the mechanism we have just described. This tendency of hiding negative research results from other people is not limited to psychology. It is dangerously common in other social sciences and in medicine as well.[5] Research shows however, that it is by far worst in psychology and psychiatry. Daniele Fanelli found that "the odds of reporting a positive result were around five times higher for papers published in Psychology and Psychiatry and Economics and Business than in Space Science."[6]

Unfortunately this is not the end of the negative consequences of publication bias. Do you recall the case of Cyril Burt? By falsifying research data, his made-up theory on the mechanisms of inheriting intelligence became widely accepted. As you remember, other published studies yielded very similar results to those published by Burt and his followers. If someone at this time conducted an inconclusive study related to the inheritance of intelligence, would he have any chance to publish his results? Or would he be classified as "unpublishable"?

Things might have looked the same when it comes to replicating the results of Watson's *Little Albert* experiment previously described in chapter five. The file drawer effect probably affected research results contradicting Watson's concept of the formation of phobias derived from Little Albert's case.

Publication bias will always focus attentively on all hypotheses and theories that have a chance to shine in printed journals, and thus temporarily become accepted. This mechanism resembles what is described by psychologists as a self-fulfilling prophecy. It allows us to easily explain why so many ridiculous ideas constantly spread in circulating scientific thoughts. Unfortunately, this is a very primitive mechanism, the same type that sustains the belief of many people in horoscopes...

The priming effect, well known and popular in psychology, serves as an excellent illustration of this phenomenon. Priming studies suggest that decisions can be influenced by apparently irrelevant actions or events that took place just before the cusp of choice. One of the most famous studies from 1998 showed that thinking about a university professor before answering general knowledge questions led to higher scores than imagining a football hooligan.[7] Media usually happily report such research results. They are easy to understand; they usually slightly oppose common intuition, and are also very promising. Wouldn't you like to ace all general knowledge quizzes by applying simple mental tricks? Research on the priming effect has rapidly spread across psychology over the past decade. Some of their insights have already made it out of the labs and into the brains of policy geeks eager to poke the masses. But in April 2013, a paper in *PLoS ONE*, a renowned journal, reported that nine separate experiments failed to reproduce the results of a famous study from 1998.[8]

Similar doubts regarding the validity of priming, especially in such simplified form, were already raised long before. Unfortunately, the amount of research papers confirming this effect overwhelms the few voices of common sense. It is worth mentioning that at the time of writing those words, scientific databases in the field of psychology report 20,878 papers that contain the word 'priming.'

The problem surfaced however not because of the few attempts to replicate original research results, but due to the intervention of one of the world's most renowned psychologists, a Nobel Prize winner in economics, Daniel Kahneman. He summarized his worries and concerns about the condition of research on priming in his open letter sent in 2012 to several dozen social psychologists. The media quickly picked this up, and the case became widely known. What was in Kahneman's letter?

> As all of you know, of course, questions have been raised about the robustness of priming results. The storm of doubts is fed by several sources, including the recent exposure of fraudulent researchers, general concerns with replicability that affect many disciplines, multiple reported failures to replicate salient results in the priming literature, and the growing belief in the existence of a pervasive file drawer problem that undermines two methodological pillars of your field: the preference for conceptual over literal replication and the use of meta-analysis. Objective observers will point out that the problem could well be more severe in your field than in other branches of experimental psychology, because every priming study involves the invention of a new experimental situation.
>
> For all these reasons, right or wrong, your field is now the poster child for doubts about the integrity of psychological research. Your problem is not with the few people who have actively challenged the validity of some priming results. It is with the much larger population of colleagues who in the past accepted your surprising results as facts when they were published. These people have now attached a question mark to the field, and it is your responsibility to remove it....
>
> My reason for writing this letter is that I see a train wreck looming. I expect the first victims to be young people in the job market. Being associated with a controversial and suspicious field will put them at a severe disadvantage in the competition for positions. Because of the high visibility of the issue, you may already expect the coming crop of graduates to encounter problems....
>
> I believe that you should collectively do something about this mess. To deal effectively with the doubts you should acknowledge their existence and confront them straight on, because a posture of defiant denial is self-defeating. Specifical-

ly, I believe that you should have an association with a board that might include prominent social psychologists from other fields. The first mission of the board would be to organize an effort to examine the replicability of priming results, following a protocol that avoids the questions that have been raised and guarantees credibility among colleagues outside the field. ...

Success (say, replication of four of the five positive priming results) would immediately rehabilitate the field. Importantly, success would also provide an effective challenge to the adequacy of outsiders' replications. A publicly announced and open effort would be credible among colleagues at large, because it would show that you are sufficiently confident in your results to take a risk.[9]

The Kahneman phrase most quoted by media was: "I see a train wreck looming." It is difficult to say whether the author's action, led by his concerns regarding the validity of a certain research area (and psychology as a whole), did more good or harm, but it definitely pointed attention to the problem of the reliability of conducted research, and the discussion is still ongoing. As we write these words, *The Head Conference* organized by the Edge Foundation and entitled *What's New in Social Sciences* is taking place. Rob Kurzban had a speech in this conference on "P-Hacking and the Replication Crisis" and he clearly declared that the current situation prevailing in psychology worldwide should be finally recognized and called a crisis.

I really wanted to take this opportunity to have a chance to speak to the people here about what's been going on in some corners of psychology, mostly in areas like social psychology and decision-making. In fact, Danny Kahneman has chimed in on this discussion, which is really what some people thought about as a crisis in certain parts of psychology, which is that insofar as replication is a hallmark of what science is about, there's not a lot of it and what there is shows that things we thought were true maybe aren't; that's really bad.[10]

The file drawer effect is ubiquitous and it virtually renders replication, one of the most precious inventions in the history of science, useless. Kurzban referred to replication as the hallmark of what science is. Repeating the results of other scientists allows us to pick-up mistakes, random occurrences and also allows us to confirm results that others achieved. However, previously described limitations imposed by scientific journals simply mean that replication can only work in one way: by confirming previous results. Unfortunately, even this statement is overly optimistic, as replication in psychology is nothing but a myth. Most editors and reviewers reject publishing research results that replicate the work of others. The rejection letter usually sounds like this: "We regret to inform authors that submitted manuscript is only a simple replication of already existing works. Our journal publishes only articles that introduce new discoveries and extend the knowledge about human behavior." And that's it; an article that would add another brick to the wall of human knowledge ends up in a drawer. Such letters from journal editors are very common. The requirement of originality of research submitted for peer-review is almost always already included in the instructions for authors.

Significant in this regard is what happened in March 2011, after Daryl Bem published his paper in the *Journal of Personality and Social Psychology* that found evidence for Psi, or extra-sensory perception. Bem stands by his work, but many psychologists question his analysis and found it hard to believe that the study was even published.

> At least one team, led by Richard Wiseman, PhD, of the University of Hertford-shire, was unable to replicate Bem's findings — but JPSP didn't publish that research because the journal does not publish replications. Psychological Science also rejected Wiseman's replication study, for the same reason. It was eventually published in the open-access journal *PLOS One*, and Wiseman set up a website for others to document their attempts to replicate Bem's study. And, in December, JPSP published a meta-analysis of the original studies, and replication attempts, by Carnegie Mellon University's Jeff Galak, PhD. But critics say that the situation exemplifies the difficulty of getting any replication work published.[11]

Such habits were bitterly criticized by authors Christopher J. Ferguson and Moritz Heene in the article "A Vast Graveyard of Undead Theories: Publication Bias and Psychological Science's Aversion to the Null." They wrote, "The aversion to the null and the persistence of publication bias and denial of the same, renders a situation in which psychological theories are virtually unkillable. Instead of rigid adherence to an objective process of replication and falsification, debates within psychology too easily degenerate into ideological snowball fights, the end result of which is to allow poor quality theories to survive indefinitely.[12]

A notable exception to the "non-publication principle," one of very few examples that we know of, is *Representative Research in Social Psychology*. This journal was founded in the late 1960s by the initiative of students from the University of North Carolina in response to a very long waiting time for publication in other journals. The main reason for its establishment was the complete lack of any journal that would actually publish methodologically correct research papers, including replications, regardless of whether their results were positive, negative or inconclusive. Since 1970, when the first issue of *Representative Research in Social Psychology* was released, the journal's editors remain faithful to the original promise. They still indicate that the magazine specializes in publishing all replicated and all negative and inconclusive results.

Another place worth mentioning that publishes replications and null hypothesis studies is the *Journal of Articles in Support of the Null Hypothesis*, established in 2002. This is an open access magazine; however, the number of published articles is very modest, reaching only few per year.[13]

When mentioning initiatives aimed at restoring the deserved honor of the process of replication, we cannot skip the Many Labs project, a group of scientists from 36 laboratories and research institutes around the world who have decided to verify and replicate 13 classic psychological experiments. The project was coordinated by Richard Klein and Kate Ratliff from the University of Florida, together with Michealangelo Vianello from University of Padova (chief statistician) and Brian Nosek from the Center for Open Science. A total of 6,344 people from 12 different countries were involved in the attempt to replicate classic experiments. The results of Many Labs replications were openly documented and will be by published in *Social Psychology*. It is worth mentioning that the project was accepted for publication before the replication started. All available data, results, analyses, methods of data collection and movies documenting the experiments are publicly available on the project's webpage.[14]

Within the same Open Science Framework, another more ambitious *Reproducibility Project* is being implemented.[15] Previously mentioned Brian Nosek and dozens of other psychologists are attempting to reproduce as many studies as possible that were published in the 2008 volumes of three prominent journals: the *Journal of Personality and Social Psychology*, *Psychological Science* and the *Journal of Experimental Psychology: Learning, Memory, and Cognition*. The group is working on about 50 studies and hopes to get 100 or more researchers to join the project. "The goal, Nosek says, is both to investigate the reproducibility of a representative sample of recent psychology studies, and to look at the factors that influence reproducibility. That's important, he says, because 'being irreproducible doesn't necessarily mean a finding is false. Something could be difficult to reproduce because there are many subtle factors necessary to obtain the results. And that's important too, because we tend to overgeneralize results.'"[16]

One of possible solutions that could allow the publication of larger numbers of repeated studies is the idea of registered replication. At this moment, the website of the Association for Psychological Science declares they will start a series of publications called Registered Replication Reports in the journal *Perspectives on Psychological Science*. The description, which we can find in their Mission Statement, precisely describes the idea of registered replications.

> A central goal of publishing Registered Replication Reports is to encourage replication studies by modifying the typical submission and review process. Authors submit a detailed description of the method and analysis plan. The submitted plan is then sent to the author(s) of the replicated study for review. Because the proposal review occurs before data collection, reviewers have an incentive to make sure that the planned replication conforms to the methods of the original study. Consequently, the review process is more constructive than combative. Once the replication plan is accepted, it is posted publicly, and other laboratories can follow the same protocol in conducting their own replications of the original result. Those additional replication proposals are vetted by the editors to make sure they conform to the approved protocol.[17]

Unfortunately, at this moment, only one research project was reported.

A website called PsychFileDrawer.org plays an important role in the attempt to save replication research, where researchers can post their unpublished negative results. It was launched by Hal Pashler from the University of California, San Diego and Barbara Spellman, a psychologist from the University of Virginia. The site has gotten some positive attention, but so far only 43 studies are posted. "We have a high ratio of Facebook likes to actual usage," Pashler says. "Everybody says that it's a good idea, but very few people use it."

> The problem, he says, is probably that researchers — especially grad students and other young researchers likely to do replication work — have little incentive to post their findings. They won't get publication credit for it, and may only annoy the authors of the original studies.
> "I'm happy we made [the site]," Pashler says "but so far it has just ended up spotlighting the incentive problem."[18]

Pashler's reflections do not sound optimistic. Don't forget that among thousands of psychological journals, all initiatives described above are nothing but a drop of water in the ocean of published research papers, more interesting, more creative and resulting in even less dependable and less replicable results. It is also very interesting to notice that most of the attempts to change the unfavorable status quo are undertaken by single individuals, small groups of researchers, or (rarely) by non-mainstream associations. Large organizations, like the American Psychological Association, responsible in fact for what is really happening in the field of psychology, seem to completely ignore the problem. Even when they seem to notice it for a moment, their actions are negligible in relation to the seriousness and scale of the problem; usually they are limited to initiating discussions.

The policy of not accepting articles for publication, that we have written so much about, is not the only reason for the reluctance to conduct replication research. "Among the top problems are that funding agencies aren't interested in giving money for direct replication studies and most journals aren't interested in publishing them. So researchers whose careers depend on winning grants and publishing studies have no incentive to spend time and effort redoing others' work."[19] Moreover, as Brian Nosek, a social psychologist at the University of Virginia says, the incentive system at work in academic psychology is heavily weighted against replication: "There are no carrots to induce researchers to reproduce others' studies, and several sticks to dissuade them."[20]

But even if we assume that institutional obstacles are removed, there are still psychological barriers that will continue to discourage scientists from reproducing studies of their colleagues. This problem was accurately summarized by Robert Kurzban:

I would go to give talks in places and, lo and behold, it turns out there's this kind of background radiation—there's the dark matter of psychology, which is a few people who fail to replicate and don't publish their work and also don't talk about it because the fact that you've failed to replicate has a reputational effect, right? The person who's in charge of this literature says, "Oh, these guys were going after me," and so maybe you don't talk about it in polite company. Right? It's sort of like sex; it's the thing that we're all doing, we're all replicating, we just don't want to talk about it too much, right?[21]

In the simplest of studies, where the strength of the linear relationship between two variables is defined as the correlation coefficient, the resulting number is usually given with the accuracy up to two decimal places. John Hunter, in his article "The Desperate Need for Replications" calculated that in order to give the exact result of such correlation, we would need a staggering number of at least 153,669 test subjects![23] For accuracies to one decimal place we "only" need to examine 1,544 subjects. In social science, it is very unusual and extremely rare to even reach second numbers of participants. The average numbers of test subjects in one study reported in the *Journal of Consumer Research* is 200. In order to be able to achieve the required accuracy for correlation to one decimal place, we should repeat the study at least eight times. If we would like to be accurate to two decimal places we would need to repeat the study at least 800 times, provided that each one involves at least 200 participants. For the above to be true, we must also assume that all of those studies would be perfectly conducted and be free of methodological errors (such as non-random selection of participants), etc. Don't forget that this is all to analyze the simplest and most basic relationship between just two variables. A test so simple… that nowadays almost nobody bothers with it. In reality, even in simple studies we analyze many more variables. Somehow we tend to forget that requirements of the sample size for more complicated studies increase exponentially.

According to Hunter, the 'novelty approach' as the main criteria for selecting papers for publication is fatal to science, hence his cries for replications. In his view, science needs hard facts and evidence good enough to accurately base our predictions on, yet we still lack those. Other authors also notice the absence of replications and the need for them, however rarely.[22] They all point to life sciences as a model to follow in this regard, but according to critics, it's primarily in the life sciences where lack of replications results in the printing and spreading many absurd and misguided concepts.[23]

Let's put some numbers to Hunter's postulates. Arif Jinha calculated that up to 2010, as this is when he was doing his computations, there were about 50 million scientific articles published around the world.[24] Today, every year over 2,000,000 papers are being published, and the number of publications per year doubles every 20 years. This is a result of rapid progress in science, but it also has some negative implications.

While brilliant and progressive research continues apace here and there, the amount of redundant, inconsequential, and outright poor research has swelled in recent decades, filling countless pages in journals and monographs. Consider this tally from *Science* two decades ago: Only 45 percent of the articles published in the 4,500 top scientific journals were cited within the first five years after publication. In recent years, the figure seems to have dropped further. In a 2009 article in *Online Information Review*, Péter Jacsó found that 40.6 percent of the articles published in the top science and social-science journals (the figures do not include the humanities) were cited in the period 2002 to 2006.[25]

Taking into account the fact that scientists should cite all articles that they have based their research on, it simply means that if an article is never cited, it had absolutely no influence on any scientist in the world. It therefore means that it had precisely zero impact on the advancement and progress of science. Even if it was read, the reader must have concluded that what he just read was worthless. This never-cited mountain of articles produced by the scientific community remains unproven, unverified, unreplicated and probably unread. If we subtract from those numbers some of the previously mentioned scientific sins, such as self-plagiarism or single citations (often a polite gesture or a need of reciprocity), this pile of scientific junk stretches even further. Wouldn't it be better if instead of "creatively" producing completely useless research papers, we properly replicated the existing ones in order to finally create solid foundations for our knowledge?

But those are not the only consequences. Think about it; someone has to devote a lot of time to read and review all this nonsense. Reviewers get increasingly more papers to review and most likely analyze them less thoroughly. Libraries have to catalogue all this junk and spend a significant amount of money to do it. Scientific databases are becoming overloaded. Young scientists find it increasingly more difficult to find valuable, relevant evidence hidden in stacks of worthless scientific "creativity." It is good to know what is happening in your own field, isn't it?

During the massive floods in Europe in 1997, one of the major problems people faced in flooded areas was the lack of drinking water. Isn't it similar, when the world of science is being flooded with unnecessary, unwanted and useless streams of new data, results and publications, but lacks scientific evidence and replication? Isn't the flood of scientific papers grotesque?

Philosophers of science proudly mention replication as a mechanism that distinguishes the system of knowledge created by science from other structures that lack self-correcting mechanisms that would allow the detection of misconduct, fraud and lies. Unfortunately in the real world, replication is a myth. It is the myth that researchers of human nature are willing to believe, a myth that psychologists are proud of, a myth whose presence in psychology was accurately summarized by Robert Kurzban: "I'm saying that in many ways, the replication crisis in psychology is a little bit like the weather, right?

We all talk about it but no one really does anything about it. We do a little about it here and there."[26]

---

[1] M. R. Goldfried and G. C. Walters, "Needed: Publication of Negative Results," *American Psychologist, 14,* (1979): 598.

[2] No Author, "How Science Goes Wrong," *The Economist* (October 2013): http://www.economist.com/news/leaders/21588069-scientific-research-has-changed-world-now-it-needs-change-itself-how-science-goes-wrong

[3] R. Rosenthal, "The 'File Drawer Problem' and Tolerance for Null Results. *Psychological Bulletin, 86,* (1979): 638-641.

[4] See I. Peterson, "Publication Bias: Looking for Missing Data," *Science News, 135,* (1989): 5.

[5] See J. Cohen, "A New Publication Bias: The Mode of Publication," *Reproductive BioMedicine Online, 13,* (2006): 754-755.

[6] D. Fanelli, "Positive Results Increase Down the Hierarchy of the Sciences," *PLoS ONE, 5,* (2010): 4.

[7] A. Dijksterhuis and A. van Knippenberg, "The Relation between Perception and Behavior, or How to Win a Game of Trivial Pursuit," *Journal of Personality and Social Psychology, 74,* (1998): 865–877.

[8] D. R. Shanks, B. R. Newell, E. H. Lee, D. Balakrishnan, L. Ekelund, L., et al. "Priming Intelligent Behavior: An Elusive Phenomenon," *PLoS ONE, 8* (2013).

[9] "Kahneman on the Storm of Doubts Surrounding Social Priming Research," *Decision Science News* (October 2012): http://www.decisionsciencenews.com/2012/10/05/kahneman-on-the-storm-of-doubts-surrounding-social-priming-research/

[10] R. Kurzban, "P-Hacking and the Replication Crisis," *The Head Conference - What's New in Social Sciences* (December 2013): http://edge.org/panel/headcon-13-part-iv

[11] L. Winerman, "Interesting Results: Can They Be Replicated? *APA Monitor, 44,* 38 (February 2013): from: http://www.apa.org/monitor/2013/02/results.aspx

[12] C. J. Ferguson and M. Heene, "A Vast Graveyard of Undead Theories Publication Bias and Psychological Science's Aversion to the Null," *Perspectives on Psychological Science, 7,* (2012): 555-561. http://pps.sagepub.com/content/7/6/555.full

[13] See, http://www.jasnh.com/

[14] See, https://openscienceframework.org/project/WX7Ck/files/

[15] See, https://openscienceframework.org/project/EZcUj/wiki/home/

[16] Winerman, "Interesting Results."

[17] See: http://www.psychologicalscience.org/index.php/replication, retrieved December 7, 2013.

[18] Winerman, "Interesting Results."

[19] Ibid.

[20] Ibid.

[21] Kurzban, "P-Hacking."

[22] See, B. Schneider, "Building a Scientific Community: The Need for Replication. *Teachers College Record, 106,* (2004): 1471-1483.

[23] W. Broad and N. Wade, *Betrayers of the Truth. Fraud and Deceit in the Halls of Science.* (New York: Simon & Shuster, 1982).

[24] A. Jinha, "Article 50 Million: An Estimate of the Number of Scholarly Articles in Existence," *Learned Publishing, 23,* (2010): 258-263.

[25] M. Bauerlein, M. Gad-el-Hak, W. Grody, B. McKelvey, and S. W. Trimble, "We Must Stop the Avalanche of Low-Quality Research," *The Chronicle of Higher Education.* (June 2010): http://chronicle.com/article/We-Must-Stop-the-Avalanche-of/65890/

[26] Kurzban, "P-Hacking."

CHAPTER 7

A BIG JAR OF COOKIES:
THE REASONS FOR CURRENT STATE OF
ACADEMIC PSYCHOLOGY

> I was alone in my fancy office at University of Groningen.... I opened the file that contained research data I had entered and changed an unexpected 2 into a 4.... I looked at the door. It was closed.... I looked at the matrix with data and clicked my mouse to execute the relevant statistical analyses. When I saw the new results, the world had returned to being logical. (p. 145)
>
> I preferred to do it at home, late in the evening, when everyone was asleep. I made myself some tea, put my computer on the table, took my notes from my bag, and used my fountain pen to write down a neat list of research projects and effects I had to produce.... Subsequently I began to enter my own data, row for row, column for column...3, 4, 6, 7, 8, 4, 5, 3, 5, 6, 7, 8, 5, 4, 3, 3, 2. When I was finished, I would do the first analyses. Often, these would not immediately produce the right results. Back to the matrix and alter data. 4, 6, 7, 5, 4, 7, 8, 2, 4, 4, 6, 5, 6, 7, 8, 5, 4. Just as long until all analyses worked out as planned. (p. 167)
>
> Nobody ever checked my work. They trusted me.... I did everything myself, and next to me was a big jar of cookies. No mother, no lock, not even a lid.... Every day, I would be working and there would be this big jar of cookies, filled with sweets, within reach, right next to me — with nobody even near. All I had to do was take it. (p. 164)[1]

As you are reading these words, thousands of scientists are alone in their fancy offices. Many others, with freshly brewed cups of tea, are starting to work on their data in their homes, when their family members are already sleeping in their beds. Most of them are never supervised. Others just trust them. And next to them is a big jar of cookies. No mother, no lock, not even a lid. They work every day and every day this big jar of cookies, filled with sweets, within reach, next to them, with nobody even near.

Diederick Staple is a criminal and any attempts to justify his actions or to put blame on external circumstances should not be taken seriously, especially a tear-inducing story that Staple spreads himself. Despite what we have just written, we should carefully analyze those stories as they point out the weaknesses of science's self-controlling mechanisms, one of the very basic reasons for pandemics of scientific misconduct, especially in psychology.

In order to better depict this pathetic condition of psychology as a science, it is worth listening to what other fraudsters have to say. We owe this opportunity to Mark S. Davis, Michelle Riske-Morris, and Sebastian R. Diaz, who examined closed cases of research misconduct (which ended with finding the accused scientists guilty), processed by the Office of Research Integrity as of December 2000.[2] Since the ORI came into existence in May 1992 as a successor to the Office of Scientific Integrity, we are referencing a period of about 8.5 years. The subjects were not a random sample of members of the scientific community. They were scientists accused and found guilty of misconduct. Of the 92 closed cases they examined, the most common offences included falsification (39%) or fabrication and falsification (37%), with plagiarism making a healthy showing as well. The charges were raised against assistant professors (12%), associate professors (13%), full professors/ department heads (9%), graduate students (12%), postdocs (13%), and technicians or research assistants/associates (24%). 17% of the sampled respondents didn't fit any of those classifications.

However, the researchers were looking for empirical data about why scientists engage in the behaviors that fall under scientific misconduct. We are not going to present details of the methodology used by authors to find answers to their questions in infinite piles of folders filled with data. We would like to refer all interested readers to the source materials, and we will focus on what Davis and his coworkers have discovered.

After collecting every possible excuse that was possible to find in the testimonies of all the scientists, they have grouped them together into 44 concepts. In order to find patterns in their data, the authors employed multidimensional scaling and cluster analysis. There was not a single case file in which all 44 of the factors implicated in research misconduct were found; at most, a single case file pointed to 15 of these factors. The researchers generated plots and matrices to identify how the various factors implicated in research misconduct coincided in these 92 case files, such as which ones seemed to frequently travel together, and which ones were hardly ever cited in the same case. Potentially, the factors that repeatedly coincide, seen as clusters, could be understood in terms of a new category that covers them (thus reducing the list of factors implicated in research misconduct to a number smaller than 44).

Researchers identified seven such clusters in their analysis of the data. Let's take a look at how these factors ended up clustering (and the labels the researchers used to describe each cluster), and then discuss the groupings:

Cluster 1 — Personal and Professional Stressors:
8. Stressful Job
9. Supervisor Expectations
12. Lack of Support System
14. Overworked/Insufficient Time
17. Insecure Position
18. Pressure on Self/Over-Committed
19. Desires to Succeed/Please
20. Personal Insecurities
22. Poor Judgment/Carelessness

29. Personal Problems
30. Psychological Problems
36. Denial of an Injury
40. Denial of Negative Intent

Cluster 2 — Organizational Climate Factors:
6. Professional Conflicts
10. Insufficient Supervision/Mentoring
11. Non-collegial Work Environment
13. Substandard Lab Procedures
15. Poor Communication/Coordination
39. Lost/Stolen/Discarded Data
41. Reliance on Others/Permission
44. Condemnation of the Condemner

Cluster 3 — Job Insecurities:
3. Inappropriate Responsibility
5. Poor Supervisor (Respondent)
16. Competition for Position
32. Language Barrier

Cluster 4 — Rationalizations A:
23. Lack of Control
25. Jumping the Gun
37. Lie to Preserve the Truth

Cluster 5 — Personal Inhibitions:
4. Difficult Job/Tasks
26. Frustrated

Cluster 6 — Rationalizations B:
21. Fear
28. Apathy/Dislike/Desire to Leave
33. Restoring Equity
35. Avoiding Degradation
42. Slippery Slope

Cluster 7 — Personality Factors:
24. Being impatient
27. Laziness
31. Character Flaw
34. Recognition
38. Public Good over Science
43. Amnesia[3]

As you can clearly see, cluster one basically describes what people in academia know as the 'publish or perish' culture. Cluster two points to factors typical for large corporations and models of interpersonal relations commonly found in such corporations. Cluster three relates to how scientists perceive their job security. Other clusters' names describe their contents fairly well.

What conclusions can be drawn from this research? Are scientists working in environments filled with irresistible temptations, just like Stapel? Maybe the reasons for misconduct actually reside within scientists? Or perhaps they start their scientific careers full of love, honesty and good intentions, only to be spoiled by the evil environment? Well, you need to remember that all of the causes presented above were actually communicated by people who were charged and found guilty in a court of law. It is quite probable that they are nothing more than their attributions and excuses. That's why it is extremely hard to draw any hard conclusions regarding those matters. Among the collected concepts and clusters we can however find clues as to where to look for leaks in the dam of scientific self-control mechanisms. We can try to find where it hurts scientists hard enough that they do not hesitate to betray the rudimentary message of science – fidelity to the truth.

It would be a huge mistake if in our attempts to analyze the causes of pathologies in psychology, we focused only on the comments of those charged with misconduct. In order to understand how we have found ourselves here, at the territory occupied by frauds, we must take a closer look at scientists. Let's look at them as ordinary human beings. Scientist are people who eat, sleep, sweat, have sex, go to the toilet. They can be envious, generous, greedy or lavish. There are virtuous people among them. There are also villains and scoundrels. When we look at scientists in the way we look at every other human being, we need to take into account four basic things that lead humans off the path of honesty:

1. Size of the temptation
2. Possible rewards and punishments
3. Chances of getting caught
4. Personal ethics

In 1920s, researchers showed that there is no such thing as a permanent trait called honesty. The hypothesis that honest people will remain honest in all circumstances (and dishonest people will always act dishonestly) turns out to be false.[4] Our honesty depends mostly on the size of the temptation, and in this regard, psychology has slightly changed over recent years. A Psychologist working as a researcher can expect a certain degree of social respect and extra money from research grants, together with fame, recognition and other privileges that come along when they make a breakthrough discovery. Research psychologists, especially when compared to scientists working in other fields, do not have too many spectacular possibilities in front of them. Tangible benefits from biotechnological, genetic, pharmaceutical or medical research are incomparably larger than what could be expected from cultivating psychological plots. True, tenured professors in America earned on average $135,000 in 2012, more than judges did.[5] However, in many European countries, their salaries are significantly lower. It is not unusual, even for department heads, to seek additional employment in order to provide for a half-decent life.

Psychologists can also rarely count on gaining worldwide fame, and even if they do, it's rather for some kind of media popularity. The Wall of Fame has been reserved for physicists, biologists and medics. The simple fact that there is no Nobel Prize awarded for research in social sciences already explains a lot. The Nobel Prize awarded to psychologist Daniel Kahneman was in fact the Nobel Memorial Prize in Economic Sciences (commonly referred to as Nobel Prize in Economics), awarded by Sweden's central bank. Though psychologists generally have good career possibilities, both in terms of making money and gaining social recognition, they are not significant enough to consider these as the main reason for their misconduct. We would even take the risk of saying that those factors actually limit the number of abuses in psychology. Having said that, they cannot be disregarded as a motivation, as we have shown in previously described cases. The temptations of research psychologists are rather insignificant, although their power may lie in their numbers: another published article, another appearance on television after "discovering" an interesting (not necessarily causative) correlation, an interview in the local newspaper, some money from a new research grant or paid lecture, applause from the audience, and so forth.

The popularity of psychology as a science is constantly changing. Over the past decades, we have witnessed the incredible growth of neuroscience. It is no longer a part of biology and it has become a truly interdisciplinary science. Numerous ambitious psychologists devote their time to research in neuroscience, and this discipline is much closer to the "hot" sciences. Many discoveries in this field could potentially be applied in medicine or education. Research results can be translated into computer science, engineering or even mathematics. Those factors certainly influence how psychologists see their possible career paths and increase the temptation to wander off the path of scientific honesty and ethical practices. It is very likely that neuroscience will continue to grow despite increasing numbers of justified criticism regarding its explanatory value and the hopes pinned to it.[6]

Another very common temptation, no matter how mundane it seems, is simply related to maintaining someone's status quo at a university. It is true that after reaching a certain position at a university (tenure), one can do nothing for the rest of his or her life, but *something* has to be done to get to this position in the first place. Many universities require constant research activity from their employees. This means that even if scientists do not even think about pursuing fame, power, recognition or money, they simply want to maintain their previously gained position and status. A desperate attempt to do so was previously mentioned by Kingsley Amis in the story of *Lucky Jim*. People need to continue researching and publishing to keep their jobs. Unfortunately, it often does not really matter how valuable the research is or what the quality of published paper is. What matters is the number of publications and their impact factor. Publish or perish! This popular saying among scientists perfectly describes the pressure of everyday life they are subjected to. Some call it an institutional pressure that motivates them to do their work. Others perceive it as a pressure 'from the bottom'—the more you publish, the higher you will climb up the corporate ladder, the further ahead you will be in the rat race and more awe and recognition you will receive. The first group avoids punishments, the other craves awards. Both might choose the evil path to achieve their goals. Let's take a look at the numbers caused by publication pressure.

> The first of the major recent scandals, that of William Summerlin at the Memorial Sloan-Kettering Cancer Center, was in the research group supervised by Robert A. Good. In the 5 years before 1975, Good had published 342 papers, an average of 68/year (16 as first author, 325 as coauthor with at least 136 coauthors). Several other frauds were associated with medical research groups that have been extremely prolific. In the period 1975 to 1980, which preceded the discovery of fraud in their laboratories, Eugene Braunwald had published 171 papers, an average of 29/year. Philip Felig had published 191 papers, an average of 31/year. Ephraim Racker, who is not on a medical faculty, had published an average of 16/year.[7]

In contrast to the presented data, we shall mention that a very active scientist publishes on average three to four articles per year at the top of his career. In the entire time of employment of the average researcher in any given university or other research institute, the average is much lower.

Another factor influencing the frequency of dishonest behaviors relates to the punishments and awards awaiting scientists in their working environments. The severity of penalties threatening dishonest researchers whose frauds are detected is so ridiculously small, that when we put them against the potential benefits on the other pan of the scale, the scale will not even twitch. The heaviest sentences against dishonest psychologists were described in Chapter 3. Despite the fact that some court judgments are a mere joke, let's not forget that they are probably the most severe sentences ever ruled against researchers in psychology. Many of them still continue to build their careers, like Milena Penkowa. Others continue to exploit their fraud, like Diederik Stapel.

When we try to look for possible benefits that would strengthen good research practices, we can hardly find any. John Ladd at the American Association for the Advancement of Science Workshop on Professional Ethics proposed in 1980 the one and only benefit that we have managed to find in available literature. He suggested that certificates might be provided to professionals for prominent display in their laboratories, offices, and such.[8] Why not? Nothing so far suggests that the academic environment considers awarding researchers for their integrity, honesty, reliability or accuracy.

The third factor, the chance of getting caught, has different significance. We have shown in previous chapters that holes in this sieve are so large that almost anything can slip through. We believe that the causes for scientific dishonesty may have three sources:

1. The nature of the research field,
2. The positioning of psychology amongst other sciences,
3. The construction of a system of public control of science.

Psychology is particular in that, due to the fact that it is a social science, it reaches into the methodology of life sciences. Psychology is no stranger to precise and controlled experiments, yet phenomena it describes are multifactorial in their nature. This means there is a high level of complexity that may lead to virtually unlimited possible interpretations of failures during replications. That is the reason why many science philosophers, including such giants as Paul Meehl or August Fridrich von Hayek did not believe that social sciences could be "scientific" enough, nor that they could become such in the foreseeable future. They claimed that there are too many variables that are difficult to measure and/or isolate, making it impossible. This favors mistakes and abuses, and the control mechanisms of obtained results very often fails.

Other scientists often have even worse opinions of the social sciences. The famous American physicist, Nobel Prize laureate, Richard Feynman, spoke about social sciences during his speech at the graduation ceremony at Caltech in 1974:

> I think the educational and psychological studies I mentioned are examples of what I would like to call cargo cult science. In the South Seas, there is a cargo cult of people. During the war, they saw airplanes with lots of good materials, and they want the same thing to happen now. So they've arranged to make things like runways, to put fires along the sides of the runways, to make a wooden hut for a man to sit in, with two wooden pieces on his head [like] headphones and bars of bamboo sticking out like antennas--he's the controller--and they wait for the airplanes to land. They're doing everything right. The form is perfect. It looks exactly the way it looked before. But it doesn't work. No airplanes land. So I call these things cargo cult science, because they follow all the apparent precepts and forms of scientific investigation, but they're missing something essential, because the planes don't land.[9]

There is a lot of truth in Feynman's comment. If we look at the social sciences from a distance, we can see the obvious relationship between the complexity of the matter of psychological studies on one hand, and on the other hand the abuses mentioned in this book. We are convinced that among thousands of "cargo cult followers," there are at least a few real scientists unraveling the mysteries of our complicated psychological lives.

In order to understand the importance of the next factor, the positioning of psychology among other sciences, we should take a closer look at the history of science. It might resemble the history of a hit parade. Perhaps only the most popular "hits" stayed on the list slightly longer than musical record-ings. Darwin's theory dominates most discussions, not only among scientists, but also in cafés for many years, and it earned its place in the top-ten. Ein-stein's theory of relativity is another example of an undeniable scientific hit. When completely unknown Gregor Mendel was counting his peas over a century-and-a-half ago, he did not even suspect that he was laying the foun-dation for a new scientific discipline that dominates today's top-ten lists of scientific achievements. There were just a few such hits in psychology. Freud's psychoanalysis was one of them; (we will discuss it in more detail later). There were some creative ideas of behaviorists and... that's about it. We are talking about scientific hits in a very wide sense, theories that an aver-age person can recall, at least, the theory's name. Currently, neuropsychology seems to be in fashion. It is also in fashion to apply it to such areas as mar-keting (where it's directly translated into money). The flow of cash seems to move "psychological recordings" higher up on the hit parade. Despite the current fashions, there are numerous doubts as to its actual applicability. Moreover, neuropsychology is complicated enough that the average consum-er of science is unable to make a sound judgment.

Social interest (or rather, the degree to which scientific theories irritate public opinion), seems to have an influence on what is considered to be "fashionable" in science at a given moment. Fashions also depend heavily on the distance between theories or particular discoveries and their practical use, (including consequences of their use). Today, even a small step forward in areas such as genetics, biotechnology or immunology immediately translates into various applications, and is immediately followed by streams of money, fame and recognition. Those are, after all, the hot sciences.

Interest in particular areas can be represented quite well with some pa-rameters calculated by the Institute of Scientific Information in Philadelphia. For example, the citation index, which is the ratio of number of citations to the total number of publications in a given field, shows huge differences between citations in various disciplines. Values showed below (from 1995-2005) represent how scientists analyze and process their colleagues' texts. Data presented below shows only the "hottest" areas of psychology, along with psychiatry and other "colder" disciplines:

– Molecular Biology and Genetics – 24:6
– Immunology – 19:6
– Biology and Biochemistry – 15:37
– Microbiology – 14:0
– Space Science – 11:6
– Psychology and psychiatry – 8:2
– Economics and Business – 4:1
– Social Sciences – 3:5

Another interesting parameter is the citation count, the average number of citations from one publication. These parameters show large differences between various disciplines. Between 2000 and 2010, the average number of citations from papers published in the field of molecular biology was around 25:5, in immunology 21:2 pharmacology averaged 12:2, economics and business only 6:2, social sciences 4:7 and mathematics 3:5.[10]

There are two main consequences of being a scientist in one of the popular disciplines. First, there is a significant threat to one's development caused by commercialization, and the resulting corruption.[11] The other is exposure to immediate criticism. Quite often, a new discovery in one of the hot fields gets verified on the day of publication. Important results are almost never left without validation and are often replicated. Even tiniest attempts to cheat will be picked up immediately and dishonest authors are publicly exposed. Jan Hendrik Shön, a German physicist working for Bell Labs can serve as a good example. He was working with a group involved in research of superconductors and electronics, and the Nobel Committee was considering the results published by this group. Unfortunately, they turned out to be faked. Another example from recent years concerns Victor Ninov from Berkeley National Laboratory. His interpretation of observational study resulted in an announcement of the discovery of a new element with the atomic number 118. In both presented cases, virtually immediate replications and analyses allowed quick exposure of fraud.

It is also worth mentioning a discovery of the researchers involved in the OPERA experiment. They claimed that the speed of neutrinos sent to an Italian INFN Gran Sasso Laboratory 730 km away exceeded the speed of light in space. The results were based on the observation of over 15,000 neutrino events. Researchers concluded that the neutrinos traveled at a velocity 20 parts per million above the speed of light, nature's cosmic speed limit. Immediately after the observation became public, a wave of criticism followed. A few scientists from the team refused to have their names published, claiming that results were announced too early. On September 22, 2011 CERN appealed to the scientific world to verify those findings. In the end, it turned out that the results obtained in this experiment were caused by an improper connection of the GPS receiver to the computer that was measuring the neutrino travel times.

Psychology, apart from large areas of neuropsychology, consistently remains outside the hot trends. This protects it from the negative consequences of commercialization, but also leaves fraudsters unpunished and unexposed, as most published results are not of much interest to their peers. Perhaps the only exceptions are in areas touching the issues of political correctness, as in the case of Helmuth Nyborg, a controversial Danish psychologist. His paper, "The Decay of Western Civilization: Double Relaxed Darwinian Selection," appeared in 2011 in *Personality and Individual Differences*, a prestigious journal in the field, and quickly aroused the fury of a group of Danish scientists. As a consequence, the Danish Committees for Scientific Dishonesty called for retraction of this article over concerns about referencing and authorship! It was not the first time Nyborg infuriated his colleagues and the public. Earlier he tried to prove that men have higher IQ than women and that there is biological explanation for those differences.[12] Apart from those extremely sensitive areas that become popular for a short while, psychology will not be able to join the hit parade for a long time. The present conditions do not encourage scrupulosity or meticulousness and certainly has an impact on the number of frauds committed.

The construction of a social control system of science is the third area that significantly impacts the chances of discovering research fraud. Weakness in these systems can be traced back to its fundaments, the educational system. We have both attended two different universities at different times, in different disciplines. Neither of us was exposed to the ethical problems in scientific work. Similarly, while working on our PhDs, nobody discussed research ethics with us. None out of a few universities we have ever worked at had research ethics as a mandatory part of their curriculum. Ethical issues were discussed with students regarding medical or psychological clinical practice, but never in regard to research. It has been changing slowly in the recent years, but workshops focused on research ethics are still sparse. It would be good if lecturers spent at least one hour during methodology courses to discuss scientific ethics and values.

The situation in the United States is significantly better. In response to the increasing numbers of misconduct since 1989, the National Institutes of Health requires that research institutions receiving grants hold regular courses in ethics. It also requires doctoral students to attend such courses, but still, the majority of courses in psychology are not subjected to those regulations. They happily hand over diplomas without ever making sure that fresh psychologists know any ethical code at all. When a young scientist does not know what norms are expected of him, how can he/she stick to them? Of course some issues, like those relating to the falsification of research results or plagiarism, are derived from general ethics, but all those other subtle rules are not at all always clear and straightforward. How can a young scientist know that citing sources after someone else, without referring to the original is a sinful practice? Do all scientists know what they are expected to do with their research results? Is it clear to them that hiding unsatisfactory results of experiments is considered as misconduct? What about holding up a publication? Or unnecessary or excessive publishing? What should he do and how should he react when he discovers dishonest behavior in his colleague or superior? Rules and regulations regarding the treatment of humans and animals quite often remain completely unknown.

Such deficiencies in education have a significant impact on scientists' behaviors. When one of us was conducting his first research, he was meticulously gathering all forms, documents, questionnaires, lists, correspondence, and all other materials created or used in his work, being mistakenly convinced that they will be used as a part of evaluation. No such thing happened. They were packed into a large box and left forgotten between old invoices somewhere in his attic. In every following research we conducted, we were both less and less scrupulous. Why? You might find it hard to believe, but raw data and basic documentation from a single study can take thousands of pages and many boxes of recyclable paper. If anyone was ever interested in the research, they usually asked for processed data, excel spreadsheets, or even only the final results of statistical analyses. One of us was very surprised when during his research at the University of Psychology in Bielefeld, Germany, he was asked to collect all questionnaires from his experiments, sort them, bind them together and send them to university's archives! You can probably still find them there. Anyone can reach for them and analyze them again. And we wonder, where are the archives containing raw data from most of the research ever conducted? Even in extremely regulated areas, such as approval for medicinal products to be used by humans, the British Medicine and Healthcare Products Regulatory Agency (MHRA) destroys all archived data after 15 years.[13]

> Scientists need well-defined and clearly written professional codes of conduct. Additionally, well-defined procedures for handling accusations of misconduct should be developed, agreed upon, and implemented. Those accused of misconduct should be afforded, at minimum, the same rights that are given to those who participate in the conventional legal system. At the beginning of their professional training, scientists should know the nature of their punishment should they fail to abide by their code of conduct. They should be frequently reminded of their professional obligations, whether by seminars, informal discussion groups, formal initiation into appropriate professional societies, or other means.[14]

The chance to create solid ethical foundations in young researchers dramatically diminishes with changes in educational systems. We talk about substituting former relations between PhD students and their supervisors with doctoral studies now aimed at the mass-production of doctors. A few years of joint cooperation and individual work under one supervisor provides far more guarantee that the work of a young scientist will be properly monitored and evaluated. Thus before, it was easier to absorb the master's values and it was easier to discuss any dilemmas with a senior researcher. Doctoral studies no longer provide such opportunities because contact with senior staff has been formalized; talks and discussions have been replaced with workshops, seminars and courses. This slow extinction of master–apprentice relations is just another element in the education system that weakens science's moral backbone. It is much easier to ignore written instructions than to betray the very values professed by your master!

We are not alone in our views. Horace F. Judson, who spends a lot of time and energy researching crimes against science says,

> To be sure, instruction in ethics can prepare otherwise naive young research scientists to recognize some of the dilemmas and plausible temptations that may take them by surprise. All to the good: yet it is not obvious that norms, to set the social conditions for psychological discomfort, is not a class once a week for a semester, but the day-to-day practice of a well-run laboratory – the example in action of a mentor. "In those days, when you came to a new lab, the first thing you did was calibrate the weights," Alfred Nisonoff said, and the remark contains not nostalgia but something much larger, however laconically stated, a stern, plain realism about the way good science proceeds.[15]

But surely creating a solid ethical foundation is not the only way to minimize the chances of unfair and dishonest behavior. Most scientists are employed by institutions, which can also guard and promote ethical values and help to detect scientific misconduct. In order to do it, they create rules and regulations that represent their policies related to abuses and rules of conduct in such cases.

> Being well aware of institutional policy prevents researchers from involving themselves in misconduct and encourages them to report suspected misconduct. Wells (2008) supported this view by reporting that scientists who read their institutional policies are more likely to make allegations. Since the 1985 Health Research Extension Act, which required universities to develop their own institutional policy to deal with scientific misconduct, almost every institution funded by the federal government has its own policy. Moreover, 78.4% of accredited medical schools in the US have their own guideline (Douglas-Vidas, Ferraro, and Reichman, 2001). These figures imply that establishing an institutional policy regarding research ethics is stabilized at the institutional level.[16]

Unfortunately, having a proper code of conduct is not enough. When confronting misconduct or abuses, institutions are reluctant to disclose them.

> Why is an institution reluctant to report misconduct of its constituents? First, it [would] damage the institution's reputation (Altman, 1997). Second, it would take away the material resources of the institution. More specifically, an institution would lose its overhead cost or it should pay for penalty under the False Claim Act.[17]

The verification system of scientific publications run by journals also leaves a lot to be desired. In today's world, nothing prevents authors from supplying raw data during the manuscript submission process. It is also hardly any challenge to verify this data, even randomly, by journal editors and reviewers. This of course will not prevent fraud, but rather will force potential cheaters to carefully falsify raw data in order to provide specific results, but this is neither as easy nor as fast as the simple fabrication of end results like Cyril Burt used to do.

At this point, we need to ask why research projects financed by public money (grants) are not externally audited? Most of them end with a final report of some sort and with a publication or two, but the entire process of deriving those results remains hidden, which encourages potential abuses. Many scientists would probably oppose such a proposal, claiming that it would be an assault on their sovereignty and independence. However, we do believe that such mechanisms should be introduced. Take a look at how Maciej Grabski, former president of the Foundation for Polish Science, argues for such a case:

> It is often stressed that it does not matter whether [the] probability of an occurrence of scientific misconduct is one per thousand, or one per hundred thousand. It also does not matter what the probability of lightning hitting a particular house is. Every building should have a lightning rod – because if such event occurs, the losses will be colossal.[18]

Areas for possible misuses expand together with progressive specialization. Specializing is necessary as it allows further progression of science. Today, we are able to look into the deepest nooks and crannies of matter. We prepare not only cells or their components, but chromosomes or even particular genes. Powerful telescopes penetrate the darkness of space, revealing stars and planets further and further away. We can track single nerve impulses in our brains and detect the slightest chemical disturbances in our bodies. The expression 'renaissance man' is nowadays only used to appreciate someone's erudition. There is no place in science for people with wide interests or broad knowledge.

On the other hand, specialization brings certain new problems and consequences to the table. Alvin Toffler summarized them by saying, "One of the most highly developed skills in contemporary Western Civilization is dissection: the split-up of problems into their smallest possible components. We are good at it. So good, we often forget to put the pieces back together again."[19] Many other scientists have a similar view of the problem. Here is another example from Roman Zawadzki, "[The] fragmentation and dissection of all sciences has reached absurd levels. The vast amount of specialist knowledge crossed the critical point of perception and arrived at a place where only two famous laws of chaos apply – the Murphy's Law and Peter's Principle. The first one, in one of its' versions, tells us that increasingly more people are proficient in increasingly smaller areas of knowledge; the second one – that more and more people reach the ceiling of incompetence."[20]

Due to the increasing levels of incompetence, specialization usually provides a safe and convenient asylum for a scientist. Specializing in one particular phobia, narrow timeframe of human development or even in a single emotion, say for example in shyness, can yield many profits for a psychologist involved in research. Primarily, it is possible to thoroughly explore and become familiar with such a narrow field. It is then also quite difficult to find another scientist specializing in the same area; therefore, such scientist automatically becomes a unique expert. The largest advantage comes from the fact that there will be only few other people able to assess the quality of his work. A reviewer will not be able to clearly appraise or evaluate his hypotheses. This opens another door to the world of scientific dishonesty: highly specialized dishonesty.

The last element in the social control system of science that we would like to point our readers to is the lack of a self-cleaning tendency in the scientific world. Even if all cases of misconduct previously described were exposed due to the vigilance of the scientific community, the consequences faced by those disclosing the crimes make an interesting case report on how whistleblowers are treated in the academic community. Most of us would not call the police if our good friend or relative got in a car under the influence, despite the fact that our phone call could possibly prevent a tragic, potentially fatal incident. The same applies to the scientific community, and for this reason alone, many frauds remain undisclosed. Chapter 4 of this book, devoted to the problem of raw data, presents research results suggesting that in the USA alone, over 33% of anonymous responders admitted to at least one of 10 unethical behaviors in past three years.

Other research has shown that 36% of observed cases of scientific misconduct were never reported. The presented results confirm our assumptions that scientists do not want to report abuses made by their colleagues to authorities. Scientists do not readily get into the role of a whistleblower. To answer why, let's take a look at the pill that Robert Sprague had to swallow when he decided to stand against the crimes of Stephen Breuning. His 31-page long report describing events that took place around Breuning's case entitled, *Whistleblowing: A Very Unpleasant Avocation*, is in reality a bitter story describing his own experiences. He wrote in the prologue,

> This article is a brief accounting of the many events surrounding this scientific fraud case with a focus on the great reluctance of universities and federal agencies to investigate vigorously an alleged case of scientific misconduct when it involves members of their own faculties or grant recipients. Even after this case was publicized in the media, there were attempts to harass and intimidate me as I continued to point out [the] delaying tactics of the organizations responsible for the oversight of Breuning's federal grant funds and research projects. Unfortunately, intimidation of whistleblowers is quite common.[21]

His own experiences authorized him to say such words. The first result of making Breuning's case public was that after 17 years of federal funding, his grants were not renewed. But Sprague was in fact lucky. Clifford Richter, a medical physicist in Columbia, Missouri, acting in accordance with his responsibilities, reported a breach of safety procedures to the Nuclear Regulatory Commission. He reported an incident of leaving radioactive isotope inside patient's body that resulted in the patient's death. Initially, Richter was boycotted by his coworkers, shifted to a position with lower responsibilities and then fired. This demonstrates how cruel the academic world may be to whistleblowers and how tolerant it is to offenders.

There are hundreds of similar examples, and the numbers confirm this grim reality. James Lubalin and his coworkers found that 69% of whistleblowers experienced negative consequences after reporting incidents or fraud. They include being pressured to drop (42.6%) or counter their allegations (39.7%), being ostracized by their colleagues (25%), experiencing material deficits such as decrease in research support (20.6%), staff support (10.3%) or denial of salary increase (11.8%), promotion (7.4%), and most seriously, being fired or being denied renewal of contract (11.8%).[22]

Problems faced by whistleblowers led to numerous discussions in academia devoted to the necessity of developing rules and regulations in order to protect people who decide to report abnormalities. Several countries and institutions have already applied certain regulations. In the United Kingdom, Germany and in the USA, similar, decentralized systems were adopted where matters related to scientific misconduct is dealt with on the local level by scientific institutions, but monitoring and appeals are handled by government agencies, usually those involved in financing research projects. The question remains whether universities or research institutions are prepared for handling such cases. Regardless of their ability to do so, such solutions seem to be better than those in the rest of the world where above-average honesty of scientists is simply assumed, while self-regulation of the scientific community and the protection of whistleblowers is ignored or belittled.

In the United States of America, after many years of endless efforts, the protection of whistleblowers working for federal scientific institutions was finally introduced as well.

The Senate's unanimous approval of comprehensive bipartisan whistleblower legislation comes after a decade-long effort to protect the federal employees who risk their careers to protect the rest of us. UCS urges the House to quickly pass the bill to make it law this year so we don't need to begin the process all over again in January. The Whistleblower Protection Enhancement Act (S. 372) makes significant progress in giving millions of federal workers the rights they need to fully serve the public interest. It restores best practice free speech rights by overturning years of flawed judicial and administrative decisions that had left current whistleblower law in tatters and federal workers unprotected.[23]

Certain modifications of the American model that significantly increase the efficiency of the whole system were introduced in Germany, where an institution of a third-party ombudsman was established. Anyone who knows about any kind of scientific misconduct can contact an ombudsman with full confidence and ask for help and support. If an ombudsman believes that a particular case represents significant misconduct, he will report it to the relevant authority and request an official investigation. This way, adequate protection of the whistleblower is also assured.

Probably ill-conceived professional solidarity is the reason for hiding many frauds and offences against science. It is also responsible for the overall decline in confidence that people have for science and scientists. To underline how important it is to protect whistleblowers and show how incredibly complicated a task that actually is, let us once again cite Robert Sprague:

> One important facet of the Breuning case must be remembered. In contrast to most other cases, I was [a] tenured full professor at a major state university with considerable research recognition when I blew the whistle on a non-tenured, new assistant professor at another university. Even with all of these important factors in my favor, I was barely able to blow the whistle successfully. I certainly did not accomplish the task without much emotional trauma, which lasted several years (Sprague, 1987), personal finance expense, hundreds of hours of time (Subcommittee of the Committee on Government Operations, 1988), and professional loneliness, with little overt support from colleagues. To potential whistleblowers, consider seriously the following advice:
>
> 1. Document as many of the events and facts as is humanly possible.
> 2. Seek support from family and trusted professional friends.
> 3. Be as truthful as you know how to be.
> 4. Do the right thing, even though such action may place you at considerable risk.[24]

In order to equip whistleblowers with tools better than just good advice, scientific commissions in many countries created their own ethical codes and good practices. Unfortunately, they usually contain declarations and rules but lack sanctions and appropriate implementing rules. "One ORI report in 2000 reviewed 156 institutional policies consistent with the ORI guideline and found that only 29% of them explicitly state the duty of reporting scientific misconduct. Also, except for two institutions, no policy clearly states if the institution will pursue anonymous allegations.[25] We can only hope that all initiatives described above will serve as a good starting point on a journey to a better system of public control in science. Postulates to further seal the system of detection of scientific misconduct can be reinforced with an analogy presented by professor Grabski:

> The need for the existence of such codes of good practice can be explained by using an analogy from the world of automobiles. When there were only a few cars on the road, it was enough to announce that drivers should act responsibly and carefully, according to general ethical rules, or even to the Ten Commandments. With the increase of traffic, such simple rules were not enough, and appropriate regulations had to be introduced, with precisely defined rules, procedures necessary to ensure [the] efficiency of those regulations, together with ap-

propriate sanctions. Otherwise the frequency and probability of accidents, including those involving responsible and careful drivers, would severely jeopardize car travel, or possibly even render it impossible.[26]

Finally, let's bring our attention the third and last factor, the ethics of scientists. Can we assume that ethics among scientists do not differ significantly from the ethics among general population? It might seem unfair, as we are used to perceiving scientists as priests called to serve the truth. Unfortunately, the reality is rather blunt. Modern day psychologist-researcher is a profession, not a divine calling, (as it actually used to be several decades ago). Its image no longer resembles the stereotypical absent-minded, distracted, emaciated enthusiast whose entire existence is focused around solving nature's puzzles, nor is finding answers to problems deemed as the ultimate reward. Today's scientists are not much different from yuppies in a rat race on career ladders. Quick, preferably brilliant dissertations, articles in renowned journals, international conferences, participation in research orchestrated by famous luminaries of the scientific world, professorship, early tenure... Now let's add some everyday dreams: perhaps a sport car, an apartment in an expensive area or a luxurious house... Those are the goals of modern scientists. If there are shortcuts that would take you to your goals, why not give it a go, just like Cyril Burt, Stephen Breuning, Karen Ruggiero, Milena Penkowa, Diederik Stapel and hundreds others have tried? Many more are still trying.

The assumption that ethical standards among scientists and other people are the same is in our opinion too optimistic. We will never know for sure, but there are good reasons to believe that scientists are actually more prone to unethical behaviors than a statistical citizen. And no, not because scientist live and work in environments less protected and less monitored than let's say banks, post offices or supermarkets. It's because scientists are usually people with above average intelligence, and this factor alone can lead to higher incidences of misconduct. Research conducted by Bella DePaulo, a psychology professor at University of Virginia, can serve as a simple illustration of the previous sentence. DePaulo has shown that people with higher IQ more often "miss the truth." She concluded that, on average, about 20% of people lie (usually white or innocent lies) during a 10-minute conversation. Among people with above average intelligence, the percentage rises, up to 33%. DePaulo tried to explain this phenomenon by richer vocabulary and above average self-confidence of people with higher IQs. It could mean that intelligence not only makes lying easier, but also makes us more likely to lie by providing higher efficiency and dexterity in our attempts to deceit. If that's the case, why would scientists be any different? The one thing that they usually have in abundance is intelligence.

Of course we are not saying that there is a simple linear relationship between intelligence and dishonesty. We both personally know many extremely intelligent people who are crystal-clear people, (and just as many who are scoundrels). But when following the paths chosen by people like Burt, Breuning or Stapel, invariably there appears a convincing thought that it was a combination of high self-confidence and their high intellectual prowess that pushed them towards wrongdoing. We believe that there are more similar villains, led by the very same temptations. We might never hear about them, even if they wandered off the path of truth just as far as those mentioned in this book.

In the previous chapter we cited the words of Robert Kurzban, who compared the crisis in psychology to weather, "we all talk about it but no one really does anything about it." So has the world of Giordano Bruno, Charles Darwin and of all other great, uncompromising seekers of the truth been replaced with a world without punishments, a corporation focused primarily on preserving our convenient existence, our bureaucratic homeostasis? Have we all conformed to never undermine the existence and homeostasis of the officials of science that surround us?

---

[1] D. Borsboom and E. J. Wagenmakers, "Derailed: The Rise and Fall of Diederik Stapel," *Observer, 26* (January 2013): http://www.psychologicalscience.org/index.php/publications/observer/2013/january-13/derailed-the-rise-and-fall-of-diederik-stapel.html

[2] M. S. Davis, M. Riske-Morris and S. R. Diaz, "Causal Factors Implicated in Research Misconduct: Evidence from ORI Case Files," *Science and Engineering Ethics, 13,* (2007): 395-414.

[3] Ibid.

[4] H. Hurstone and M. May, *Studies in the Nature of Character: Studies in Deceit.* (New York: McMillan, 1928).

[5] No Author, "How Science Goes."

[6] See S. Satel and S. O. Lilienfeld, *Brainwashed: The Seductive Appeal of Mindless Neuroscience,* (New York: Basic Books, 2013); W. Uttal, *The New Phrenology: The Limits of Localizing Cognitive Processes in the Brain.* (Cmbridge: MIT Press, 2001).

[7] P. K. Woolf, "Pressure to Publish and Fraud in Science," *Annals of Internal Medicine, 104,* (1986): 254-256. The example presented here describes William Summerlin, a dermatologist working in Sloan-Kettering Cancer Center in New York in 1974 who announced a revolutionary skin transplantation technique, free of risks associated with graft rejection by host. Soon one of his lab technicians revealed that Summerlin simply painted the white fur of lab mice with black marker. It is one of the most infamous scandals in cancer research history.

[8] R. Chalk, M. S. Frankel and S. B. Chafer, *AAAS Professional Ethics Project,* (Washington: American Association for the Advancement of Science Publication *8* 1980).

[9] R. P. Feynman, "Cargo Cult Science," *Engineering and Science, 37* (June 1974): 10-13.

[10] Thomson-Reuters, Web of Science (thomsonreuters.com)

[11] See S. Krimsky, *Science in the Private Interest: Has the Lure of Profits Corrupted Biomedical Research?* (Lanham, MD: Rowman & Littlefield, 2004).

[12] A. Marcus, "Citing 'Scientific Dishonesty,' Danish Board Calls for Retraction of Controversial Paper on Decline of Western Civilization," *Retraction Watch* (November

2013): http://retractionwatch.com/2013/11/13/citing-scientific-dishonesty-danish-board-calls-for-retraction-of-controversial-paper-on-decline-of-western-civilization/#more-16534

[13] P. C. Gøtzsche and A. J. Jørgensen, "Opening Up Data at the European Medicines Agency," *British Medical Journal, 342,* d2686. (2011): http://www.bmj.com/ content/342/bmj.d2686

[14] V. N. Hamner, "Misconduct in Science: Do Scientists Need a Professional Code of Ethics?" (1992): http://www.files.chem.vt.edu/chem-ed/ethics/vinny/www_ethx.html

[15] H. F. Judson, *The Great Betrayal. Fraud in Science.* (Orlando: Harcourt, 2004): 370-371.

[16] J. Lee, "The Past, Present, and Future of Scientific Misconduct Research: What Has Been Done? What Needs to be Done?," *The Journal of the Professoriate, 6,* (2011): 67-83.

[17] Ibid.

[18] M. W. Grabski, Uczciwość i wiarygodność nauk. *Osiągnięcia Nauki i Techniki Kierunki Rozwoju i Metody, Konwersatorium Politechniki Warszawskiej, 7,* (2006): 1-11.

[19] A. Toffler, "Introduction," in Order Out of Chaos, I. Prigogine and I. Stengers, (New York: Bantam Books, 1984), p. xi.

[20] R. Zawadzki, *Magia i mitologia psychologii.* (Warszawa: Wydawnictwa Uniwersytetu Warszawskiego, 2008), 37.

[21] Sprague, "Whistleblowing."

[22] J. S. Lubalin, M. E. Ardini and J. L. Matheson, *Consequences of Whistleblowing for the Whistleblower in Misconduct in Science Cases* (Rockville MD: The Office of Research Integrity, 1995).

[23] Union of Concerned Scientists, "Protecting Scientist Whistleblower," (December 2010): http://www.ucsusa.org/scientific_integrity/solutions/big_picture_solutions/protect-scientist.html

[24] Ibid, 131.

[25] Center for Health Policy Studies, *Final Report: Analysis of Institutional Policies for Responding to Allegations of Scientific Misconduct,* (Rockville, MD: The Office of Research Integrity, 2000).

[26] Grabski, *Uczciwość i wiarygodność nauk.*

PART II

CONQUERING PATIENTS' SOULS: SINS OF PSYCHOTHERAPISTS

Whatever houses I may visit, I will come for the benefit of the sick, remaining free of all intentional injustice, of all mischief and in particular of sexual relations with both female and male persons, be they free or slaves.

From the *Hippocratic Oath*

Psychology doesn't only attract seekers of ultimate truth who focus on empirical evidence behind academic walls. It is not only the constant process of pushing the boundaries of our understanding of human beings, the process of constant challenging of what we know. When one of us taught first-year psychology students and inquired about the motives that prompted them to undertake such studies, more than half of them invariably declared a desire to help others. Many psychologists–therapists are driven by similarly noble motives for helping others. You would expect noble and pure souls among them. Unfortunately, charlatans, megalomaniacs and even sadists lurk among therapists in great numbers. We even suspect that they are more common in this group than among any other profession. Further, when we come across noble and generous people, they are often the victims of rogues and of their own ignorance. Can, however, noble motives and good will justify their behavior? Should therapists be skeptical of the knowledge they acquire? Have they not studied psychology for many years to be able to separate the wheat from the chaff? Unfortunately, facts show that they don't necessarily think in that way.

Looking at therapies, we will repeatedly use the results of empirical research as the principal foundation of our judgments. An attentive reader might accuse us of hypocrisy. At one point, we try to convince them, using a whole section of the book, about the poor quality of research and then we squander it in order to prove our own theses. To avoid such allegations, we should, therefore, explain the grounds for our behavior. There are several of them. First of all, looking at the results of research describing any phenomenon, it is always worth asking ourselves, in what way can they be adulterated? For example, if we have data on corruption, we can be pretty sure that they are understated rather than overstated. There are so many reasons as to why people hide corruption and they are so obvious that it would be difficult to expect realistic results. Thus, with the result of such a study in hand, we can be almost certain that it tells us only that a certain number of people were involved in acts of corruption, but probably there were many more. Similarly, that will occur in the results of research which reports the number of acts of marital betrayal or of improper sexual relations between therapists and their patients. There are very many reasons for this. Many people are interested in glorifying various therapies, and have little desire to criticize them.

Psychotherapy is, beyond all, a market-oriented activity. The main players involved are interested in maintaining its positive image, as demonstrated by the descriptions of numerous scientific frauds. We have not found a single case of fabricated research results used for a critical purpose. Most frauds found were conceived rather to confirm some hypothesis. For this reason, major research studies do not take into account, for example, the results of research conducted by therapists actively engaged in any kind of therapy. There is an overwhelming number of studies showing the positive effects of therapy and a modest number of critical works. Additionally, there are many grounds for such criticism to be avoided, because this usually brings only trouble. We have experienced this ourselves, presented in examples in the third part of this book. Because nobody wants to fake "negative" results for their own glory, there are far less publications that show unfavorable outcomes of therapies. These are usually much more methodologically robust. Therefore, the results of critical studies can usually be trusted with very high probability (as we will show later). Conversely, we should treat results that show surprisingly positive effects with great skepticism. We will confidently approach research conducted by independent institutions not involved (in any way) in market activities, and whose purpose is solely to issue an objective opinion. We shall strongly distrust research, for example, on the effectiveness of psychoanalysis, conducted by declared psychoanalysts. We think that that is an adequate rationale for the careful use of research results.

Taking into account all of the concerns regarding the quality of research conducted by psychologists, we still personally believe that to produce a picture of some part of reality, we have, unfortunately, nothing better than the results of research and our own self-critical mind. As Albert Einstein stated, "All our science, measured against reality, is primitive and childlike — and yet it is the most precious thing we have."

Armed therefore with a critical filter of our own reasoning—the most precious thing we have—let's look through another keyhole, this time into the temple of knowledge, a place where the mysterious academic formulas turn into methodologies of influencing people. Let us have a look at some of the therapeutic systems.

CHAPTER 8

PSYCHOANALYSIS: CASTLE BUILT ON SAND

After a long search, the Australian reporter Karin Obholer finally finds Sergei Pankejev, the 70 year old aristocrat who escaped from Russia at the beginning of the century. She manages to conduct an interview with him. Who is this man after all? Why did the journalist invest tremendous amounts of effort, time and money to reach him? He's none other than the famous "Wolf Man," the greatest therapeutic success of Sigmund Freud, and at least partly, "living proof" of the validity of the assumptions that psychoanalysis was based on.

Sergei Pankejev suffered from depression since early childhood; he experienced anxieties related to an obsessive fear of wolves. They started when Sergei was 4 years old. He'd had a nightmare that a pack of wolves were lurking just outside his home. Starting from 1910, Freud intensively subjected him to his psychoanalysis for 5 years, and in 1918, he finally published his "Wolf Man" case in the article, "From the History of Infantile Neurosis." He claimed that Sergei's obsessions and fears were cured and declared his patient completely free of them.

Freud claimed the cause of Sergei's problems lay in a single event from his childhood. Apparently, as a little child, Pankejev witnessed his parents having sex. According to Freud, this event triggered the Oedipus complex and induced a fear against castration. When Freud was taking care of this patient, he was developing his great theory. His analysis of Sergei's case was essential for key elements of his theory. From this work, Freud derived the concept of the structural model of the psyche and described three parts of the psychic apparatus: id, ego and superego. Freud classified the Oedipus complex and fear of castration not only as the root cause of neurotic disorders, but as the universal cause for forming each individual's superego. This, he maintained, played an important role in suppressing and controlling our primitive impulses, and therefore, it was responsible for preserving the basics of our civilization.

Unfortunately, the reality is a bit disappointing because Freud's alleged success in treating Pankejev was... a lie. The aristocrat still suffered from the very same fears and obsessions that made him visit Freud in the first place; he suffered from them for the rest of his life. He used the services of many psychoanalysts and was hospitalized multiple times, without any effect. When

Karin Obholzer finally reached him, she discovered that Pankejev was receiving a regular monthly payment from an association of psychoanalysts—salary for keeping his mouth shut! Yet, in a spurt of desperation, he revealed to Karin: "The whole thing looks like a catastrophe. I am in the same state as when I came to Freud, and Freud is no more."[1]

The health condition of Pankejev is only one side of the coin. The other facts and assumptions, which were fundamental for Freud's theories, were also false. Freud imposed all major "memories" from Pankejev's childhood onto him. The patient never saw his parents having sex. It was simply impossible because at that time, children in Russia used to sleep in separate rooms with their minders, never with their parents. This rule was strongly enforced in Pankejev's home. Also, his wolves turned out to be dogs. Wolves simply fit better the poetic description of a dark unconsciousness than a few mutts do.

Case analysis is the foundation and the only "basis" for psychoanalysis. It wouldn't be honest to judge Sigmund Freud solely on a single case, so let's take a closer look at some other, slightly less known cases. Here we need to introduce Freud's friend and trustee, Wilhelm Fliess, who had a significant influence on Freud's beliefs and on the way he interacted with his patients. The letters they exchanged contain many scandalous facts that are best compared to a dynamite stick placed below the sand castle of psychoanalysis and its father. Complete publication of that correspondence emerged in 1985, and it greatly undermined Freud's authority. Actually, the first attempt to publish them was undertaken in Germany in 1950. However this edition was heavily censored by Freud's successors and by his own daughter Anna. It has been sufficiently edited to ensure that the "greatest psychoanalyst of our times" was shown in glorious light.

Emma Eckstein's case, carefully hidden by Freud's successors for 90 years, was eventually revealed in published letters and reads like a good horror story from the darkest times of inquisition. Fliess created a theory in which the nose was an equivalent of genitalia; thus, any disorder of the nose could be directly responsible for psychosexual problems. The treatment of this 'nasal reflex neurosis' as diagnosed by Fliess, involved thermal coagulation of internal nasal bones, followed by the application of cocaine. Freud fully accepted the eccentric, and as it turned out, also dangerous, dogmas of his friend. In 1895 Freud asked his friend to treat an attractive young patient Emma Eckstein. Fliess diagnosed her with nasal reflex neurosis and performed the abovementioned procedure, removing parts of her nasal bones. Following the surgery, a serious infection developed which led to massive bleeding that nearly cost Emma her life. Freud diagnosed the cause himself: she was "bleeding her love" for... him. Her life was saved when it was discovered that Fliess accidentally left almost 2-feet long piece of gauze in her body! This "therapy" left Emma glued to her bed for the rest of her life. It also permanently deformed her face.

The above case is often described by modern-day supporters of psycho-analysis as merely the juvenile, "pre-psychoanalytical mistake" of a young doctor. Let's take a look at other cases then. The first one was described and published by Freud in 1905☐ (although it dates back to 1900). It is the story of Dora. The title of the article was "Fragment of an Analysis of a Case of Hysteria." Dora was considered by orthodox psychoanalysts as a classical illustration of the structure and genesis of hysteria. Today Dora's case seems more like a classical example of unethical malpractice. Dora's real name was Ida Bauer; she lost herself in the strange network of relationships among the adults in her surroundings. Her father, despite being diagnosed with symp-tomatic late stage syphilis, engaged for many years in a romance with a wom-an referred to as Lady K. Ida adored Lady K., even though she knew about her father's affair for years. Mr. K. was a friend of the Bauer family. He often showered Ida with gifts. His sexual interest in the girl was more than obvious. During her analysis, she spoke with a great disgust about an event that took place when she was barely 14 years old where Mr. K. forced her to kiss him; two years later he sexually harassed her. Ida slapped Mr. K. in his face and ran away. She told her father about the incident. When her dad confronted Mr. K., he simply denied everything. She kept pressing her father to abandon all relationships with the K. family. Her father decided to side with Mr. K. and continued the affair with his wife. The girl developed a series of somatic symptoms, including coughs lasting many weeks, vaginal discharges, bed-wetting and suicidal thoughts. Her father sent her to Freud for treatment of both her symptoms and her rebellious attitude. She started the therapy against her will.

Her therapy was intense but lasted only three months. During this time, Freud analyzed her and performed a bold (to say the least) interpretation of her dreams. The diagnosis was that Ida Bauer was subconsciously in love with Mr. K. The girl denied it, but Freud interpreted her objections as sub-consciously transferred feelings for her father onto Mr. K. (which in fact became part of another of Freud's fundamental theses, the theory of trans-ference). He claimed that bedwetting and vaginal discharge were proof of frequent masturbation, a practice he condemned. The disgust the girl felt during the forced kiss was supposedly masking sexual arousal. According to Freud, this was a completely healthy reaction in an adolescent girl. The de-termination to stop all relations with the K. family was caused by her at-tempts to hide homosexual attraction to her mother and to Lady K, who somehow acted as her mother's substitute. She was also apparently jealous of her father's romance and so forth.

Ida decided to stop her therapy. Freud mentioned this fact in his letter to Fliess with great grief, as he was sure that the girl's conceptions were entirely wrong. Malcolm Macmillan described this case in great detail on 15 full pages of his work.[2] The analysis shows Freud's arrogant tyranny towards Ida; it justified Mr. K's pedophilic behavior; it favored Ida's father. This scientific fraud, derived from unethical, sloppy practice, originated in Freud's systematic and extreme ignorance of the facts presented to him by Ida Bauer. As Macmillan demonstrated, Freud completely ignored Ida's real feelings, reactions and associations. Freud always favored his own theories and preconceptions during the analysis of his patients. He then presented his biased and fake results as a triumph of his new discipline.

Freud cheated since the very beginning of his career. During his 1896 speech to neurologists and psychiatrists in Vienna, he claimed that his ideas had been confirmed in 18 cases of hysteria which he had successfully treated. Meanwhile, in his letter to Fliess, he complained that he had worked with 13 women with hysteria and none of his patients were ever cured. Most of the cases described by Freud were fabricated in exactly the same manner.

In fact, there are only few real cases on which Freud based his theory of psychoanalysis and his treatment recommendations. Frank Sulloway was able to find only eight of them. Even when we assume that eight cases are enough for a creation of an entirely new discipline and relevant treatment procedures (even though it is not), bear in mind that every single case used by Freud was completely useless. They were all falsified or fabricated, just as in the three examples described above[3].

We are aware that our style of disproving Freud's conceptions could be considered as demagogic to a certain degree. History teaches us that they were numerous scientists or philosophers whose intuition went far beyond the framework of available evidence at their time. George Mendel created the foundations for modern genetics without knowing anything about genes, chromosomes or DNA. Perhaps Freud was ahead of his time and accusing him of fabricating or falsifying evidence is simply a sign of our timidity and blind trust in empirics? Let's take a look then at his intellectual legacy.

In March 1999, *Time* published an article called "The Century's Greatest Minds." Einstein and Freud appeared on the front cover of the magazine and Freud's name was the first on the list of greatest minds of the century, followed by the Wright brothers and Albert Einstein. The justification for Freud's merits read, "Sigmund Freud, more than any other explorer of the psyche, has shaped the mind of the 20th century. The very fierceness and persistence of his detractors are a wry tribute to the staying power of Freud's ideas. … He opened a window on the unconscious — where, he said lust, rage and repression battle for supremacy — and changed the way we view ourselves."[4]

The tenth issue of *Newsweek Poland* from 2005 included an insert add-on titled "79 People Who Changed History." The top four most outstanding names and discoveries according to the journal editors were, "Evolution by natural selection, subconsciousness [*sic*], theory of relativity, freedom. Darwin, Freud, Einstein, Mill. Four pillars, on which our vision of the world has been based."

Oh, Holy Ignorance! Of course we were told at our universities that Freud discovered sub-consciousness and that humans can be guided by unconscious impulses, but this is simply untrue! Even Freud himself did not aspire to be acclaimed as the "discoverer" of this idea. The idea of the dynamic sub-consciousness—the unaware mind and the battle of urges and impulses within it—was the mainstream thought of German philosophy and culture from the early nineteenth century. Some researchers even derive its sources from Plato! It's worthwhile reading the following story by Plato to have one's own opinion about it:

As I said at the beginning of this tale, I divided each soul into three--two horses and a charioteer; and one of the horses was good and the other bad: the division may remain, but I have not yet explained in what the goodness or badness of either consists, and to that I will now proceed. The right-hand horse is upright and cleanly made; he has a lofty neck and an aquiline nose; his color is white, and his eyes dark; he is a lover of honor and modesty and temperance, and the follower of true glory; he needs no touch of the whip, but is guided by word and admonition only. The other is a crooked lumbering animal, put together anyhow; he has a short thick neck; he is flat-faced and of a dark color, with grey eyes and blood-red complexion ... the mate of insolence and pride, shag-eared and deaf, hardly yielding to whip and spur. Now when the charioteer beholds the vision of love, and has his whole soul warmed through sense, and is full of the prickings and ticklings of desire, the obedient steed, then as always under the government of shame, refrains from leaping on the beloved; but the other, heedless of the pricks and of the blows of the whip, plunges and runs away, giving all manner of trouble to his companion and the charioteer, whom he forces to approach the beloved and to remember the joys of love. They at first indignantly oppose him and will not be urged on to do terrible and unlawful deeds; but at last, when he persists in plaguing them, they yield and agree to do as he bids them. And now they are at the spot and behold the flashing beauty of the beloved; which when the charioteer sees, his memory is carried to the true beauty, whom he beholds in company with Modesty like an image placed upon a holy pedestal. He sees her, but he is afraid and falls backwards in adoration, and by his fall is compelled to pull back the reins with such violence as to bring both the steeds on their haunches, the one willing and unresisting, the unruly one very unwilling; and when they have gone back a little, the one is overcome with shame and wonder, and his whole soul is bathed in perspiration; the other, when the pain is over which the bridle and the fall had given him, having with difficulty taken breath, is full of wrath and reproaches, which he heaps upon the charioteer and his fellow-steed, for want of courage and manhood, declaring that they have been false to their agreement and guilty of desertion. Again they refuse, and again he urges them on, and will scarce yield to their prayer that he would wait until another time. When the appointed hour comes, they make as if they had forgotten, and he reminds them, fighting and neighing and dragging them on, until at length he on the same thoughts intent, forces them to draw near again. And when they are near, he stoops his head and puts up his tail, and takes the bit in his teeth and pulls shamelessly. Then the charioteer is worse off than ever; he falls back like a racer at the barrier, and with a still more violent wrench drags the bit out of the teeth of the wild steed and covers his abusive tongue and jaws with blood, and forces his legs and haunches to the ground and punishes him sorely.[5]

Doesn't it look like a psychoanalytical description of conflicting motives? And the three parts that the souls are divided into? Doesn't it resemble id, ego and superego? We can leave those questions to more inquisitive historians searching for the source of psychoanalytic theory. Meanwhile we will move back to nineteenth century, where we have already faced (this time for sure) the subconscious mind and the 'battle of the urges.'

Frederick Nietzche was most popular and renowned for developing the afore mentioned concept. He described it in great detail in his work called *Beyond Good and Evil.* He described individuality ("I") as an autonomously created construct, while in reality, the structure of our minds consists of many souls. He also devoted a lot of attention to the sublimation of impulses into other, potentially unconnected urges. For example, the sublimation of sexual impulses drive into art. Andrzej Śliwerski has found in his analysis that we can find most of the ideas and thoughts that are nowadays commonly associated with Freud in the works of Nietzsche:

> The father of psychoanalysis copied philosophers' theories; usually he did not even bother to change examples used to explain them. Furthermore, Freud repeatedly claimed that he never had an opportunity to read Nietzsche's work. ... Freud said that he never had a chance to study Nietzsche, partially due to the reason of tremendous similarity of his intuitive theories to Freud's laboratory findings and partially because of the frivolous, light and easy way in which he intended to present it. It stopped him from reading more than half a page, every time he tried to read Nietzsche's books... This statement can be verified by following Freud's early career and searching for answer to the question: Could works of Frederick Nietzsche disturb the "lightness" of thought of the future father of psychoanalysis? Freud however, as every perfect criminal [does], purposely got rid of all possible notes, letters, abstracts, manuscripts and all other possible traces of his early "scientific" work.[6]

Why then was Freud considered the author of those ideas? How did he become so popular? The answer to the first question is ignorance, especially in the English-speaking parts of the world. At the time Freud was revealing "his" ideas, the theories of nineteenth century German philosophers were mostly unknown to the rest of the world. Freud was therefore able to easily take credit for them. Much more interesting is the issue of Freud's popularity. One particular historic event had a significant impact on the popularity of psychoanalysis.

It the afternoon of May 10, 1933, on a square between the Berlin Opera and the university, students wearing swastikas (Nazi symbols) on their sleeves began piling up books. They gathered over 30,000 of them and shortly before midnight, thousands of students with torches in their hands paraded around the square. Then, they set fire to the piles of books. The minister of propaganda, Josef Goebbels, explained that the book burning was part of the battle against decadence and moral corruption. Books of Thomas Man, Albert Einstein, Erich Maria Remarque, Upton Sinclair, Emil Zola and of many other great scientists and authors were burned. Freud's books were on the list. Well, if you are about to be hanged, you'd better make sure you are surrounded by appropriate company.

Nothing better could have happened to Freud. The act of Nazi book burning caused an immediate resentment, especially in the United Kingdom and in the USA, countries with a long tradition of freedom of speech. The prestige of every author whose books were burned on that day skyrocketed immediately. All burned books rapidly became bestsellers. Publishers were ordering reprints as everyone wanted to have their copy. Freud's books were no exception. Even today, we tend to praise certain books only because they were forbidden. Paul Halpern, professor of physics at the University of the Sciences in Philadelphia, equates Freud and Einstein. Yes, this is not a misprint! Einstein next to Freud, "five years after Freud published his bold attempt to decipher the patterns of dreams and thereby map the hidden mechanisms of the mind, Einstein introduced his *equally ambitious* [emphasis ours] scheme to reveal the behavior of near-light speed objects."[7] Well, the bright light of burning stakes shines equally on both theories...

Nazis disapproved of Freud, and called him an idiot and a pervert. Where some saw degenerated fantasies, others regarded every book burned on the other side of the ocean as a brilliant masterpiece. In fact, today many historians agree that the Nazi's aversion to Freud was the key reason for his popularity in the USA, although it wasn't the only one.

Historians point out other factors that had an impact on Freud's recognition, but before we discuss them, let's deal with the issue of his intellectual input on science. Apart from the ideas of other German philosophers that are mistakenly attributed to him, did Freud actually bring new or original ideas into the temple of knowledge? This is how Leonard Sax commented on it: "Freud's unique contribution can be summarized in one sentence: Everything that happens in your mind is ultimately sexual. Everybody is bisexual, every boy wants to kill his father and have sex with his mother, every six-month-old baby is sexually fixated, and every woman subconsciously wishes she had a penis. All these notions were invented by Freud, and almost every one of them has been abandoned, even among those who still believe that Freudian techniques retain some utility."[8]

There is more. Those "original" theories of Freud were constructed on rather bizarre observations. How did Freud, "discover" the Oedipus complex if none of his patients ever suggested they had such a problem? The answer, again, can be found in Freud's letters to Fliess. On October 15, 1897 he wrote: "I found in myself a constant love for my mother, and jealousy of my father. I now consider this to be a universal event in early childhood."[9]

Unfortunately this is the one and only evidence documenting the existence of Oedipus complex that historians all over the world managed to find in Freud's works. His remaining theories are "supported" by similarly "reliable" evidence.

Perhaps some more comprehensive evidence could be obtained from Freud's works. Unfortunately, we only know a small part of his work—the part that successfully passed psychoanalytical censorship. Freud's heirs and supporters locked a huge number of documents, letters, notes and other such materials in the US Library of Congress. They will not be revealed until the twenty-second century! This time capsule, deposited in a renowned institution, was created "for the benefit of science." Well, neither us, nor our readers will live long enough to witness the opening of the capsule.

Why then, despite extremely weak evidence and the significant amount of time that has passed since the memorable events in front of Berlin University, is psychoanalysis is still so popular? Mostly because of the constant efforts of psychoanalysts themselves to nurture the good image of psychoanalysis. They carefully protect their pseudoscience and the entire therapeutic system based on it. After all, it still is an excellent money-making machine. There are also other reasons from different sources.

Psychoanalysis caused significant turmoil among literary critics and other specialists who delve into minds of artists and try to understand their motives. Thanks to all complexes, repressions and sublimations made up by Freud, characters in popular literature suddenly gained a secondary meaning. All of the sudden, the entire literature could now be reanalyzed in search of another level of depth. A paradise for critics, but an excellent pretext for writers to create even weirder, more degenerated characters suffering from their unconscious desires and torn by internal discrepancies. Literary critics are hardly ever concerned with evidence undermining the scientific value of psychoanalysis. Neither are writers. Psychoanalysis continues to inspire them and allows them to create extremely rich worlds of mysterious secrets and unrealized urges. While such inspirations may enrich literature, they also greatly contribute to popularizing an idealized image of psychoanalysis to the general public.

The last reason for Freud's continued attractiveness that we would like to mention here, (although there are many more), is its emphasis on sex. Many people, especially the male part of the population, sooner or later discover a powerful desire for sexual gratification. Sexual fantasies, intrusive thoughts, the inability to concentrate and masturbation are just a few typical symptoms that normally occur in young, healthy, adolescent people. Often, these continue and persist after puberty. At the time when Freud announced his "discoveries," the symptoms described above were considered inappropriate, sinful or even equal to succumbing to Satan. Suddenly, thanks to Freud, millions of people were freed from those sins. Not only natural sexual desires, but also all possible deviations were placed in the sub-consciousness, which effectively meant it was outside of the individual's control. Freud turned active sinners into the passive slaves of unrealized impulses. Isn't that convenient? Any remaining problems could be now brought to a trusted psychoanalyst. Just as a car mechanic can fix a faulty engine, psychoanalysts were now supposed to fix these issues. And don't forget, sex is an excellent commodity. It sells well. It has always sold well. Puritan, Christian Europe stigmatized sex and turned it into something negative and sinful. So we shouldn't be surprised that lifting the sexual bans under the banners of science was enthusiastically accepted.

So if it was proven that most of Freud's evidence was made up, fabricated or intentionally misrepresented, his contribution to science virtually nonexistent, and his popularity merely due to a lucky chain of events, what is actually left of Freud? There's another important rampart of the besieged castle, so fiercely defended by psychoanalysts: therapy. Freud created the foundations for an entire therapeutic system that is widely used by psychoanalysts to help others. Doesn't that count?

Well, it does not. Freud himself was already using therapy predominantly to achieve his personal goals. He was doing it in a way that would be considered today as extremely unethical. To illustrate this, let's take a closer look at an interesting story that surfaced recently after Freud's letters were discovered. It focuses on Freud's relations with Horace Frink, his patient and protégé. This married American had a romantic affair with Angelica Bijur, heiress to a bank. Freud suggested to Frink that he was a homosexual and that there was a significant danger that his hidden tendencies could be publically revealed. To avoid public disgrace, Freud convinced Frink to divorce his wife and marry Bijur. At the same time he was advising Bijur to leave her husband. It would be also worth noting that Freud had never met their spouses. His motives became clear when we read fragments of Freud's letters to Frink: "Your complaint that you cannot grasp your homosexuality implies that you are not yet aware of your fantasy of making me a rich man. If matters turn out alright, let us change this imaginary gift into a real contribution to the Psychoanalytic Funds."[10]

The tone of the letter might seem frivolous at the first glance. We can only realize how cynical they were when we take to the account that the divorces and the set up marriage actually happened. Further consequences included the death of both abandoned and shattered spouses. The arranged marriage did not last. Frink's newly wed wife soon requested a divorce and Frink, steeped in guilt, soon developed severe depression, which led to multiple suicide attempts.

In a very similar way, Freud led his former protégé Victor Tausk to suicide. Victor was suffering from chronic severe depression and Freud instructed him to abandon his treatment. There is no evidence that Freud had ever shown any grief or sorrow for the suffering or death of his patients. On the contrary, in his letter to Carl Jung from 1912, he wrote about a woman who had been his patient since 1908, "there's no way she can possibly benefit in any way from my therapy, but it's her duty to dedicate herself for science."[11]

Freud's disregard for his patients can be found in a diary of his closest friend, Sandor Ferenchi. The dairy is still, after 53 years from its completion, awaiting publication. "I remember certain statements Freud made to me. Obviously he was relying on my discretion. He said that patients are only riffraff. The only thing patients were good for is to help the analyst make a living and to provide material for theory. It is clear we cannot help them. This is therapeutic nihilism. Nevertheless, we entice patients by concealing these doubts and by arousing their hopes of being cured."[12] Even if we assume, despite many similar and well-documented events, that Freud could create a complete therapeutic system that benefited patients, we should carefully confirm its efficacy in an empirical way. Unfortunately, such evaluation of psychoanalysis does not leave any doubts about its real value.

Technically, Hans Eysenck already dealt a devastating blow to psychoanalysis in the 1950s. He asked a very simple, yet accurate, research question: Is the Freudian psychoanalysis more effective than any other face-to-face interaction, for example a meeting with a preacher who talks to his patients about the Bible? Eysenck compared psychoanalysis to various talk-based therapies and discovered that 44% of patients with symptoms of depression or anxiety declared improvement. Patients treated with other therapies managed to improve in 64% of cases over two years of treatment. Patients who were left untreated for two years reported improvement of their symptoms in 60% of cases.[13] Well, the difference between talk-based non-Freudian therapies is probably not statistically significant. We can therefore conclude that therapists do not significantly harm their patients; perhaps even sometimes they can provide limited (statistically insignificant) benefit. But when patients are subjected to psychoanalysis, it seems that it actually prevents their natural recovery! This has been confirmed in numerous subsequent research in which Eysenck was also involved. According to this evidence, patients with genetic predispositions to cancers or with severe cardiovascular diseases (such as thrombosis) lived much shorter if they were subjected to psychoanalytical therapy in comparison to those who did not receive any psychological support.[14]

As we never give up, let's give psychoanalysts one more chance. After all, a research conducted by devoted opponents and critics of a certain theory might not be entirely objective or robust. Let's analyze an attempt to measure the effectiveness of psychoanalysis conducted by the American Psychoanalytic Association. Certified psychoanalysts were asked to provide their own data on the efficacy of psychoanalysis. This "study" had a fatal design flaw in its conception, as nobody bothered to ensure that submitted results would be complete, that is, would cover every treated patient, (or was it just an "honest" omission?). This is important, as such methodological mishap could significantly boost the efficacy of treatment; psychoanalysts could only report cases where their therapy was successful. Instead, APA declared, "due to potentially misleading statistical analysis, it is recommended to consider this report as confidential." 3,000 reports were submitted and classified, but since then, not a single publication of those results was ever published or made otherwise available to public.[15]

Multiple attempts of empirical verification or confirmation of efficacy of psychoanalysis were undertaken by Lester Luborsky from the University of Pennsylvania. Unfortunately, even superficial analysis of his works shows that if there are any results of psychoanalysis, they only appear in comparison to patients who were not treated at all. The global standard for empirical research requires comparing results from treatment groups to the best available treatment, or (if such treatment is not available) to compare results obtained from treatment groups to results obtained from control groups treated with placebo. In such comparisons, psychoanalysis always proves to be not only less effective than placebo (!) but also much more time consuming, and therefore more expensive.

Freud repeatedly encouraged his patients to use high doses of cocaine and morphine, exposing them to significant harm. Psychoanalysts however, consequently refuse to prescribe any medications to their patients. This included medications developed in the 1950s that could rapidly and significantly improve patients' symptoms. Why? Perhaps the reason was linked to the need to prolong and continue the profitable, never-ending therapy, rather than to provide relief to their patients. Actually, medications developed in the middle of twentieth century very quickly exposed numerous mistakes, misconceptions and frauds in the very foundations of psychoanalysis. Here are a few of many examples:

In 1954 chlorpromazine, also known under a trade name Thorazine, became available for the treatment of psychotic disorders, including schizophrenia. This was the first modern pharmaceutical proven to actually be effective in the treatment of these conditions. Numerous patients, previously forced into straightjackets could finally be freed from restraints. In many cases, patients could return to their everyday lives, friends, families and get back to work. The transformations of patients' lives were often surprising to psychiatrists themselves. The efficacy of the new medication was shocking, especially in comparison to patients' previous helplessness and the mediocre results of available treatments. In 1956, a new breakthrough medication, imipramine, the first effective antidepressant, was introduced. This was not good news for psychoanalysts. They considered any pharmaceutical products unnecessary and believed they interfered with the process of analysis. These novel pharmaceuticals, primarily due to their high efficacy and low costs, not only interfered with long (many years) of expensive and extremely profitable psychoanalysis therapy, but they have also provided definite proof that the principles on which Freud based his theory were completely wrong.

According to the theory of psychoanalysis, psychiatric disorders were caused by "internal conflicts." When the alleged conflicts were relatively weak, patients suffered from anxieties. When the conflicts were significant, psychoses were supposed to develop. Thorazine had shown to be extremely effective in the treatment of psychoses, but it was completely useless in patients with anxieties. Therefore, Freud's theory, which assumed continuity of mental processes and considered psychoses as quantitative intensity of same processes that form anxiety disorders, turned out to be false.

Unfortunately for the patients suffering from manic-depressive disorder, (currently known as bipolar disorder), most psychiatric hospitals and other mental health institutions in the United States were dominated by psychiatrists with psychoanalytical orientations. In 1954 it was discovered that lithium salts are an effective treatment of manic-depressive disorders. The misfortune of American patients was based on the fact that prominent and influential psychiatrists in the USA declared lithium salts as "a dangerous nonsense."[16] Why? Simply because Freud's theory could not explain the efficacy of new medication. The result? Until 1970, treatment with lithium salts was prohibited in the USA, and patients who could have been treated effectively, quickly and cheaply, were forced into useless, ineffective, tiring and extremely expensive psychoanalytical séances for many years.

The evidence for the fragility of psychoanalytical theories presented here is partial and incomplete. Their presentation is also limited by the form of this book. We do believe however that it is appropriate to briefly mention a few critical works, written reliably, with great attention to detail, and according to highest scientific standards. These works originated in the last 30 years, when scientists have finally and clearly proven Freud's psychoanalytical assumptions to be outright false.

Perhaps the most credible critic of Freud's theories is Frank Sulloway. In 1979, he wrote his book *Freud, Biologist of the Mind: Beyond the Psychoanalytic Legend*,[17] in which he convincingly defended young Freud. He was awarded by delighted Freudians and also received the scientific *History of Science Society* prize. But Sulloway turned out to be a true, born and bred scientist. Relentlessly pursuing new facts and knowledge, he renounced his previous work that defended Freud. By the end of the 1980s, he published a series of critical articles and an updated version of his book.[18]

Another critical work was devoted to the epistemological analysis of the clinical foundations of Freudian theories. Its author, Adolf Grünbaum, was a professor of philosophy and psychiatric researcher working at the University in Pittsburgh. In his work called, *The Foundations of Psychoanalysis: A Philosophical Critique*,[19] Grünbaum shows that the epistemological foundations of Freud's clinical theories are entirely false.

Malcolm Macmillan's work has been devoted mainly to the analysis of the methodology of Freud's practice throughout his entire career. This scrupulous and erudite masterpiece, written on over 600 pages, was summarized by its author as follows: "Should we therefore conclude that psychoanalysis is a science? My evaluation shows that at none of the different stages through which it evolved could Freud's theory generate adequate explanations. From the very beginning, much of what passed as theory was description, and poor description at that. In every one of the later key developmental theses, Freud assumed what had to be explained.[20]

We would like to mention one more book, a collection of all the articles written in 1993 and 1994 by Frederick Crews in *The New York Review of Books*. The author attached his correspondence, often quite harsh, with the editor of the journal who initiated the ideas of those works. In a book titled *The Memory Wars. Freud's Legacy in Dispute. Frederick Crews and his Critics*, Crews exposes the falsehood of the entire Freudian project by presenting a detailed review of complete literature, both critical and in favor of psychoanalysis. In a polemic and ironic manner, he shows how Freud falsified and fabricated all the evidence he needed and rejected or destroyed all the inconvenient proof that didn't favor his theories.[21]

There are many similar books, articles and publications like those presented above. Today, it's hard to resist the impression that we are dealing with ruins of the castle of psychoanalysis—a castle built on sand, without foundation, using inferior materials, joined with mediocre grout, but with obsessive attention to gloss and glare. The castle did not last a century. No reasonable scientist or practitioner today takes psychoanalysis seriously. Ironically, this pseudoscientific monster, based on made up evidence, falsified cases and on fabricated, fraudulent claims, continues to exist in popular opinion as one of primary canons of practicing psychology.

---

[1] Judson, *The Great Betrayal*, 89.

[2] M. Macmillan, *Freud Evaluated: The Completed Arc* (Cambridge, MA: MIT Press, 1997), 249-263.

3 E. Sulloway, "Reassessing Freud's Case Histories: The Social Construction of Psychoanalysis," *Isis*, (1991): 245-274.

[4] "The Great Minds of the Century," *Time*, (1999): 105.

[5] Plato, *Phaedrus*, trans. B. Jowett, (The Project Gutenberg, 2008), EBook #1636.

[6] A. Śliwerski, "Freud nie wymyślił Nietzschego," [Freud Did Not Invent Nietzsche]. *Studia Psychologica, 6*, (2006): 261-282.

[7] P. Halpern, *The Pursuit of Destiny. A History of Prediction.* (New York: Basic Books, 2000), 63.

[8] L. Sax, "The Future of Freud's Illusion," *World & I*, 15 (8), (2000): 263-283.

[9] J. M. Masson, ed., *Sigmunt Freud: Briefe an Wilhelm Fliess* (Frankfurt: S. Fischer, 1999), 285.

[10] L. Edmunds, "His Master's Choice," *John Hopkins Magazine, 40*, (1988): 40-49.

11 W. McGuire, ed., *The Freud-Jung Letters: The Correspondence between Sigmund Freud and C. G. Jung*, (Princeton University Press, 1974): 473-474.

12 S. Ferenczi, *Journal Clinique*, (Payot: Paris, 1985). As cited in J. M. Masson, *Against Therapy* (Kindle version: 2012) Retrieved from Amazon.com.

13 E. Shorter, *A History of Psychiatry* (New York: Wiley, 1996), 312.

14 R. Grossarth-Maticek and H. Eysenck, "Prophylactic Effects of Psychoanalysis on Cancer-prone and Coronary Heart Disease-prone Probands, as Compared with Control Groups and Behavior Therapy Groups," *Journal of Behavior Therapy and Experimental Psychiatry*, *21*(2), (1990): 91-99.

15 Ibid.

16 Ibid.

17 F. Sulloway, *Freud, Biologist of the Mind: Beyond the Psychoanalytic Legend* (London: Burnett Books, 1979).

18 Sulloway, "Reassessing Freud's Case Histories," 245-274.

19 A. Grünbaum, *The Foundations of Psychoanalysis: A Philosophical Critique* (Oakland: University of California Press, 1985).

20 Macmillan, *Freud Evaluated*, 625.

21 F. Crews, *The Memory Wars*, (The New York Review of Books, 1997).

CHAPTER 9

THE MYTH OF CHILDHOOD:
FOUNDATION OF THERAPIES EXPLORING THE PAST

In 1944, Dr. Walter Langer, a psychologist from Harvard University, was approached by the Office of Strategic Studies (the predecessor of today's Central Intelligence Agency) and asked to create a psychological profile of Adolf Hitler. Langer proceeded according to the recognized methods of the day, and focused on analyzing events from the Third Reich's leader's childhood. Although getting hold of witnesses of this period was not easy, Langer and his team overcame these difficulties; he contacted Hitler's ex-doctor and two of his former collaborators who had immigrated to the United States. Langer's search became the subject of David Steward's documentary film *Inside the Mind of Adolf Hitler*. Here are some excerpts from the script, which will give us a good idea of what Hitler's profilers found as interesting and relevant:

> Hitler's Mother's obsessive cleanliness was significant for Langer. It indicated that the toilet training phase of Hitler's childhood might have gone wrong. ...
> From what we know about his mother's excessive cleanliness and tidiness, we may assume she employed rather stringent measures during the toilet training period of the children. We know that this usually results in a residual tension in this area. It is regarded by the child as a severe frustration that arouses feelings of hostility. Now, this facilitates an alliance with his infantile aggression that finds an avenue for expression through anal activities and fantasies. These usually center around soiling, humiliation and destruction and form the basis of a sadistic character...Langer felt that he had a potential fit for Freud's Oedipus complex and extreme difficulties with the toilet training stage of childhood, and he suspected that this would be most clearly seen in Hitler's attitudes to sex.[1]

Having identified the cause of Hitler's sadistic predispositions, the team went on with further research along their chosen path. Another discovery was made during an interview with Otto Strasser, a former collaborator of the Third Reich's leader, who spoke about Hitler's relationship with his niece, Geli Raubal.

> She said that Hitler made her undress while he lay on the floor. Then she would have to squat down over his face while he examined her at close range. This

made him very excited. When his excitement reached its peak, he demanded that she urinate on him, and that gave him his sexual pleasure. Geli said that she found the whole performance extremely disgusting. And although it was sexually stimulating for Hitler, it have her no gratification whatsoever.[2]

Dr. Jerrold Post, a world-class expert and an authority on psychological profiling of politicians who worked for the CIA in the 1970s and founded the first psychological profiling unit in the world, commented as follows:

> In the act of being subject, in his perversion, to being really sexually humiliated by a woman, that represented the unmasked wish to surrender, capitulate, be seen as a weak man, against which he was forcefully quarreling psychologically, and in this power, the power of the will, this was central for him…This was the highly potent, powerful leader, but underneath—[there] was this man who was desperately weak and desperately afraid of, and yet seeking, submission and capitulation…Langer believed that Hitler found a way to deal with the terrible psychological consequences of his perversion by adapting a political ideology, which was a prevalent and powerful part of mainstream European culture: anti-Semitism.[3]

The findings of Langer's team fit so well in psychoanalytical hypotheses that nobody even troubled to verify the witnesses' credibility. Their statements were accepted as true and Adolf Hitler's psychological profile was pieced together. A number of hypotheses predicting future actions of the Führer were based solely on this profile. The predictions were actually quite often correct, but the question emerges whether it was really necessary to know about Hitler's problems using the toilet in his childhood to predict how much more neurotic Hitler would become if the war took the wrong turn for him? Would it not have been easier to predict such behavior by analyzing his reactions to the recent events? Were Langer's suggestions about increasingly more frequent outbursts of Hitler's fury possible to predict because of his analysis of Hitler's sexual behavior or rather his reactions to current events? And last but not least, was the likelihood of the Third Reich's leader's suicide in the face of defeat based on the analysis of his childhood? Or was it rather based on Hitler's well-known announcement already made in fall of 1939 that he would commit suicide if he lost the war?

The history of Hitler's profiling shows we are ready to believe that everything, including the most atrocious crimes, might be explained in terms of childhood experiences. It would be great if that were possible, wouldn't it? Unfortunately, it is not possible. There are hundreds of thousands or even hundreds of millions of children who had experiences similar to Hitler's or Stalin's, but only a few of them grew up to be monsters. The opposite is also true: there are many people who never experienced a lack of love or warmth in their childhoods, but committed hideous, hateful crimes as adults.

One of us came across such a story in of 2010 during a summer holiday on the Indonesian island of Bali. This exotic place is fascinating, not only because of its beautiful landscape, colorful nature, culture and religion, but also, or perhaps primarily because of the inhabitants' lifestyle. The Balinese are always smiling and, at least in areas not yet swarming with tourists, they have a consistently positive attitude towards other people and towards all living creatures, an attitude deeply rooted in their culture and religion. Common belief in reincarnation means to locals that every living creature might be an incarnation of their loved ones' spirits. They especially adore children, believing that souls of deceased members of their families return to them in their children's bodies. Guided by this religious devotion toward children, the Balinese will not let them touch the ground with their feet in infancy. The process of raising children, full of care and affection, puts equal emphasis on self-reliance and resourcefulness. No wonder then that the Balinese are not warrior-types and their history books do not mention any aggressive conquests. With one exception, however, which took place in the 1960s…

At that time, the majority of Balinese people supported the Communist Party of Indonesia (Indonesian: Partai Komunis Indonesia, PKI), primarily because it favored land reforms. The Communists encouraged people to take over the bulk of the arable land, which then belonged to only several hundred great landowners. The PKI was chiefly supported by about eighteen thousand islanders who had suffered a decimating famine in 1964. They believed they would have avoided the disaster if the land had been more justly distributed. The PKI was largely seen as a challenge to the old feudal and caste systems. A split developed between its supporters and supporters of the Nationalist Party of Indonesia (Indonesian: Partai Nasional Indonesia, PNI). All over the island, demonstrations calling for radical changes were organized. The propaganda campaign was joined by folk artists, such as dancing groups and puppeteers, who spread proletarian ideas. In 1965, the communists were blamed for a failed coup attempt in Jakarta, the capital of Indonesia. The coup claimed the lives of six generals, and communists were blamed for this conspiracy. Afterwards, events took a quick turn. The PNI and other anti-communist parties decided to take matters into their own hands in order to finally deal with the PKI. A bloody massacre followed. The carnage soon lost its political character and became a way to settle all local scores. Entire villages were butchered and then razed to the ground. The extent of the slaughter reached such magnitudes that General Sarwo Edhy, nicknamed the "Butcher of Java" on account of his cruelty, described it in the following way: "In Java, we had to egg the people on to kill Communists. In Bali, we had to restrain them; make sure they don't [sic] go too far."[4]

It is estimated that between 80,000 and 100,000 people lost their lives in Bali, which in those days accounted for about five percent of the island's entire population. The number of those killed was proportionally larger than on any other Indonesian island; (there were massacres on virtually all of them). It is notable that the victims in Bali were not strangers to each other, not followers of different religions, not the members of a different race or nationality. The Balinese simply murdered each other.[5] Even today, almost half a century later, the inhabitants of this lovely island refuse to talk about those events.

Could this behavior of the Balinese in 1965 be explained in terms of their early childhood experiences? We doubt it. Neither can we explain why the African Tutsi and Hutu tribes, otherwise very gentle toward their children, decided to massacre each other. We also doubt if it is possible to use the analysis of childhood to explain acts of genocide in Bosnia and in many other similar conflicts from the grim history of mankind. Most criminals and mass murderers probably had caring, sensitive and loving parents.

However, for many people, and unfortunately for many psychologists as well, the childhood of Hitler, Stalin or other criminals prove their thesis about the significance of early childhood experiences for adult life. From the scientific point of view, they are nothing more than anecdotes. This is mere confirmation bias, so often pointed out by Karl Popper. Instances that confirm our hypothesis reinforce our belief. Not only do we stop looking for examples that might disprove it, but if we do come across them, we activate defense mechanisms to uphold the conclusions we have already reached. In fact, if we take a closer look, we will find that:

> Though serial killers like Charles Manson were abused and neglected as children, the list of serial killers with a normal childhood is long. Famous serial killers such as Ted Bundy, Jeff Dahmer and Dennis Rader grew up in healthy households with supportive family members.[6]

The belief that the analysis of early childhood experiences might enable us to understand the behavior of adults is one of the most deeply rooted, but also one of the most dangerous myths in psychology. In his book *Myths of Childhood*, Joel Paris, professor of psychiatry at the McGill University in Canada, makes a thorough analysis of this claim, distinguishing three more detailed ones:

> Myth 1: Personality is formed by early childhood experiences.
> Myth 2: Mental disorders are caused by early childhood experiences.
> Myth 3: Effective psychotherapy depends on the reconstruction of childhood experiences.[7]

None of the above statements is true, as we will attempt to show further in this part. Somehow however, they are all deeply embedded, not only in ordinary people's minds, but also in the minds of well-educated psychologists. This was confirmed by the results of our questionnaire survey carried out in 2011 and 2012 among a group of 185 respondents between the ages of 20 to 68 (the average age was 33). The group was further split into two subgroups. The first group was comprised only of 34 trained psychologists. The second group consisted of 151 representatives from a broad professional spectrum (the majority with a higher education degree). We asked the study participants three questions (among others) related to their opinions about the childhood myth:

1. Human personality is mainly formed by early childhood experiences. Please indicate, in percentage terms, how much you agree with this statement.
2. Mental disorders are caused by early childhood experiences. Please indicate, in percentage terms, how much you agree with this statement.
3. Effective psychotherapy depends on how thoroughly one's childhood issues were dealt with. Please indicate, in percentage terms, how much you agree with this statement.

The participants used a percentage scale. On average, they gave the first statement the score of 60.8%, thereby confirming the prevalence of popular notions about personality formation. The psychologists fared a little better with the average of 44.1%. The respondents were also firmly convinced that mental disorders originate from childhood experiences since the average score for the second statement was 57.3%. The psychologists were less stereotypical in their thinking with the average score of 40.6%. Such thinking is still far off from currently available evidence. Similarly, the prevalence of popular ideas about what happens during psychotherapy was confirmed by the average score of 55.8% in regard to the third statement. The psychologists did much better with the average score of only 26%.

The prevalence of these stereotypical views is also found in public statements of famous psychologists. For instance, Professor Jerzy Mellibruda from the University of Social Sciences and Humanities in Warsaw, a well-known and recognized Polish expert in the treatment of addictions, said in an interview for a major Polish newspaper:

In order to break free from the influence of sufferings experienced in childhood, you have to go back to them in your memories and "work through them." In the process of healing, it is also necessary to emotionally separate yourself from your parents, forgive them for the harm they did to you, and free yourself from the irrational sense of guilt for their evil actions. It's only then that you will reach autonomy and maturity.[8]

As we will show further in this part, none of the above statements is substantiated by research. Instead, more questions emerge, which we will discuss more thoroughly further in this book. Who decides whether a person is autonomous and mature? Professor Mellibruda? A therapeutic support group? Who gave those authorities the power to decide what is mature and autonomous and what is not?

Such and similar claims in media greatly contribute to dissemination of popular psychological myths. Psychology textbooks do not do much to limit the spread either. The truth is that statements about the influence of childhood experiences on adult life are not supported by scientific research. This has been shown very clearly by Jerome Kagan, among others, in the book *Three Seductive Ideas*.[9] Let us then take a closer look at the particular elements of the childhood myth to see what they look like from the vantage point of research.

1. Personality is formed by early childhood experiences.

This popular claim is so deeply rooted in our culture that it will probably take many years before empirical findings will be able to debunk it, or even to create a small opening for rational thinking about what really shapes our personalities. The first foundation of this view is the Myth of the Blank Slate. Steven Pinker devoted a seven hundred-page long work to debunk it.[10] This myth, however, is quite well entrenched even in the minds of educated people, and arguments against it seem to reach them with significant delay.

It is no wonder, however, because this myth was further reinforced by the results of research conducted in the 20th century by zoologists. As a matter of fact, it is for such discoveries that Konrad Lorenz (along with Niko Tinbergen and Karl von Frisch) received the Nobel Prize in physiology and medicine in 1973. Lorenz's research was focused on *imprinting*. He discovered this phenomenon as he observed geese, which, having no other objects in their environment, began to perceive the researcher as their mother. Imprinting occurs in young animals during a very precise phase of their ontogenetic development (the so-called critical period, sometimes only several hours long), when a pattern of parent and sibling recognition is established together with behaviors typical for a given species. Once imprinted, the pattern remains practically unmodified throughout animals' lives. During this critical period, in certain species, every moving object or organism will be "considered" (or imprinted) as a mother, provided it does not resemble congenital patterns typical of the predators of the given species. Other imprinted traits that usually last for animal's entire life include mother-child or partner-partner relationships. Also, at least to a certain extent, individuals' sexual preferences are imprinted in adolescence.

Over time, however, it turned out that imprinting is a phenomenon typical only for precocial birds and selected herding mammals, such as sheep or deer, where newborn animals immediately start to follow their mothers after birth. In ornithological terms, humans are a typical representative of altricial animals, in which imprinting does not occur. This phenomenon fit in so well with prevalent myths and with psychoanalytical hypotheses that it was uncritically adopted to explain human behavior and has persisted in psychology to current day in the form of such terms as a "critical" or "sensitive" period.

Belief in the existence of a critical period was also boosted by works of David Hubel and Torsten Wiesel. During their experiments, they covered one eye of new-born kittens so that visual stimuli could only reach one side of their cerebral cortex. This resulted in permanent functional blindness of the covered eye. This happened due to the specific neurochemistry of cats' cerebral cortex during a short critical period between the 14th and 30th day of life, when a lack of stimulation leads to permanent changes in the brain's anatomy.[11] The same experiments conducted on 30-day old cats caused no changes in their brains. The researchers concluded that there is a critical period in the development of cats' cerebral cortex, which requires constant visual stimulation.[12]

Similar reports from animal research stirred the imagination of psychologists who speculated about critical periods in man connected with speech development, forming relationships and so forth. Since it is not possible to carry out the same experiments on children, the zoologists' claims were accepted by analogy without evidence, and because they fit well within existing framework of opinions about human nature, they never met any opposition.

Unfortunately, what is true for birds does not have to be true for mammals. The behavior of one species does not have to occur in another, even a closely related one. Nature is extremely varied and reasoning by analogy is merely indicative of intellectual laziness. It turns out, for instance, that young *Homo sapiens* have such strong developmental potential that most of them will overcome virtually any obstacles in their way and will eventually develop according to their genetic potential.

Stories of children orphaned in World War II and during the Korean war show that even if deprived of adult affection during, as psychologists claim, the most "critical" period of their lives, they developed absolutely normally after being adopted by caring foster parents, and no dramatic effects (as predicted due to 'affective deprivation in the critical period') were ever reported.[13] There are many similar examples. "More recently, a group of children who had spent their first year in orphanages in Romania were adopted by nurturing British parents. When they arrived in London, they were emaciated and psychologically retarded, as one would expect, given their harsh experiences. However, when they were evaluated several years after adoption, a majority, though not all, were similar in their intellectual profile to the average British child."[14]

Other interesting conclusions came from analyses of children from certain parts of Holland, where children in the first year of their life are treated in a rather odd way. Many middle-class mothers from these areas believe that a harsh environment helps to develop strong character. For this reason, they keep their babies tightly wrapped up, in closed bedrooms, and do not give them any toys. A comparison of their developmental stage with that of other children who were raised in a more conventional way shows that even though they initially lag behind a little, the differences disappear by the age of five, and there aren't any side effects of such early childhood experiences.[15]

At this juncture, it is worth quoting Jerome Kagan, who aptly describes early childhood experiences and their effects on later life:

> The thousands of infants who will be born today across the world will experience very different environments in their first two years. Some will be raised by surrogate caretakers on kibbutzim; some will be cared for by grandmothers or older sisters; some will attend day care centers; some will remain in home with their mothers. Some will have many toys; some will have not one. Some will spend the first year in a dark, quiet hut wrapped in old rags; some will crawl in brightly lit rooms full of toys, picture books, and television images. But despite this extraordinary variation in early experience, excluding the small proportion with serious brain damage or a genetic defect, most will speak before they are two years old, become self-conscious by the third birthday, and be able to assume some family responsibilities by age seven. The psychological differences among these children are trivial when compared with the long list of similarities. The prevalence of serious mental disorders like schizophrenia and depression, as well as the less impairing anxiety disorders, is surprisingly similar around the world, even though children are being reared in different environments. This fact is not consistent with the awarding of significance to the first two years.[16]

The above is true because each of us makes use of the huge genetic potential we all have when we come into this world. This characteristic is typical of all living creatures; it manifests itself even in the development of mentally impaired children. This potential will be utilized under any conditions children might find themselves in. An attempt to create circumstances that might counteract these predispositions would be an exceedingly difficult task. While research on the influence of early childhood experiences on adult personality show its negligible significance, genetic research clearly shows that our genetic material is the main driving force of our development.

Nowadays, with a high degree of certainty, we can tell to what extent the environment forms personality and how much it is determined genetically. This is possible due to research on monozygotic twins, that is, siblings who are born with the absolutely identical genetic material. Additionally, finding a sufficient number of twins who were separated in early childhood and grew up in different environments makes it possible to come very close to answering the question about the role of genetic factors in the process of forming of our personality.

One of the biggest pieces of research into this subject was carried out by the team of Thomas Bouchard from the University of Minnesota, who managed to reach more than sixty pairs of identical twins separated shortly after birth and growing up in different homes.

Bouchard and his colleagues, including Auke Tellegen and David Lukken, found that these twins often exhibited eerie similarities in personality and habits. In one case of male twins raised in different countries, both flushed the toilet both before and after using it, read magazines from back to front, and got a kick out of startling others by sneezing loudly in elevators. Another pair consisted of two male twins who unknowingly lived only 50 miles apart in New Jersey. To their mutual astonishment, they discovered that they were both volunteer firefighters, big fans of John Wayne westerns and, although fond of beer, drinkers of only Budweiser. While attending college in different states, one installed fire detection devices, the other fire sprinkler devices.[17]

This striking and obviously anecdotal material from the research is used equally readily by proponents of identical twins' supernatural powers of communicating with each other. Much more important were the findings of Bouchard's team about personality. The analysis of such characteristics as anxiety levels, tendency to take risks, motivation for achievement, hostility, traditionalism or impulsiveness showed that identical twins brought up separately are as similar to each other as those who grew up together. Thus, different environments influencing them in early childhood did not have an impact on their adult personality.[18] Other pieces of research yielded similar results, namely that it is genetic rather than environmental factors that are more instrumental in shaping personality.[19]

There is also a different way of verifying the hypothesis about the relationship between personalities and childhood experiences on the one hand and (particularly) upbringing on the other. Nancy Segal called this methodological approach "virtual twins technique." It is based on the study of unrelated children adopted and raised by the same family. Segal's research, involving pairs of virtual twins, demonstrates that they bear very little resemblance to each other as far as personality traits are concerned, and they differ significantly in many respects, for instance in anxiety levels. Also, they do not present with similar behavioral problems.[20]

In 2000, Erick Turkheimer set forth three laws of behavioral genetics, which in his view, once and for all ended the nature versus nurture debate concerning the development of humans. These laws are the following:

First Law. All human behavioral traits are heritable.
Second Law. The effect of being raised in the same family is smaller than the effect of genes.
Third Law. A substantial portion of the variation in complex human behavioral traits is not accounted for by the effects of genes or families.[21]

Unfortunately, as Steven Pinker points out,

The three laws of behavior genetics may be the most important discoveries in the history of psychology. Yet most psychologists [have] not come to grips with them, and most intellectuals do not understand them, even when they have been explained in the cover stories of newsmagazines. It is not because the laws are abstruse: each can be stated in a sentence, without mathematical paraphernalia. Rather, it is because the laws run roughshod over the Blank Slate, and the Blank Slate is so entrenched that many intellectuals cannot comprehend an alternative to it, let alone argue about whether it is right or wrong.[22]

2. Mental disorders are caused by early childhood experiences.

Only once in the history of science has the Congress of the United States held a debate about a scientific study, and then put it to the vote, condemning it by 355 votes to 0. Matt Salmon, a congressman from Arizona, called the paper a "proclamation of pedophiles' emancipation." The American Psychological Association was given a rap over the knuckles for publishing it; the media, in their turn, called it "junk science at its worst" and accused its author of a "not-so-veiled attempt to normalize pedophilia."[23]

Who is that *enfant terrible* of American psychology and what did he do to incur the anger of state institutions and God-fearing citizens? The article was actually penned by three authors: Bruce Rind, Philip Tromovitch and Robert Bauserman. It was a meta-analysis of research into the correlates of child sexual abuse among college students. What most annoyed the representatives of the American nation was the conclusion drawn from the study, namely that there was no empirical evidence pointing to the permanent negative consequences of child sexual abuse. The authors looked for correlations between sexual abuse in childhood and 18 categories of mental disorders in adulthood, including depression, anxiety, eating disorders, and so forth. The analysis showed that the average correlation between these variables was only 0.09, which means that sexual abuse is not linked with these disorders at all. Interestingly, the correlation between child sexual abuse and later serious psychological problems was not higher even in those cases where acts of abuse were frequent or took extreme forms.[24]

It is clear that the article of Rind and his collaborators struck at the foundations of one of the most powerful pseudo-sciences in psychology. Not only did it disprove the assumptions of psychoanalysis, dispelled popular notions about human development and took the wind out of hosts of militant feminists' sails, but it also shattered the assumptions of an ideology on which thousands of therapists had relied upon; and so did their jobs, their profits, their sense of self-importance, their mission and the very need for their existence. That hysterical reaction to a scientific paper showed once more how deeply misunderstood the function of science actually is, and how often it is used in the service of ideology.

The duty of scientists is to discover the laws behind the reality in which we live, regardless of whether we will like them or not. They should bring us closer to the truth about reality. Neither the United States Congress nor the Supreme Court—of any country—have the power over the results of scientific research and their implications. Yes, self-correcting mechanisms in social sciences are not working well and not all scientific publications should be blindly trusted, but one apt methodological comment means more that the opinion of 355 congressmen who do not understand that reality might be different from their notions about it. This, among other things, is the strength and beauty of science.

The whole uproar over Rind's paper reveals yet another grotesque notion about human nature. Apparently, it would seem that people can only be prevented from raping and hurting children as long as they are convinced that this will have significant consequences in their adult lives; otherwise, everybody would abuse, beat and take advantage of kids in every imaginable way. That seems the only reasonable explanation of the accusations that publishing the results of Rind's meta-analysis constituted the legalization of those "natural" urges. Clearly, this assumption is totally absurd. Most people love children and do all they can for their best. These are sufficient reasons for them not to harm them. No one needs to threaten people with the consequences of abuse in order to force them to take care of their own children! Constantly cautioning parents, babysitters and teachers against the consequences of violence must come, to put it mildly, from a mind of a simpleton.

Rind and his collaborators did not make up their thesis out of thin air. As we have already written, they conducted a meta-analysis of many published studies. Earlier on, Bruce Rind and Philip Tromovitch published a meta-analysis of other studies, which yielded similar results.[25] Perhaps the only reason why it did not attract as much publicity as their subsequent study was because it was not published in such reputable journal as the *Psychological Bulletin*. Replications of the research that followed, further confirmed the initial conclusion.[26] Despite intensive research, psychologists did not manage to find any proof of what had long been given much publicity: a specific personality profile of child abuse victims.[27] Subsequent studies also challenged the view about the existence of the so-called 'cycle of child sexual abuse,' – the hypothesis that victims turn into abusers in adulthood.[28]

Now, as the debate over Rind's paper has slightly simmered down, it is worth taking a closer look at the even farther-reaching conclusions of numerous pieces of research on sexual abuse. As Scott Lilienfeld and his colleagues concluded: "The most telling finding in the research literature on the apparent long-term consequences of child sexual abuse is the absence of findings. Numerous investigations demonstrate that the typical reaction to a history of child sexual abuse is not psychopathology, but resilience.[29]

Let's make it clear here: in no way whatsoever are the above statements an excuse or a justification for child sexual abuse, pedophilia, or any other harm done to children. Neither are they an incentive to "resilience training." They might be thus interpreted as such only by those exceptionally cruel or extremely limited intellectually. All research conclusions should describe reality, as it is, whether or not this description fits in with someone's ideology. The sufficient reason for not harming children is that they are, like us, human beings gifted with consciousness and the ability to feel pain, and what is more, they are frailer than ourselves.

It is a good time to recall one tremendously important fact: psychology is a branch of knowledge that formulates general conclusions based on the probabilities at which certain events occur. In this particular case, it means that child sexual abuse, especially that which takes extreme forms, may (but does not have to) have harmful effects in some situations. This is another reason why any attempts to justify cruel deeds with research results is purely ideological.

It is in a similar vein that one should comment on the main theses of a 2010 book by Susan Clancy debunking the myth of trauma being the result of sexual abuse. The author, referring to the results discussed above, but also drawing on the findings of her own research, resolutely dispels another myth of pop psychology; as she contends: "In fact, less than 10 percent of the participants reported experiencing their abuse as traumatic, terrifying, overwhelming, life-threatening, or shocking at the time it happened."[30] What is more, the mindless assumption that sexual abuse victims are deeply traumatized does them more harm than good, argues Clancy. She claims that the victims deserve to be treated in a way reflecting well-formulated research questions and solid investigation.

Naturally, the facts presented above do not entirely disprove the claims that mental disorders are caused by early childhood experiences. Alleged sex abuse-induced traumas are given much publicity, but they are not the only problems. In order to obtain a full picture of this element of the childhood myth, let us analyze other situations. Let us begin with an event described by Terr, which most therapists would surely call traumatic:

> For example, on July 15, 1976, 26 schoolchildren ranging in age from 5 to 14 were victims of a nightmarish kidnapping in Chowchilla, California. Along with their bus driver, they were taken hostage on a school bus for 11 hours and buried underground in a van for 16 hours. There, they managed to breathe through a few small air vents. Remarkably, the children and driver managed to escape, and all survived without injury. When found, most of children were in shock, and some had soiled themselves. Two years later, although most were haunted by memories of the incident, virtually all were well adjusted.[31]

This is unthinkable for those psychologists who have been arguing for generations that events of this kind are not only bound to leave a permanent psychological scar, but that they should also be thoroughly 'worked through.' The above example, however, is merely anecdotal. George Bonano from Columbia University, who examined the reactions of children and adults to bereavement, war, terror and disease, systematically analyzed similar cases for many years. These people not only adapted themselves surprisingly quickly to most of these situations, but as soon as several months later, they regained their mental balance from before the traumatic events. After half a year, not more than 10% of the participants were still suffering from some effects of a trauma.[32]

Other systematic attempts to trace mental disorders back to childhood events yielded similar results. Granted, there have been just a few longitudinal studies observing the development from infancy to adulthood, but those that exist provide very important information: in most cases, the results directly contradict the view that we are entirely determined by early childhood experiences. One of such studies investigated the lives of more than six hundred children from the day they were born on the Hawaiian island of Kauai for over thirty years. It turned out that mental disorders in adulthood could be predicted only on the basis of extreme poverty of parents coupled with birth pathologies (premature births, etc.). But even these indices gave poor predictability of future symptoms. For this reason, the authors of the study concluded that the most characteristic feature of all the investigated children's development was self-correcting mechanisms leading to normal development. It was only in highly adverse circumstances that these tendencies did not work.[33]

Many researchers into human development arrived at similar conclusions. As the child psychiatrist Michael Rutter claims: "The ill effects of early traumata are by no means inevitable or irrevocable ... the evidence runs strongly counter to views that early experiences irrevocably change personal development."[34]

It is also worth quoting Joel Paris again: "The outcome of adversity in childhood is highly variable. If possible, it is better for children to avoid trauma, family discord, family breakdown, or serious poverty. Each of these difficulties may increase the risk for psychiatric disorder, at least in *some* children and in *some* adults. The more adversities that are present, the more likely pathology is to develop. Yet even when exposed to multiple adversities, most children manage to escape any form of mental disorder.[35]

3. Effective psychotherapy depends on the reconstruction of childhood experiences.

The previous two myths were actually bound to crystallize into the above statement, which is another element of the early childhood myth. Thus, arguments that challenge them should be quite sufficient to disprove any therapies that focus on analyzing childhood experiences. However, to be on the safe side and for the sake of those readers who were not convinced by the empirical evidence undermining the first two elements of the childhood myth discussed above, we will offer additional arguments for the uselessness of the therapeutic approach based on early childhood determinism.

The first argument draws on the limited usefulness of memory as material for therapy. Despite many commonly accepted notions (and this is yet another psychological myth), memory does not work like audiotape, mechanically recording everything that the microphone picks up. If it were so, the task of the therapist might indeed be to competently retrieve the appropriate parts of the recording and listen to them.[36] But if someone does want to stick to the audio tape metaphor, he should assume that it is a tape on which recordings were made many times; additionally, some parts of it became distorted and others got erased, so that after a few years, no tape recorder, even the best one, could accurately play it back. Another myth is that such a "playback" might be possible under hypnosis, which is more likely rather to create memories than retrieve them.[37] Psychologists specializing in memory processes and research into court witnesses' credibility have many times pointed out what an unreliable tool the analysis of memory is,[38] and no studies indicate that psychotherapists acquired some supernatural ability to make use of their patients' memory.

On the contrary, research demonstrates that insight into childhood experiences is not necessary to achieve therapeutic results even in the classical Freudian psychoanalysis. Patients show improvement without gaining insight into their 'source problems' and without 'working through' their traumatic childhood events, but only due to the emotional support of the therapist.[39] The view that insight into the emotional history of the patient is not necessary to achieve therapeutic effects has also been confirmed by other, far more extensive studies.[40]

One of the most striking arguments for rejecting the analysis of childhood events as a therapeutic tool is the fact that therapies which were rated highest in efficiency rankings drawn up on the basis of Evidence Based Practice (EBP) do not trace patients' problems back to their past, but rather try to change factors causing the disorders.[41] Unfortunately, most psychotherapists still dabble in amateurish archeology, delving through layer after layer of patients' memory with infinite patience in the hope of finding some glimpses of the past, provided they fit in well within the jigsaw puzzle.

Some very powerful arguments against the persistent delving into patients' past are furnished by patients themselves and by therapists involved in memory recovery practices. These practices usually result in serious, real problems for patients. Problems are caused by vague or completely unreal memories, such as when patient and therapist thoroughly work through indistinct memories perhaps related to sexual abuse which may bring back the trauma. Digging with ideological enthusiasm into the family past in order to identify the cause of a phobia may lead to a serious crisis and a break-up of the family or to lawsuits dragging on for years, which will be discussed in greater detail in the next chapter. It is not only the alleged victim but also the alleged perpetrator that suffers the consequences of such a "therapy." The stories of those who were wrongly accused of pedophilia could fill more than one book and the stigma resulting from such an accusation might be a real nightmare.

The last and probably the most shocking reason why we should be extremely vigilant about past life regression therapy is the fact that it was (and in some cases still is) instrumental in influencing people in therapeutic and religious groups that were deemed as cults (some of which were banned). Primal therapy, as it was called, was used by the Moonies and Scientologists, in Synanon, and at the Center for Feeling Therapy, where the biggest psycho-therapeutic scandal in human history took place, of which we will write more in Part III. Primal therapy argued that neurosis is caused by the repressed pain of childhood trauma that could be resolved through re-experiencing and expressing the resulting pain during therapy (the primal scream). It criticized the 'talking' therapies as they dealt with reasoning and not the emotional source of pain.

Social sciences shape the reality they investigate to a much higher degree than other branches of science do. Releasing figures on the population of wood sandpipers in Europe will in no way influence the behavior of this bird, but publishing data indicating a low detection rate of crime in a particular area serves as an open invitation to theft or some other offence. Similarly, an airing of the *March of the Penguins* movie will not change the behavior of these beautiful birds in even a slightest way, but a public debate about the preva-lence of domestic violence could cause many people to discard their inhibi-tions. They might start asking themselves: "If it's such a common phenome-non, why should I be restraining myself all the time?" and even statistics, which were initially exaggerated might, with time, turn out to be a grim reali-ty. Likewise, human behavior could be influenced by irresponsible publica-tions advocating the necessity (and benefits of) 'relieving one's suppressed energy' or by movies and other products of mass culture indicative of such attitudes.

One of the ways people learn is by imitating others, as was proved by the research of Albert Bandura,[42] culminating in his theory of social learning. If someone were to daily watch movies which show that relieving one's aggres-sion promotes relaxation, he would probably begin to put this rule into prac-tice. And even if this kind of behavior did not initially bring him the relief for which he hoped, his expectation might be so intense that after some time he would take some of his reactions as clearly indicative of achieving the desired effect.

Nowadays, psychotherapy has become an inseparable element of mass culture, and its image is largely shaped by the movie industry. *Dr. Dippy's Sanitarium*, the first (silent) film about the work of a psychiatrist and a psycho-therapist was made as early as 1906. In a cult 1955 film, *Rebel without a Cause*, a boy nicknamed Plato says,

- I used to lie in my crib at night and I'd listen to them fight.
- Can you remember back that far? I can't even remember what happened yes-terday. I can't. How do you do it?
- I went to a head shrinker. He made me remember.

With this short dialogue, the scriptwriter Stewart Stern plants two all-powerful psychological myths in the viewers' minds, one about the significance of childhood and the other about the possibility of recalling everything (with the help of a therapist). In the comedy *Analyze This*, a therapist (played by Billy Crystal) persuades a mafia boss (Robert de Niro) to confide in him and open up about his childhood. Similarly, therapists in *Good Will Hunting* or *The Sopranos* TV show encourage their patients to delve into their past.

Movies featuring mental disorders and psychotherapy are so numerous that they have themselves become the subject of scholarly studies. For instance, Paul Gordon wrote a paper about "the celluloid couch" in which he attempted to analyze the image of psychotherapy in contemporary cinematography,[43] while Steven Hyler even tried to classify the cinematic portraits of mentally ill people according to DSM III.[44]

Encouraging patients on the couch to open up about their childhood and the sudden emotional change that results from unearthing their experiences is an almost inseparable element of therapy scenes; it is endlessly repeated by movie-makers. No wonder then that a patient coming to a therapist is full of preconceived notions about what is going to happen. The most stereotypical scenes are the couch: conversations about childhood, looking for parents' mistakes, and finally, a sudden breakthrough inspired by the discovery of a forgotten childhood event or by its brilliant interpretation.

This might have two consequences for therapy. If the patient comes across a therapist who believes in the same therapeutic rituals that are so readily popularized by Hollywood, they will both celebrate them with equal enthusiasm, and perhaps they will even take something that happened during therapy as proof of their ceremonies' effectiveness. But if the patient does not fully share the therapist's views, it is a much more probable scenario that they will repeat the rituals ad infinitum without any effect. Let us notice, however, that in both cases, it is therapists that benefit because, firstly, it is patients that pay for the repeated, though ineffective ritual practices, and secondly, if there are no results, therapists can always attribute this to patients' resistance or some other "indisputable" mechanism; (after all, what are psychotherapeutic dogmas for?) This saves the therapist from doubt and enables them to continue their ceremonies with a clear conscience even to the end of their days.

---

[1] D. Stewards (Director), *Inside the Mind of Adolf Hitler* [Motion picture] (United Kingdom: Discovery Channel, TF1, British Broadcasting Corporation, 2005).

[2] Ibid.

[3] Ibid.

[4] J. Hughes, *The End of Sukarno: A Coup that Misfired: A Purge that Ran Wild* (London: Angus & Robertson, 1968), 181.

[5] G. Robinson, *The Dark Side of Paradise: Political Violence in Bali* (Cornell University Press, 1995): 1-4.

[6] B. Brogaard, "The Making of a Serial Killer: Possible Social Causes of Psychopathology," (2012): http://www.psychologytoday.com/blog/the-superhuman-mind/201212/the-making-serial-killer

[7] J. Paris, *Myths of Childhood*. (London: Routledge, 2000): xi-xii.

[8] D. Frączak, Pożegnanie z rodzicami. *Gazeta.pl*. D. (2010): http://wyborcza.pl/dda/1,110613,8705870,Pozegnanie_z_rodzicami.html.

[9] J. Kagan, *Three Seductive Ideas* (Cambridge, MA: Harvard University Press, 1998): 83-150.

[10] S. Pinker, *The Blank Slate: The Modern Denial of Human Nature* (London: Penguin Books, 2003).

[11] Most likely to allow other parts of the cortex to take over the function of a disabled eye and develop other areas of the brain to compensate for the primary deficiency.

[12] D. H. Hubel and T. N. Wiesel, "Receptive Fields and Functional Architecture in Two Nonstriate Visual Areas (18 and 19) of the Cat," *Journal of Neurophysiology, 28*, (1965): 229-289.

[13] C. Rathbun, L. DiVirgilio and S. Waldfogel, "A Restitutive Process in Children following Radical Separation from Family and Culture," *American Journal of Orthopsychiatry, 28*, (1958): 408-415.

[14] Kagan, *Three Seductive Ideas,* 108.

[15] F. Rebelsky, "Infancy in Two Cultures," *Nederlands Tijdschrift voor de Psychologie, 22*, (1967): 379-385.

[16] Kagan, *Three Seductive Ideas,* 108-109.

[17] S. O. Lilienfeld, S. J. Lynn, J. Ruscio, and B. L. Beyerstein, *50 Great Myths of Popular Psychology* (Chichester, UK: Wiley-Blackwell, 2010), 154.

[18] A. Tellegen, D. T. Lykken, T. J. Bouchard, K. J. Wilcox, N. L. Segal and S. Rich, "Personality Similarity in Twins Reared Apart and Together," *Journal of Personality and Social Psychology, 54* (1988):1031-1039.

[19] J. C. Loehlin, *Genes and Environment in Personality Development* (Newbury Park, CA: Sage, 1992).

[20] N. Segal, *Entwined Lives: Twins and What They Tell Us about Human Behavior.* New York: Dutton, 1999).

[21] E. Turckheimer, "Three Laws of Behavior Genetics and What They Mean," *Current Directions in Psychological Science, 5* (2000): 160-164.

[22] Pinker, *The Blank Slate,* 372-373.

[23] Lilienfeld et al., *50 Great Myths,* 168.

[24] B. Rind, R. Bauserman and P. Tromovitch, "A Meta-analytic Examination of Assumed Properties of Child Sexual Abuse Using College Samples," *Psychological Bulletin, 124* (1998): 22-53.

[25] B. Rind and P. Tromovitch, "A Meta-analytic Review of Findings from National Samples on Psychological Correlates of Child Sexual Abuse," *Journal of Sex Research, 34* (1997): 237-255.

[26] H. Ulrich, M. Randolph and S. Acheson, "Child Sexual Abuse: Replication of the Meta-analytic Examination of Child Sexual Abuse by Rind, Tromovitch, and Bauserman," *Scientific Review of Mental Health Practice, 4* (2006): 37-51.

[27] K. A. Kendall-Tackett, L. M. Williams and D. Finkelhor, "Impact of Sexual Abuse on Children: A Review and Synthesis of Recent Empirical Studies," *Psychological Bulletin, 113* (1993): 164-180.

[28] D. Salter, D. McMillan, M. Richards, T. Talbot, J. Hodges, A. Bentovim A. et al., "Development of Sexually Abusive Behavior in Sexually Victimized Males: A Longitudinal Study," *Lancet, 361* (1993): 471-476.

[29] Lilienfeld et al, *50 Great Myths,* 167.

[30] S. A. Clancy, *The Trauma Myth. The Truth about Sexual Abuse of Children – and its Aftermath* (New York: Basic Books, 2010), 37.

[31] L. C. Terr, "Chowchilla Revisited: The Effects of Psychic Trauma Four Years after a School-bus Kidnapping," *American Journal of Psychiatry, 140,* (1983): 1543-1550.

[32] G. A. Bonano, *The Other Sides of Sadness: What the New Science of Bereavement Tells Us about Life after Loss* (New York: Basic Books, 2009).

[33] E. E. Werner and R. S. Smith, *Vulnerable but Invincible.* (New York: McGraw Hill, 1982).

[34] M. Rutter, "Continuities and Discontinuities from Infancy," In *Handbook of Infant Development,* ed. J. D. Osofsky (New York: John Wiley, 1987), 1256-1296.

[35] Paris, *The Myth of Childhood,* 36.

[36] Lilienfeld et al, *50 Great Myths,* 117-124.

[37] Ibid.

[38] See D. L. Schacter, *The Seven Sins of Memory: How the Mind Forgets and Remembers* (Boston: Houghton Mifflin, 2001).

[39] H. Bachrach, R. Galatzer-Levy, A. Skolnikoff, and S. Waldron, "On the Efficacy of Psychoanalysis," *Journal of the American Psychoanalytic Association, 39,* (1991): 871-916.

[40] P. B. Bloom, "Is insight necessary for successful treatment? *American Journal of Clinical Hypnosis, 36,* (1994): 172-174; J. R. Weisz, G.R. Donenberg, S. S. Han, and B. Weiss, "Bridging the Gap between Laboratory and Clinic in Child and Adolescent Psychotherapy," *Journal of Consulting and Clinical Psychology, 63,* B. (1995): 542-549.

[41] E.g., M. M. Linehan, *Cognitive-behavioral Treatment of Borderline Personality Disorder* (New York: The Guilford Press, 1993); M. M. Weisman, J. C. Markowitz, and G. L. Klerman, *Comprehensive Guide to Interpersonal Psychotherapy* (New York: Basic Books, 2000); A. Wells, *Cognitive Therapy of Anxiety Disorders: A Practice Manual and Conceptual Guide* (Chichester: Wiley & Sons, 1997).

[42] A. Bandura, *Social Learning through Imitation* (Lincoln: University of Nebraska Press, 1962).

[43] P. Gordon, "The Celluloid Couch: Representations of Psychotherapy in Recent Cinema," *British Journal of Psychotherapy, 11,* (1994): 142-145.

[44] S. E. Hyler, "DSM-III at the Cinema. Madness in the Movies," *Comprehensive Psychiatry, 29,* (1988): 195-206. DSM stands for *Diagnostic and Statistical Manual of Mental Disorders,* a classification of mental disorders drawn up by the American Psychiatric Association. In the United States, the DSM serves as a universal authority for psychiatric diagnosis. The Roman numerals following the abbreviation refer to its subsequent versions. DSM-IV is still officially recommended for clinical diagnosis. The fifth edition (published in 2013) was criticized by various authorities – both before and after it was formally published. General criticism of the DSM-V ultimately resulted in a petition signed by 13,000, and sponsored by many mental health organizations, which called for outside review of the document.

CHAPTER 10

THE NIGHTMARE OF RECOVERED MEMORIES

Memory is a mirror that scandalously lies.
—Julio Cortázar

One in four women was raped at least once in her life. A third or even up to half of all women were sexually harassed in childhood, usually by someone from their family. Approximately 50-60% of all patients in psychiatric wards in hospitals were physically or sexually molested in childhood. Over 50,000 out of 255,000 therapists are deeply convinced that their patients were sexually harassed in childhood, but somehow they managed to "displace" or deny their memories and experiences. This is the image of American society from the 1980s based on questionnaires and reports filed by psychotherapists.[1]

It is even worse in Canada: "98 percent of Canadian women have personally experienced sexual violation, 51 percent of women (sixteen and over) have been victims of rape or attempted rape, and 40 per cent of women reported at least one experience of rape. The Canadian national news carried the headline: 'Two Out of Three Canadian Women have been Sexually Assaulted.'"[2] But this dark image is not specific to North America. J. Owen, a British journalist for the Independent, reports:

One in three women in Europe - 62 million people - have been a victim of violence during their lifetime, according to the Violence Against Women report by the European Union Agency for Fundamental Rights (FRA), which is released today. It is an "extensive but widely under-reported human rights abuse" faced by women at home, work, in public and online, warns the report. The findings are from face to face interviews with 42,000 women in the 28 countries that make up the European Union. Participants, aged 18-74, were asked about their experiences of physical, sexual and psychological violence, as well as issues such as sexual harassment and stalking.[3]

According to the same report, the situation in the United Kingdom is among the worst:

Britain ranks among the worst countries in Europe when it comes to women being violently abused, coming fifth worst in terms of violence suffered by women in their lifetime (defined as being since the age of 15). The proportion of women affected, at 44 per cent, is far above the European average of 33 per cent. The only countries worse than Britain are the Netherlands, Sweden, Finland, and Denmark.[4]

In Eastern Europe, things are just as bad. It turns out that in Poland 35% of women and 29% of men were sexually harassed in childhood; almost 17% of women and 9% of men were harmed by their closest relatives; over 10% of women and 3% of men were raped in early childhood. Almost none of them revealed the facts of being abused, never asked for medical, psychological or legal help. The data above was collected and published by Polish Society of Sexology in 1991.

The problem of sexual abuse and harassment became a hot topic in specialist press. In 1984 in the USA, only 46 scientific papers on sexual child abuse were published among 241 other research papers dealing with other forms of harassment. Six years later, this proportion was 500 to 349 respectively. In Sweden between 1974 and 1986, the total number of publications dealing with sexual abuse in children was 66 and it quickly rose to 757 between 1987 and 1992. Analysis of Polish press has shown that the numbers of articles doubled in just 2 years between 2000 and 2002, while exposure indicators have risen significantly, meaning that those articles began to hit the headlines.[5]

Those numbers forced their way into our reality suddenly, without warning, as if someone had ripped the veil of silence and hypocrisy. But let's think about it for a while. Is it really possible that every second or every third of our neighbors sexually abuses their children? Are we living among creatures whose sole purpose is to satisfy their degenerated sexual lusts?

According to Ellen Bass and Laura Davis, authors of the book published in 1998 and entitled *The Courage to Heal*, the answer is yes.[6] According to the thousands of women who brought their relatives, usually parents, to courts and accused them of sexual harassment, the answer is also yes. This claim is also widely accepted by proponents of so-called "recovered memories movements." They believe that those facts have now come to light due to new possibilities and innovative methods of recovering lost memories.

Unfortunately, actual sexual abuse and the social stigma associated with it are real problems that were underestimated for many years. However, the scale in which it is presented today is nothing but a creative invention of therapists. Not only it is based on wrong conceptions, but they have also managed to contaminate their work with ideology. Only few of them however er withdrew from those false conceptions, because "discovering" sexual abuse in patients' memories turned out to be a very profitable business. And it still is so in many countries. But how did this happen?

Previously mentioned activists of the women's liberation movement, Bass and Davis, noticed while leading group meetings for women, that some of them confided their unpleasant sexual experiences or sexual assaults from the past. Such confessions invariably met with sympathy and support from other women in the groups. This encouraged other women to similar confessions. When therapists suggested the idea that, according to Freud's conceptions (or rather misconceptions), similar experiences could have been repressed, it usually helped other women to "restore" their memories. It also helped to justify its inaccuracies.[7] On top of that, there was also a need to meet the expectations of patients (by therapists and by the group). Finally they were "finding" the source of most of their problems in a very convenient way: it was not their fault. It was these traumatic experiences that have made them incapable of coping with their lives.

Let's take a look at what such a "memory recovery process" looks like:

> We might use an age regression technique like holding a ribbon or a rope that goes to the past. … Another technique that works well for a lot of people is imagining they are watching a portable TV. … It has a tape that covers the traumatic experience, and we can use it in slow-motion, we can fast-forward it, we can reverse it. … Then I will suggest that the tape or the dream is going to tell us something about the trauma. I will count and then they will begin to report to me. I watch very closely for changes in facial expression, body movements. If a memory is going to come up, it comes at this time. We work with whatever comes up. Sometimes when it's an image of a very young child being abused, I will check whether it's all right to continue. … People come out of trance with a lot of affect but also with some distance. A lot of the affect is sadness, and feeling appalled and stunned by the brutality. On coming out of trance they frequently will begin to make connections for themselves. There are suggestions to help them do that: they will remember only what they are ready to remember, they will have thoughts, images, feelings, and dreams that will help them understand it better over time, they will be able to talk about it in therapy.[8]

There are a few elements in this example that are typical for various forms of pseudo-therapies. First, take a look at the claim that "there are suggestions to help them" in drawing conclusions—after all, someone could arrive at different conclusions than the therapists! Also note the statement "they will remember only what they are ready to remember." There is a well-hidden, implicit assumption, characteristic for many therapies, that the patient defends himself from the truth…a truth that his therapist can, of course, easily see. He already knows it all. Therapists, especially those with psychodynamic orientation, usually describe such mechanisms in patients as psychological resistance. In this case, we are dealing with the expression "ready to remember." It is to be expected that when the "victim," with the help of his/her therapist, is ready to remember, the resistances will become unblocked and then surely some nightmarish memories from childhood will inevitably surface.

If the patient is too resistant in his ability to "restore" the trauma, therapists have a few more tricks up their sleeve:

In the majority of cases, an adequate narrative can be constructed without resort to formal induction of altered states of consciousness. Occasionally, however, major amnesiac gaps in the story remain even after careful and painstaking exploration. At these times, the judicious use of powerful techniques such as hypnotherapy is warranted. …In addition to hypnosis, many other techniques can be used to produce an altered state of consciousness in which dissociated traumatic memories are more readily accessible. These range from social methods, such as intensive group therapy or psychodrama, to biological methods, such as the use of sodium amytal.[9] In skilled hands, any of these methods can be effective.[10]

Well, this is not the end. There is more. Therapists who failed to "recover" memories can go a step further in their attempts:

Maltz directs people who can't remember to "spend some time *imagining that you were sexually abused*, without worry about accuracy or having your ideas make sense." Others give clients the instruction to "ground the experience or event in as much knowledge as you have and then let yourself *imagine what actually might have happened*."[72] …On other occasions, psychologists rely on "body memories" as the means to access lost information. Despite the lack of any research evidence to support the theory, Smith reports that in her survey of psychologists specializing in sexual abuse recovery, 59 percent claimed that their clients experienced body memories, and 95 per cent of the therapists said it was common for memories to surface via body memories. Operating from this belief, many psychologists touch, massage or move their clients' bodies, watching for any reactions, memories or feelings that come to the client. Others practice "somatic bridging," by asking clients to notice a feeling in the body and then let their minds follow that feeling back in time to another occasion when that feeling was stronger. One woman, who felt physically uncomfortable sitting in the chair, was instructed to "follow that body feeling back in time." Eventually she described a pain in her bottom that was interpreted as an early act of anal penetration.[11]

Courts of law in the United States of America where flooded with thousands of accusations based solely on "recovered memories" obtained with methods such as those described above. Thousands of guilty and innocent people were sentenced. It is estimated that in the 1980s alone, more than 1,000,000 people believed that they were molested in childhood. Therapists who specialized in recovering repressed memories still influence relations between parents and children, grandparents and grandchildren in hundreds of thousands of families around the world. The changes in those relations are usually impossible to reverse. Accusing fathers of sexually abusing their children became an effective and readily available weapon that allowed divorced women to take revenge on former partners. It also left children with permanent significant psychological trauma.

Can memory recovery therapy actually be used to find out about previously committed crimes? Perhaps in our society, children are commonly molested by their parents, and all the tricks that memory plays allow them to get away with their crimes? What is the actual scale of this problem? And what allows us to say that "recovering" memories is nothing but a prosperous business?

We can get the answers to those questions when we analyze several factors that led to this situation in the first place: misinterpreted and simplified understanding (or rather misunderstanding) of Freud's assumptions, feminist ideology and a focus on meeting customers' expectations.

In 1970s and 1980s, when feminist movements raised the problem of sexual harassment, this actually was a very hurtful, severe and real problem, often pushed past the margins of public awareness. Thanks to feminist movements, this problem was identified and feminists managed to draw public attention to it. This deserves the most praise, as for people who experienced such horrors in childhood, therapists ignoring their problems was nothing more than a continuation of their nightmares. By the way, feminist therapy objected to Freud's assumptions. Freudians initially believed their patients' relations in regard to molesting and harassment, but later took the stance that they were nothing but sexual fantasies. It is easy to imagine how people trying to share their traumas must have felt, when they met with the haughty disbelief of psychoanalysts.

Feminists managed to reject one of the psychoanalytic prejudices: a categorical disbelief in the relations of abused patients. Unfortunately, they have also managed to introduce another psychological monster in place of the previous one – repression. Although traditionally associated with Freud, the term and the concept of repression was introduced into psychology before the development of psychoanalysis, by Johann Herbart, to designate the inhibition of ideas by other ideas.[12] Herbart applied repression to the classic problem of the limited span of consciousness: ideas compete for entry into consciousness and the stronger or dominant ideas inhibit, that is, repress, weaker ideas, which are not destroyed, but remain in a "state of tendency."[13]

The concept was popularized by Freud, who noted that the purpose of repression was a defensive one. It was his claim that when ideas or memories produce anxiety, the individual tends to repress the material from consciousness to get rid of the intolerable effect that they produce. It is also worth mentioning the process of suppression and to explain the differences between those two terms. Suppression is a conscious attempt to remove certain contents from memory; when we try not to forget or not to think about something, we are aware of what we are suppressing as well as how and when we have done so. Repression is supposed to be an unconscious process, a certain type of a defense mechanism that we cannot control. We are not aware of what or when we repressed a feeling. Only a trained psychoanalyst is able to access those memories and experiences. The psychoanalyst has enough knowledge and experience to see repressed contents and he is skilled at using previously described "tools" to access the vault of lost memories. Can you see where this is going?

Advocates of recovering repressed memories conveniently assumed that if a patient (usually a woman) has problems, they must have originated from events repressed from their memories. The first sentence of the book *Trauma and Recovery* by one of the most ardent and eager supporters of recovering memories, Judith Herman, states, "the ordinary response to atrocities is to banish them from consciousness."[14] Therefore, memories of violence must be freed in order to achieve full insight and patients will feel immediate relief and improvement. While this assumption at first glance seems a logical one, the assumptions regarding the scale of the problem were rooted in feminist ideology, i.e., that most women are oppressed, discriminated, molested, harassed and so forth. Therapists with such a view of the world were no longer asking the question "was this patient molested or not?" He asked himself "how do I get to the repressed content of absolutely molested victim?" instead. Not a very objective nor scientific approach...

Such a model of working with patients is extremely convenient for the customer and is still extremely common in the most profitable therapies. Let us think about it, a tormented patient comes to see a therapist who is able to offer him a good explanation of all his problems. This explanation frees the patient (customer) from absolutely any responsibility, for example by suggesting that the cause of his problems is rooted in the behavior of other people: perhaps parents' wrongdoings in early childhood, harassment from other family members, his uncontrollable libido, or even the impact of some cosmic forces or any other more or less likely explanation, but always beyond the patient's control. At the same time, such therapeutic systems also contain a promise, such as 'working though the problems with parents,' revealing repressed memories, getting insight into the sub-consciousness, cathartic experiences, understanding or even neutralization of cosmic or divine forces, virtually anything that will allegedly help the patient to get rid of his problems. Would such model of therapy be chosen more often than those systems where the therapist point out the patient's lack of motivation, poor willpower or insufficient effort? This is a rhetorical question; the market of therapeutic services provides all the answers.

Perhaps the biggest fraud in recovering repressed memories is not the ideology, not the completely misrepresented scale of the problem, not even its marketing structure, what makes this model a scam are its false theoretical assumptions. We have said previously that the assumption may appear logical at a first glance. Why aren't they? First of all, the alleged mechanism of repression, despite being tested, experimented on and analyzed prospectively and retrospectively for over a hundred years, was never shown or proven to be true. Secondly, the therapeutic procedure described above leads straight into the so called induction of memories, not to their recovery. Induction of memories on the other hand is a proven phenomenon— dozens of therapists learned about it in a rather painful way:

The case of Patricia Burgus, whose family had accepted a $10.6 million settlement, the biggest to date in any lawsuit accusing mental health professionals of implanting false memories into the minds of patients. Burgus had been referred to a large Chicago hospital for severe postpartum depression. Burgus alleged that while she was undergoing psychiatric therapy from 1986 to 1992, she was persuaded that she had been part of a satanic cult, had been abused by numerous men, had abused her own children, and had had sex with John F. Kennedy. Drugs and hypnosis helped her "recall" that she had cannibalized people.[15]

In 1986 Nadean Cool, a nurse's aide in Wisconsin, sought therapy from a psychiatrist to help her cope with her reaction to a traumatic event experienced by her daughter. During therapy, the psychiatrist used hypnosis and other suggestive techniques to dig out buried memories of abuse that Cool herself had allegedly experienced. In the process, Cool became convinced that she had repressed memories of having been in a satanic cult, of eating babies, of being raped, of having sex with animals and of being forced to watch the murder of her eight-year-old friend. She came to believe that she had more than 120 personalities-children, adults, angels and even a duck – all because, Cool was told, she had experienced severe childhood sexual and physical abuse. The psychiatrist also performed exorcisms on her, one of which lasted for five hours and included the sprinkling of holy water and screams for Satan to leave Cool's body. When Cool finally realized that false memories had been planted, she sued the psychiatrist for malpractice. In March 1997, after five weeks of trial, her case was settled out of court for $2.4 million.

In Missouri in 1992 a church counselor helped Beth Rutherford to remember during therapy that her father, a clergyman, had regularly raped her between the ages of seven and 14 and that her mother sometimes helped him by holding her down. Under her therapist's guidance, Rutherford developed memories of her father twice impregnating her and forcing her to abort the fetus herself with a coat hanger. The father had to resign from his post as a clergyman when the allegations were made public. Later medical examination of the daughter revealed, however, that she was still a virgin at age 22 and had never been pregnant. The daughter sued the therapist and received a $1-million settlement in 1996.[16]

If you think that those examples are horrifying, keep in mind that those very courts have sentenced innocent people for made-up crimes implanted in memories of ordinary people. The empirical evidence, not the court of law, should decide which theoretical concepts (including therapies based on those concepts) are appropriate. And the evidence is rather clear in this matter: "Despite over sixty years of research involving numerous approaches by many thoughtful and clever investigators, at the present time, there is no controlled laboratory evidence supporting the concept of repression."[17]

Although some evidence suggests that adults who have been through overwhelming trauma can suffer a psychic numbing, blocking out memory of or feeling about the catastrophe, it appears that the trauma more often strengthens memories due to heightened emotional or physical sensations. Repression and suppression are today regarded as cases of limited access to previously remembered information. Both phenomena have a quantitative character: we can talk more or less about restricted access. It is rather not the case that some information can be completely displaced, lost or repressed. This a clear departure from psychoanalytic concepts, as they did not allow the possibility of partial repression.

The cases above should not be qualified as malpractice. The false memories planted in thousands of patients were a result of false theoretical assumption of the therapeutic system. Today, advanced neurological examinations allow us to have much clearer understanding of memory processes which could have something to do with repression. Our ability to remember certain events, names or ways of performing certain tasks is coordinated in different parts of our brain, and each part is responsible for different types of memories. Without indulging into neuroanatomy (more than necessary), we can conclude that most memory centers are located in the cortex. However parts of limbic system such as the amygdala and hippocampus, located in the medial part of temporal lobe, are among the structures that play an important role in memorizing and recalling stored information.

The amygdala is primarily responsible for determining the motivational meaning of a stimulus and allows for fear conditioning. Information collected from the environment is transferred into the thalamus, (think of it as a kind of a switchboard), and then to the amygdala and into the cerebral cortex. The amygdala can immediately stimulate the nervous system, and the cortex is responsible for cognitive analysis of the stimulus. The results of those analyses are sent back to amygdala, allowing the proper reaction to stimuli.[18]

The existence of the mechanism for double processing of information has particular benefits to living organisms. First of all, due to the existence of the connection between thalamus and amygdala, organisms can react immediately to threatening stimuli without having them processed first in the cortex. The amygdala receives information from the thalamus and can instantly react to stimuli, before feedback from the cortex reaches it. This process is much faster when compared to what happens when information comes from the cortex only. The third benefit of having this double processing mechanism is the fact that the amygdala, by picking up threats, can act as a kind of alarm system that allows an individual's attention to be immediately redirected to stimuli outside of an individual's attention or focus. Therefore, the amygdala is a structure responsible for the emotional reaction to external stimuli before it gets the chance to be processed and analyzed in the cortex. Famous brain researcher, Joseph E. LeDoux also believes that the amygdala is responsible for the generation of emotional memories, referred to as implicit memory. Those experiences remain subconscious and it is impossible to recall them. They can be only observed in behavior (physiological reactions). This mechanism also leads to the so called "illusion-of-truth effect," which suggests that subjects are likely to rate a statement as true if they have previously heard it, regardless of its veracity.[19]

The hippocampus is a structure responsible for the consolidation of information and the creation of a context for memorized information. It is responsible for so-called explicit memory, allowing for the conscious, intentional recollection of previous experience and information. Neuroscientists generally agree that the hippocampus plays an important role in the formation of new, recallable memories. Damage to the hippocampus affects some types of memory only. It may not be possible to learn new skills (like playing piano) or to acquire a conditional reaction to certain contexts, but it would be still possible to create new conditional reactions to physical stimuli (as they depend on the function of amygdala). This suggests a certain level of independence between those two structures, including the types of memories that they represent: explicit and implicit.[20]

We already know that the amygdala is responsible for unconscious emotional reactions, and he hippocampus directs conscious memories. Let's also take a look how a stressful situation influences both of those structures. This will help us understand whether regression is possible or not. Stress or stressful events lead to the release of stress hormones (such as cortisol and epinephrine). Cortisol acts on both the hippocampus and amygdala, but it has an opposing effect on them. Corticosteroids inhibit processes in the hippocampus and intensify those that depend on the amygdala. Therefore, it is possible that extremely stressful situations can distort memory processes in the hippocampus (the creation of conscious memories and associations) and improve efficacy of collection and storage of emotional and subconscious experiences formed in the amygdala. This would favor repression.

But is it possible that during extreme stress, only emotional experiences, revealing themselves in physiological reactions, are stored in our brains, as the repression model suggests? LeDoux believes that when we analyze the processes in our brains responsible for storing memories, it could be theoretically possible for an individual to have a subconscious memory of a traumatic event without being able to recall what actually happened to him or her. There is only one problem. It is also impossible to prove that such individuals exist. And even if they do, we will never be able to find out, because there is absolutely no way (including any possible psychological procedures, including hypnosis), to recall or access information stored by the amygdala. Unfortunately memory does not work as a recording equipment (contrary to popular belief). For this very simple reason, we can safely conclude that every attempt to access or recall memories is impossible from the neuropsychological point of view. Alleged attempts of their use in therapy are an absolute nonsense, not to mention fraud.

Paradoxically, those absurd claims of lost memory therapies were an impulse into extended empirical research on repression, suppression and other defensive mechanisms that allowed us to better understand them and to reveal their actual real-life values. Furthermore, analyses of the actions of numerous therapists allowed for the discovery of a very powerful manipulative mechanism: the implantation of false memories, sometimes called imagination inflation by academics. This is an interesting process:

> In 1968, the psychiatrist Herbert Spiegel gave a live demonstration designed to show how easily such false memories can be implanted. In front of colleagues and cameras, he hypnotized a man and told him that communists intended to take over radio and television stations. He gave no details but did suggest to the man that he would remember specific information later. When the session ended, the subject began to talk about the plot, providing an elaborate story complete with details such as a description of the furnishings in the room where he first heard about the planned takeover. He also accepted and incorporated further suggestions given to him by others he talked to, even though no longer in hypnosis. Subsequently, he was rehypnotized and the suggestions were "removed." Later, when shown the videotape of the event, he reacted with shock and surprise.[21]

More systematic experiments were conducted on a wave of recovered memories that precisely described the nature of this phenomenon. These were done as an intervention to a dramatic increase in the number of court cases. In order to better understand how it worked, let's take a closer look at a typical experiment in this area.

People at the age between 18 and 53 were invited to take part in an experiment where they were asked to recall various events from their childhoods that they might have heard of from their parents or other relatives. Prior to the experiment, Elizabeth Loftus and Jacquie Pickrell, authors of the experiment, collected descriptions of such events from subjects' relatives, making sure that none of the examined subject got ever lost in a supermarket in childhood. All subjects then received four short stories. Three of them described real events from their past. One was fictional. For all test subjects the fictional story was exactly the same and described a five-year old child lost in a large shopping center. The story described a small child unable to their find parents for a long time and crying for that reason. Finally,an elderly woman took care of them before the family finally found the missing child.

After reading all four stories, the subjects were asked to write details to those stories. In case they could not remember a particular event or situation, they were asked to simply write "I don't remember." Later the authors conducted two talks with participants. Instructions given to study participants stated that those conversations would be testing how well they can recall particular details and how well those details correlate with information provided by their families. Immediately after reading a factual story from their pasts, 68% recalled that such event took place. What is more interesting is the fact that 29% of participants recalled and remembered events that never happened! The second talk did not change this picture; still 25% of participants recalled a fictional event.[22]

There were many similar studies. For example Hyman, Husband and Billings managed to get 20-25% of participants to recall a night spent in a hospital due to high fever or a birthday party with a clown, when no such event ever happened in subject's lives. What is interesting however is the size of imagination inflation effect in most experiments. It is quite stable: imagination inflation affects 20-30% of all subjects.

Now let's summarize a few facts. First of all, don't forget that people taking part in experiments were healthy subjects, without psychological problems, without a strong need to free themselves from everyday problems, the needs that often accompany people looking for professional help. Researchers did not put any pressure on the study participants (neither time pressure, nor social pressure, nor any other kind of pressure at all). Induction of false memories happened as a result of one, single, simple procedure. We are not aware of studies that would test what happens, for example, after ten such sessions, and whether repeated session increase chances of induction of false memories. Intuition tells us that repeated procedures would significantly increase subjects' susceptibility. All the empirical studies were carried out in accordance with ethical regulations, and induced memories were of a neutral character. Having said that, let's now try to imagine how different the study environment is from a therapeutic setting where a patient faces a therapist convinced of his hypothesis, prepared and ready to fight with patient's resistance and denial. Let's look at Freud's statements to help us discover the difference between innocent experimental situations and "real" therapy:

> The work keeps on coming to stop and they [the patients] keep on maintaining that this time [no memory] has occurred to them. We must not believe what they say, we must always assume, and tell them too, that they have kept something back... We must insist on this, we must repeat the pressure and represent ourselves as infallible, till at last we are really told something. ... "Something has occurred to me now, but you obviously put it to my head" ... In all such cases, I remain unshakably firm ... I explain [to the] patient that [these distinctions] are only forms of his resistance and pretexts raised by it against reproducing this particular memory, which must [be] recognized in spite of all this.[23] ... Quite often we do not succeed in bringing the patient to recollect what has been repressed. Instead of that, if the analysis is carried out correctly, we produce in him an assured conviction of the truth of the construction, which achieves the same therapeutic results as a recaptured memory.[24]

Judging by the effects obtained by memory recovery therapists, such as implanted memories of cannibalism, abusing one's own children and similar horrors, Freud's teachings were applied quite zealously. They still are today.

We dare to suspect that the tenacity of therapists and the psychological and mental condition of their patients allowed them to recover false memories during repeated sessions in significantly higher numbers than the 20-30% mentioned in research studies. How surprisingly low seems the real number of actual cases of sexual offences when compared to data resulting from proponents of recovering memories.

Memory recovery therapy is applied with similar eagerness in children. In addition, a few other tricks and effects are used here, such as the importance of authority, guilt, modeling and so forth. Let's take a look at the words of S. E. Draheim, a researcher who is an expert in the field of recovering memories:

"The work necessary to uncover abuses" usually starts with inducing in a child the need to clarify the causes of his anxiety, poor performance at school, nightmares, low self-esteem and other symptoms that will be used to make the "diagnosis". The therapist – an adult, a figure of authority – then suggests answers to the question of why those problems appeared. In order to do it, therapists could use additional help in the form of audiovisual materials, which explains what sexual abuse is, to which consequences it leads and how to cope with the trauma of such experience. The child would then receive "homework," for example, "Think about it again at home; think whether nothing has really happened". The child's imagination was stimulated by asking questions such as "What would happen, if…?" or by presenting them with dolls with exposed genitals. Then any statement from a child implanted with abuse was accepted without criticism and reinforced. Leading, suggesting questions were asked repeatedly, which, especially for small children, might be a signal that their answers were not correct and should be changed. Silence or denial were treated as a "lack of readiness" to verbalize abuses. If your child withdrew his previous statement, it was interpreted as "secondary denial". Gradually, at least in some children, subjective belief of abuse appeared. Younger children were particularly vulnerable to such manipulations, especially if they did not know or understand the concepts of sexual abuse in the first place. An additional source of reinforcement of pseudomemories was the fact that children could be spending time during therapy with other children who were actually abused. People who have exercised such "recovery" techniques were convinced that it was a completely safe method, because it is not thought possible to suggest to a child an event like sexual abuse.[25]

In case of failures in attempts to recover memories, there are still many psychologists (at least in Poland) ready to unequivocally declare a case of abuse based solely on results of so-called projective techniques.[26] We need to stress here, that such techniques originate from Freudian psychoanalysis. Making conclusions based on projective tests is done in a similar fashion and with similar reliability as reasoning based on reading tea leaves.[27] It completely lacks empirical evidence.

Symptom analysis used in the process of diagnosis produces similarly unreliable results as projective techniques,[28] but despite this fact, the wheel of justice keeps rolling and even due to lack of proper evidence, rules on such unreliable grounds. Psychologists acting as expert witnesses believe that it is better to make a 'false alarm' type of mistake, than to make an error of omission, because we should primarily focus on the child's wellbeing. In their view, an adult, even if mistakenly convicted by the court, will somehow manage to deal with it.

It would be worthwhile to ask those troubled child psychologists, how a child's personality develops when it is falsely convicted of sexual abuse? How do those convicted adults "somehow managing to deal with it," often labeled as pedophiles, cope? How do they cope in prison when they end up behind bars thanks to a concerned psychologist? What about the families of children implanted and stigmatized with abuse? How do they cope? What happens to them? Members of the False Memory Syndrome Foundation could provide us with at least few answers.[29] Most of them are relatives of "victims" of alleged sexual abuse. Since 1992, when the Foundation was established, until 2001, over 4,400 people, for whom false memories are a horrible reality turned to the Foundation for support.[30] Unfortunately, this is the only report published by the Foundation. Elisabeth Loftus, memory researcher, presented a shocking statement:

In 1990, Washington State permitted individuals to seek treatment under the Crime Victim Act if they claimed previously repressed memory for childhood sexual abuse. From 1991-1995, 670 repressed memory claims were filed. Of these, 325 (49%) were allowed.

In the study, a nurse consultant reviewed 183 of these claims. Of these, 30 were "randomly selected for a preliminary profile." Some of the findings of this analysis are reported here. The sample was almost exclusively female (29/30 = 97%) and Caucasian (29/30 = 97%), with ages ranging from 15 to 67 yrs. with a mean of 43 yrs.

The women (and one man) saw primarily Masters level therapists (26/30 = 87%), although 2 saw a Ph.D., 2 saw an MD, and 6 saw a Master's level therapist in conjunction with an MD. The first memory surfaced during therapy in 26 cases (26/30 = 87%).

All 30 were still in therapy three years after their first memory surfaced. Over half were still in therapy five years after the first memory surfaced (18/30 = 60%).

Prior to memories, only 3 (10%) exhibited suicidal ideation or attempts; after memories, 20 (67%) exhibited suicidal ideation or attempts. Prior to memories, only 2 (7%) had been hospitalized; after memories, 11 (37%) had been hospitalized. Prior to memories, only 1 (3%) had engaged in self-mutilation; after memories 8 (27%) had engaged in self-mutilation.

Virtually all the patients (29/30 = 97%) contended they had been abused in satanic rituals. They claimed their abuse began when they were, on average, 7 months old. Parents and other family members were allegedly involved in the ritual abuse in all cases (29/29); Most remembered birth and infant cannibalism (22/29 = 76%) and consuming body parts (22/29 = 76%); the majority remembered being tortured with spiders (20/29 = 69%). All remembered torture or mutilation (29/29). There were no medical exams corroborating the torture or mutilation.

The sample of (mostly) women was fairly well educated, and most had been employed before entering therapy (25/30 = 83%), many of them in the health-care industry (15/30). Three years into therapy, only 3 of 30 (10%) were still employed. Of the 30, 23 (77%) were married before they entered therapy and got their first memory; within three years of this time, 11/23 (48%) were separated or divorced. Seven (23%) lost custody of minor children; all (30/30) were estranged from their extended families.

Whereas the average cost of a mental health claim in the Crime Victim Compensation Program that did not involve repressed memory was $2,672, the average cost for the 183 repressed memory claims was dramatically higher, $12,296.[31]

The wave of fascination with the problem reached Eastern Europe some time ago. The media willingly published statements just like the one found by us below:

> I am 52 years old. For six years now, I know that I have been molested in childhood. This happened for a very long time, over 10 years. I did not feel harassed. I was sure I was loved. I don't remember any of it. Now I know that I was saved by this defense mechanism. It was necessary, as my life with such a burden on my shoulders proved to be too difficult. I did not manage to put together my personal life; I couldn't do it. Today I am not surprised why. My marriage fell apart. I do not have any children. ...I have a therapist. Since 2 years now. Thanks to her hard work I am starting to change. I am starting to grow emotionally. I do not feel guilty about what happened anymore. Now I can and want to talk, I want to scream: STOP MOLESTING ME![32]

It seems that the therapist's hard work was "effective," the patient feels better. Finally someone helped her to understand the reasons why she was unable to cope with her life. It is by no means her own fault. It's the fault of the person who molested her. Isn't that a beautiful conclusion?

This wave has already brought misfortune, unhappiness and misery to thousands of people. The worst thing is that those people do not know where to look for help. Here is an example of such desperate attempts to find a solution:

> Are you aware of anything like the False Memory Syndrome Foundation, or any similar organization here in Poland? My daughter together with my husband became fascinated with Hellinger, Jung, integral psychology, and so forth. For many hours, every day, they spend vivisecting their minds. I warned them, that this is very dangerous – I am a psychologist. It all started four years ago. Two years ago she broke contact with us. Last year she accused us of sexual harassment, rape, lies. ...I offered any possible help and support, moderation, all possible procedures: psychological, psychiatric and legal evaluations. She rejected everything. She set a condition: first we admit, then apologize, and then maybe

we can talk. I believe I have done everything I could. I don't know where to go next. I don't know anybody experiencing similar drama. Perhaps you could put us in touch with other people with similar problems? I am worried about my daughter. We were very close. She fell into her own trap. Her husband believes her (I didn't even know when she got married, I have found out about it from the Internet). So does her mother-in-law. She might be convinced that this is all true. I have never "sent my husband to her bed to rape her". I asked her to google "Loftus implanted memory", "two-factor theory of emotion", so she could at least consider a possibility of being wrong. All for nothing. Perhaps your project can give our family a hope?[33]

Is there any way to remain objective and retain common sense in psychological evaluations without compromising the real problem of child abuse? Of course, we can find the answers, for example in the book by Rita Carter *The Human Brain Book*. The author describes research on the human brain, conducted using various methods of imaging, where researchers try to demonstrate the possibility of assessing the legitimacy of conscious memories (both real and implanted), using functional magnetic resonance imagining (fMRI). If the memories are real, images from fMRI will highlight areas corresponding to increased activity in the cerebral cortex, which is responsible for confirming that a particular situation is remembered and comes from one's own experience. If the memories are the products of "creative thinking," then different parts of the brain, areas responsible for signaling the detection of false content and for initiating emotions such as consternation or confusion will be highlighted.

In other research, Daniel Schacter from the University of Harvard, examined twelve women with positron emission tomography (PET) while showing them lists of words that were previously seen by them, and some lists containing new, previously unseen words. Lists of words seen previously induced activity in hippocampus and speech areas of the brain. The lists that the women considered as known, but that contained new items, activated the same areas in the brain plus additionally orbitofrontal cortex (OFC), a part of the brain responsible for inducing reactions like "Ooops, something is wrong in here!" Its activation when we are mistakenly recalling facts suggests that, despite happening outside of someone's consciousness, the brain registers defective memory recalls and continues to generate internal question marks.[34]

Schacter's research was focused on short-term memory, but if his results can be confirmed in regard to long-term memory, after including results conducted by others, brain imaging could be successfully introduced in courtrooms, replacing certified charlatans in psychology and finally providing clear answers in regard to the authenticity of someone's memories.

Readers who have traced our previous arguments about the therapies can be convinced of our negative attitude towards previously described therapies, and psychoanalytic therapy in particular. This is indeed true. We are not enthusiastic about them at all. We accuse psychoanalysis not only of feeding on human misery, but also of creating the miseries in the first place, and then reaping the profits from it. But apart from accusing psychoanalysts and therapies tinted with feminist ideologies, we would also like to add to the list the horde of psychologists in academic centers who create junk science in their warm offices, usually never read by anyone but themselves. Their creative approach to irrelevant problems gives unwritten consent to the creation of monstrosities, such as recovered memory therapy, in the walls of universities built for the purpose of seeking the truth. For nearly one hundred years, empiricists did not care about the problems and mechanisms of repression or other commonly recognized 'defense mechanisms,' while clinicians strongly discouraged them from doing so. The very first research papers on induced and implanted memories started to appear after the waves of people with "recovered" memories flooded American courts in order to accuse their loved ones of the greatest crimes. In spite of conclusive, clear results, representatives of psychological society in different countries preferred to sweep the problem under the rug.

> It is clear that these professional associations and licensing boards cannot be relied upon to carry out their mandate to protect the public. With only a few exceptions, they have been notably inert even in expressing concern about recovered memory therapy. For instance, in 1993, the APA established a working group to address the issue of memories of childhood abuse. Three years later, it submitted a report that acknowledged a split between the clinicians who believe in recovered memories, and the researchers who study human memory. Conceding failure, the group stated that it was "important to acknowledge frankly that we differ markedly on a wide range of issues." Other organizations, such as the Canadian Psychological Association and the British Association for Counselling, have been equally equivocal in their position, making it clear that professional associations and boards cannot be relied upon when issues of serious public concern conflict with those of the Psychology Industry.[35]

We are deeply convinced that the attitude of academic psychologists, based on simply ignoring all new trends in psychotherapy without subjecting them to empirical verification is very convenient for all those fraudulent psychotherapists. This silent acquiescence stretches like an umbrella over the hatching of new little monsters. Academics are not yet confident enough to unanimously discredit the current therapeutic plague, yet another one is already closing in upon us:

> The Psychology Industry's induced public hysteria about sexual abuse is creating yet another new specialization: "children who molest." ... Coined by psychologist Toni Cavanaugh Johnson in 1988, this diagnostic description has been applied to children as young as two, for "inappropriate" behaviors such as diddling, licking, flashing, mooning, masturbating compulsively, looking up under girls' skirts, lying on top of a girl in bed; even using sexual language or asking endless questions about sex. This has led to siblings, cousins and

playmates being diagnosed with "sexual behavior problems," charged with assault and removed from their families. And members of the Psychology Industry are seeking the inclusion of "juvenile sex offending" into the DSM, the catalogue of psychopathologies.[36]

---

[1] The numbers are collected from: Crews, *The Memory Wars*,.and M. Pendergrast, *Victims of Memory: Sex Abuse Accusations and Shattered Lives* (Hinesburg: Upper Access, 1996).

[2] T. Dineen, *Manufacturing Victims: What Psychology Industry is Doing to People* (2007), 48. Retrieved from http://tanadineen.com/documents/MV3.pdf

[3] J. Owen, "The Violence Epidemic: Half of Women in Britain Admit They Have Been Physically or Sexually Assaulted according to Shocking New Figures," *The Independent*, (2014): http://www.independent.co.uk/news/uk/home-news/the-violence-epidemic-half-of-women-in-britain-admit-they-have-been-physically-or-sexually-assaulted-according-to-shocking-new-figures-9169143.html

[4] Ibid.

[5] M. Sajkowska, "Medialny obraz wykorzystywania seksualnego dzieci," In *Dziecko we współczesnej kulturze medialnej*, ed., B. Łaciak, (Warszawa: Instytut Spraw Publicznych, 2003): 71-95.

[6] E. Bass and L. Davis, *The Courage to Heal: A Guide for Women Survivors of Child Sexual Abuse* (Harper Perennial, 1988).

[7] Ibid. 60-110

[8] As cited in J. L. Herman, *Trauma and Recovery* (New York: Pandora – Basic Books, 2001), 186.

[9] Amobarbital (sodium amytal), pentothal (sodium pentothal) and other short-acting barbiturates are often called "truth serums." These substances can alter higher cognitive functions and induce a "need" to tell the truth. There are no reliable clinical studies to support their effectiveness. They equally encourage subjects to disclose facts as well they induce daydreaming and hallucinations. Courts usually reject testimonies obtained under the influence of such medicines.

[10] Herman, *Trauma*, 185-186

[11] Dineen, *Manufacturing Victims*, 106.

[12] M.H. Erdelyi, "The Unified Theory of Repression," *Behavioral and Brain Sciences, 5*, (2006): 499-511.

[13] M.H. Erdelyi, "Repression: The Mechanism and the Defense," In *Handbook of Mental Control*, ed., D.M. Wegner, (Upper Saddle River, NJ: Prentice Hall, 1993), 126-148.

[14] J. Herman, *Trauma and Recovery*, (New York: Basic Books, 1992), 1.

[15] E. F. Loftus, "The Price of Bad Memories," *Skeptical Inquirer, 22*, (1998): 23 – 24.

[16] ———, "Creating False Memories," *Scientific American, 3*, (1997): 70-75.

[17] D.S. Holmes, "The Evidence for Repression: An Examination of Sixty Years of Research," In *Repression and Dissociation: Implications for Personality Theory, Psychopathology and Health*, ed., J. L. Singer, (University of Chicago Press, 1995), 96.

[18] J. LeDoux and E. A. Phelps, "Emotional Networks in the Brain," In *Handbook of Emotions* eds., M. Lewis and J.M. Haviland-Jones (New York: Guilford, 2008), 159-179).

[19] See: J. LeDoux, *The Emotional Brain. The Mysterious Underpinnings of Emotional Life* (New York: Simon and Schuster, 1996).

[20] LeDoux and Phelps, "Emotional Networks," 159-179.

[21] Dineen, *Manufacturing Victims*, 35

[22] E. F.Loftus and J. Pickrell, "The Formation of False Memories," *Psychiatric Annals, 25*, (1995): 720-725.

[23] S. Freud, *The Standard Edition of the Complete Psychological Works of Sigmund Freud*, vol. 2, (London: Hogarth Press, 1995), 279-80.

[24] S. Freud, *Therapy and Technique* (Middlesex, UK: Collier Books, 1963), 282.

[25] S. E. Draheim, "O wszczepianiu dzieciom pseudopamięci, czyli o manipulacji w dobrej wierze." In *Wokół psychomanipulacji*, eds., E. Zdankiewicz-Ścigała and T. Maruszewski (Warszawa: Wydawnictwo SWPS Academica, 2003). 45-55.

[26] A. Sikorska-Koza, "Ocena psychologicznych aspektów wiarygodności zeznań małoletnich świadków w praktyce biegłych sądowych psychologów na podstawie analizy spraw karnych dotyczących wykorzystania seksualnego dzieci," *Dziecko krzywdzone, 1*, (2010): 70-79.

[27] S. O. Lilienfeld, J. M. Wood, and H. N. Garb, "The Scientific Status of Projective Techniques," *Psychological Science in the Public Interest, 1*, (2000): 227-266.

[28] Draheim, "O wszczepianiu dzieciom," 45-55.

[29] http://www.fmsonline.org/

[30] P. R. McHugh, H. I. Lief, P. P. Fryed, and J. M. Fetkiewicz, "From Refusal to Reconciliation. Family Relationships after an Accusation Based on Recovered Memories," *The Journal of Nervous and Mental Disease, 192*, (2004): 525-531.

[31] E. Loftus, B. Grant, G. Franklin, L. Parr and R. Brown, "Crime victims' compensation and repressed memory. A letter to the Mental Health Subcommittee, Crime Victims Compensation Program, Department of Labor and Industries, State of Washington," (Revised Version January 5, 1996,). The original letter has never been published. As cited in Dineen, *Manufacturing Victims*, 55.

[32] "Listy do redakcji," [Letters to the editor]. (2006, May 13). *Gazeta Wyborcza - Wysokie Obcasy*.

[33] Private correspondence.

[34] Schacter, *The Seven Sins of Memory*.

[35] Dineen, *Manufacturing Victims*, 134.

[36] Ibid, 86.

CHAPTER 11

OF THE NEED FOR...PROSTITUTION:
A DIFFERENT PERSPECTIVE ON PSYCHOTHERAPY

Momo, there's nothing wrong with going to the professionals. The first few times you should always go to professionals, women who know their job really well. Later on, when you get involved and it gets to be more complicated, when feelings are added, then you can make do with amateurs.

—Eric-Emmanuel Schmitt

Hang on a minute!—many readers may cry out in exasperation—surely, leaving aside psychoanalysis, past exploring therapies, restoration of memories, and other nonsense, there *must* be some honest therapists who actually help people? Why are the authors of this book presenting the issue in the worst imaginable way, shocking us with the most perverted forms of psychotherapy? True, the authors would reply, there are many other therapeutic modalities, and even though the ones we described have to a large extent dominated psychotherapy, it would be unfair to shape the image of therapists' work solely on the basis of these phenomena. In 1959, Robert A. Harper identified 36 various therapies.[1] By the end of the 1970s, Herink reported that the number of name-brand psychotherapies exceeded 250.[2] By as early as 1986, Daniel Goleman mentioned more than 460 various kinds of psychotherapy in his guidebook[3], and by the turn of the twenty-first century, revised estimates reached 500.[4] Nowadays, they are certainly much more numerous. That's why we will now immediately set about completing the picture of therapies as they are today. However, if we try to find objective data from research results, we will encounter considerable difficulties. These are inherent in the very nature of the subject and, as we will soon see, not only originate from its complexity, but are also indicative of psychotherapy's weaknesses.

The *first difficulty*, resulting in the relative scarcity of data or their incompleteness, is the unwillingness of inventors of therapies and of therapists themselves to submit their ideas or practices to empirical verification. As Maruszewski wrote, "Freud himself was opposed to doing empirical research on defense mechanisms, including repression. In reply to the letters of Sears, who had written to him about his idea of investigating projection, Freud said that conducting experimental research on projection and other defense mechanisms was pointless, as these were so vague and delicate phenomena that they might only be observed during clinical interviews and not during experimental research, which imposes a certain level of standardization.[5]

Thus, it is clear that ever since its creation (with which Freud is commonly credited), psychotherapy has deterred empirical verification and scientific confirmation of its efficacy.

But Freud was not the only one to systematically discourage it. His successor, Neville Symington – member of the Middle Group of British Psychoanalysts, claims that therapy is "a deep experience that occurs between two people and can only be communicated very inadequately to another person."[6] Proponents of Gestalt therapy also admit they are not interested in statistical evaluation of its effectiveness. Since Gestalt therapists begin by assuming that patient behavior will not change in any particular way, the only measureable outcome might be patients' subjective conviction that their condition has improved, a conviction which cannot be accurately gauged. Similarly, Richard Bandler, co-originator of neuro-linguistic programming, went on record repeatedly expressing disdain for any scientific testing of NLP hypotheses; he argued that his system was an art, not a science, and hence the probing of its assumptions was pointless or even impossible, since art cannot be examined with the methodology used in psychology.

Hans Strotzka analyzed the reasons for this reluctance to submit psychotherapy to empirical confirmation. He described them in 1983 in his paper meaningfully entitled "The Psychotherapist's Fear of Empirical Research," where he cited the following arguments of therapists:

> The reasons for the reluctance of psychotherapists to submit psychotherapeutic sessions to empirical research are the following: (1) Psychotherapy is a "hermeneutic" process that cannot be measured (basic argument); (2) empirical research interferes in the process (technical argument); (3) empirical research contradicts the specific private working contract of psychotherapy (ethical argument); (4) in comparative psychotherapy, individual therapists, not systems, are compared (personal argument); and (5) in many institutions, care has a higher priority than documentation (pragmatic argument). It is asserted that these arguments are mainly rationalizations and that empirical research is necessary for the survival of psychotherapy as a respected discipline in medicine.[7]

However, the problem was fully brought to light only in 2009 due to a paper by Timothy B. Baker, Richard M. McFall and Varda Shoham, as well as its review published in *Newsweek*. Sharon Begley, the science editor wrote:

> It's a good thing couches are too heavy to throw, because the fight brewing among therapists is getting ugly. For years, psychologists who conduct research have lamented what they see as an anti-science bias among clinicians who treat patients. But now the gloves have come off. In a two-years-in-the-making analysis to be published in November in Psychological Science in the Public Interest, psychologists led by Timothy B. Baker of the University of Wisconsin charge that many clinicians fail to "use the interventions for which there is the strongest evidence of efficacy" and "give more weight to their personal experiences than to science." As a result, patients have no assurance that their "treatment will be

informed by science." Walter Mischel of Columbia University, who wrote an accompanying editorial, is even more scathing. "The disconnect between what clinicians do and what science has discovered is an unconscionable embarrassment," he told me, and there is a "widening gulf between clinical practice and science."[8]

As might have been expected, such strong statements set off a fierce public debate during which the author of these words found herself at the receiving end of criticism. The main attack, naturally, came from the camp of clinician-practitioners, but the arguments were not new. They might as well be classified according to the Hans Strotzka's proposal quoted above; as in his research, so in this debate, the most common argument was the one about the specificity of the psychotherapeutic process and the difficulty of measuring it.

Indeed, variables involved in psychotherapy are hard to measure; they are numerous, and mutually interact on many levels. Perhaps this field does require a specific approach. But if this is so, then the burden is on the originators of a therapy (just like on the inventors of a drug) to develop a methodology or approach that would unambiguously demonstrate the positive outcome of the therapy (or the lack of such) and the differences among modalities (or the lack thereof). This simple truth is expressed in the words of D. Myers: "If at seven feet and age fifty-nine, I claim to be able to dunk a basketball, the burden of proof would be on me to show that I can do it, not on you to prove that I couldn't."[9] Complaints about methodological inadequacies in measuring therapeutic outcome have for decades accompanied every criticism leveled at it. However, with no possibility of evaluating the effects of therapy, practicing it might have far worse consequences than bragging about being able to do a slam-dunk without as much as stepping onto the court. Unfortunately, this state of affairs seems to be rather firmly established, and hence almost impossible to change. Dr William M. Estein wrote:

> In any enterprise that invokes science in order to establish its authority, the responsibility rests with the new treatment, invention, or product to prove effectiveness and safety; the burden of proof rests with the engineer, the physician, the drug maker, and the psychotherapist. As a scientific enterprise, psychotherapy clearly carries the burden of its clinical ambitions to certify the achievement of its clinical goals – cure, prevention, and rehabilitation. Just as clearly, the burden has been shifted by cultural influences to the shoulders of the skeptic. The long-standing tolerance for the ambiguities of psychotherapy's outcomes speaks volumes about social attitudes, about the quiet but deep meaning of psychotherapy in the United States as a secular religion – a social ideology and a series of rituals that justify and dramatize embedded culture preferences.[10]

In order to make the process of evaluating therapeutic outcomes more real, and to steer clear of the argument between scientists and clinicians, let us assume for the sake of our analysis that it is not aimed at settling academic disputes. Let's assume that we would like to find a therapy, which we might, with a clear conscience, recommend to someone dear to us. Could we, on the basis of the above statements, recommend to our loved one a method of having a "deep experience" that cannot be verified? We don't know about you, dear reader, but we, for now, will hold on and keep on looking for the objective indices of effectiveness.

A rather similar approach was adopted in this regard by the British Psychological Society as it explained its stance on the compulsory registration of psychotherapists: "Existing psychotherapeutic techniques are insufficiently grounded in formal evaluative research ... until this is done, the claims of the different psychotherapeutic methods rest on the clinical experience of practitioners rather than on public evidence capable of withstanding critical scrutiny."[11]

Allowing subjectivity in the evaluation of therapeutic outcome causes considerable confusion. Who should assess it? The patient? The therapist? The patient's family and friends? At least several research studies employed many perspectives in outcome evaluation: that of the patient, of his or her nearest and dearest, of competent judges, and of the therapist. Many of the studies revealed divergences in the assessment of the same therapy's effectiveness.[12] The most striking were those that showed that therapists *were less accurate in assessing changes in patients' behavior than other people*, such as outside observers or even patients themselves.[13] Practitioners warn— in similar manner to an Australian psychologist and expert on depression Dorothy Rowe— that, "You cannot be sure that when patients say they are better, they really are. Patients, who are always trying to be good, kind people, are likely to tell the therapist what he wants to hear. After all, if this nice young chap has gone to all this trouble, it's a pity to disappoint him."[14]

Even if therapies rated high by patients were subjected to detailed scrutiny and only the perspective of patients was taken into consideration, the accuracy of such assessment can be easily questioned. Why? This has to do with the **second difficulty**, which stems from the fact that within the framework of most therapeutic approaches, no initial discussion is made with patients about their expected effects and how these effects will be measured and evaluated. (One of the few notable exceptions is Haley and Madanes's strategic family therapy, which precisely defines its purpose and the way of measuring it, as well as cognitive behavioral therapy, where, for example, the purpose of therapy with a person suffering from agoraphobia will be to bring him or her to the point of being able to leave a room without fear). The purpose of many other therapies is defined vaguely and might change throughout; moreover, more often than not, the therapist leads the patient straight toward his or her own vision of the world. Thus, therapy might seem successful as soon as the patient has adopted the therapist's point of view, and that, as research demonstrates, is easy to achieve.

As early as 1960, Bandura and his collaborators proved that psychotherapists, depending on their personality traits, are prone to expressing approval or disapproval for what patients are saying. Twelve therapists were examined for their ability to directly express hostility and for the extent to which they looked for others' approval. Then they were monitored during therapeutic sessions with patients. It turned out that those who were unable to show hostility were also prone to discouraging patients from doing so. Therapists who had been found to have a strong need for social approval were opposed to patients' expressions of hostility. Therapists who were able to show hostility encouraged similar reactions in patients. Thanks to therapists' reactions, patients were given hints on how to behave. When their hostility met with some sort of disapproval, patients showed it much more rarely by the end of therapy; the opposite was true in cases of approved hostility.[15]

Murray conducted an even more telling experiment.[16] It involved therapists specializing in the non-directive therapy developed by Carl Rogers, one during which therapists do not express their views at all. They also do not react in any way to what patients are saying. Their task is only to listen to their customers in a friendly manner. The role of therapists in non-directive therapy is to show patients unconditional acceptance, warmth and commitment, no matter what they say or do. Therapists are given intensive training to be able to follow these guidelines.

However, Murray's experiment demonstrated that it is not that easy to put Rogers' principles into practice. Recordings of therapeutic sessions showed that the experimenters were able to accurately predict what the therapists approved and disapproved of on the basis of their reactions to the information they had heard or based on the feelings displayed by the patients. Their attitude was quite clear—even though the therapists thought they were abstaining from any judgments. It also turned out that indications of approval or disapproval influenced patients. Successive sessions taught them to avoid behaviors which were frowned upon. This is an example of exceptionally powerful influence, because patients, unaware of the pressure exerted upon them, are not motivated in any way to put up any resistance. This influence affects even patients' moral attitudes, making them fall in with therapists', as shown in successive studies by Morris Parloff.[17] Also, Joan Welkowitz and her colleagues have demonstrated the ability of therapists to influence the values of their patients to come in line with their own. They arbitrarily assigned clients to therapists and subsequently found that the values of the therapists resembled those of their own patients more than those seen by other therapists, and that the similarity of values tended to increase over time or length of treatment.[18] Similarly, a study by Rosenthal found a positive relationship between ratings of improvement and a change of clients' moral values towards those of the psychologists with respect to sex, aggression and authority.[19] The above experiments point to unintentional effects only. It is easy to imagine what a powerful tool for indoctrination is possessed by those therapists who might purposely want to bring certain changes in patients' worldview or to modify their systems of values!

The subjective assessment of therapy's effectiveness is best illustrated by an old joke about the meeting of two therapists: one behavioristically oriented, and the other adhering to principles of humanist psychology. After a brief conversation, it turned out that both therapists struggled with a similar case of an adult patient who was still wetting the bed. They talked about it for a moment, exchanged remarks on the methods they were going to apply and then went their respective ways. Two months later they chanced upon each other again on the street. Having passed the time of day, they went back to their previous talk and asked each other about the effects of their therapies. "Well," replied the behaviorist, "I managed to reduce the patient's bed-wetting by 80%. It happens sporadically now. And how's your patient?" he asked. "Mine," answered the humanist, "keeps bed-wetting, but he's now proud of it!"

The subjective evaluation of therapeutic outcome does not make this picture any brighter, and difficulties connected with measuring its efficacy multiply as we go deeper into the subject. The situation will get even worse because it turns out that, as Myerson said, "The neuroses are 'cured' by Christian Science, osteopathy, chiropractic, nux vomica and bromides, benzedrine sulfate, a change of scene, a blow on the head and psychoanalysis, which probably means that none of these has yet established its real worth in the matter ... moreover since many neuroses are self-limited, anyone who spends two years with a patient gets credit for the operation of nature.[20] This is not an isolated view. Let's quote Dorothy Rowe once again:

> You can take a ward full of patients of whatever diagnosis, age and sex, and you can give them all a new drug, or a new kind of therapy, or simply a change in their routine, and a third of them will get better, a third will stay the same, and a third will get worse, give or take a few each way. Of course, a few weeks or months later, some of those who got better will get worse, and some of those who got worse will get better. But you cannot be sure that when patients say they are better they really are.[21]

The above quotes introduce us to **another difficulty** and immediately provides us with its explanation. Following their line of thought, we can in one breath add the following items to the list of "efficient" therapies: neuro-linguistic programming, facilitated communication, neurokinesiology, neuro-transgression, regression, numerology and tarot. The only variable which accounts for the efficacy of all those "therapies" is... time. There is also the natural tendency of an organism to regain a state of balance, a tendency which developmental psychopathologists call resilience. In their view, it is an ability to recover one's lost or weakened strength as well as resistance to harmful factors. This individual ability to revert to the initial state of balance depends on the proper functioning of neurohormonal structures, including the pituitary gland, renal cortex, and hypothalamus.[22]

It is because of this tendency that 60% of those examined in Eysenck's research, described in the chapter about psychoanalysis, showed an improvement without any therapy. Whenever research is carried out without reliable comparison with control groups not subjected to any therapy, obtained results might be due to the influence of the variable of time, (sometimes described as the regression to the mean). Sooner or later, the organism recovers its balance, and if someone "assisted" in this natural process, he can easily claim the laurels of being a healer, but is he really one?

This credit certainly cannot go to a psychoanalyst mentioned by Symington, who, upon terminating years-long psychotherapy, was told by his patient that throughout all those years, he had told him nothing that he couldn't learn from his friends in a pub.[23] Another woman, described by Colin Feltham, drew similar conclusions, but before she completed her therapy. Having noted that that the professional psychotherapy she had gone into did not reach to her expectations, she hit upon an excellent idea. She began to regularly invite her friend out to a restaurant. In return for her advice and the possibility of confiding in her, she agreed to pay for food and drinks. Those meetings were probably more profitable to her than visits to the psychotherapist and certainly more pleasant because accompanied by eating.[24]

There are many similar cases and anecdotes. This is how Raj Persaud describes this issue:

> The tendency for large numbers of people to get better without help, but with the passage of time, is so well known that some psychiatrists have suggested that the waiting list to see a psychiatrist could be seen as form of treatment in itself. Sometimes if you managed to see clinician too soon after an upset, when you were in a terrible state, he or she might prescribe treatment which in the end could prove unnecessary because recovery was possible without it – given enough time.[25]

Maybe it *is* mostly time, but perhaps some other factors may cause an improvement in patients in similar situations. Without going too deep into this phenomenon, scientists have referred to it as placebo therapy, sometimes also calling it "talk therapy." To make this notion clearer, let us once more give the floor to Eysenck:

> "'Placebo treatment' is a pseudotreatment" which has no rationale or meaning, and is not intended to benefit the patient; it is simply instituted to make him believe that he is being treated, while in actual fact, he is receiving no kind of effective treatment whatsoever. A placebo treatment is a control for non-specific effects, such as a patient's going to see a therapist, believing that something is being done for him, and possibly talking to the psychiatrist or psychologist. It should therefore be a control.[26]

Therefore, if a comparison of a given therapy with a placebo therapy does not point to statistically relevant differences, we can confidently assume that it is worth as much as a chat with friends in a bar or a restaurant.[27] The converse is also true: if such differences do appear, they may be a rationale for recommending such a therapy to a person who is looking for it. Do psychotherapies emerge victorious from such comparisons?

The first study of this kind, conducted by Eysenck, was described in the chapter on psychoanalysis. Its findings were shocking because they demonstrated differences that spoke against psychoanalysis. It turned out that more patients had made an improvement when they were left to themselves than those who were subjected to psychoanalysis.[28] The publishing of the results provoked a stormy debate and Eysenck found himself on the receiving end of criticism. Granted, this contributed to identifying certain flaws in his methodology, but his research was later replicated many times and in various ways. Many years afterward, as late as 1985, Eysenck had already realized that psychoanalysis did not necessarily produce negative results as compared with placebo therapy, but he also knew that it was not worth much more. Here is how he summed up the state of knowledge on the subject:

> Even now, thirty years after the article in which I pointed out the lack of evidence for therapeutic effectiveness, and some five hundred extensive investigations later, the conclusion must still be that there is no substantial evidence that psychoanalysis or psychotherapy have any positive effect on the course of neurotic disorders over and above what is contributed by meaningless placebo treatment. Treatment or no treatment, we get rid of our colds, and treatment or no treatment, we tend to get rid of our neuroses, although much less quickly and much less surely.[29]

As a well-known and intransigent opponent of psychoanalysis, Eysenck was always suspected of one-sidedness. But he was not the only one to prove that there are no differences between psychotherapy and placebo treatment, and not only with regard to psychoanalysis. Gavin Andrews demonstrated that good psychiatric care has better effects, poses fewer risks, is cheaper than dynamic psychotherapy and is only slightly less effective than cognitive-behavioral therapy.[30] Eysenck and Andrews's views are not isolated. As Richard Stuart claims in his analysis of therapeutic failures based on a review of 21 empirical studies: "In view of this evidence, it can be said that persons who enter psychotherapy do so with a modest chance of marked improvement, a much greater chance of experiencing little or no change, and a modest chance of experiencing a deterioration in their functioning.[31]

Similarly in reviewing the literature on therapist effectiveness, Robyn Dawes concluded that, "there is no positive evidence supporting the efficacy of professional psychology. There are anecdotes, there is plausibility, there are common beliefs, yes - but there is no good evidence."[32] William Epstein concurs with these findings. In his book *Psychotherapy as*

Religion, he contends that, "Psychotherapeutic intervention has not demonstrated any benefit to any patient group under any circumstances."[33] A similar view is presented by Tana Dineen: "Outcome evaluations and cost-benefit studies suggested, at best, that psychotherapy was somewhat effective with some of the clients some of the time."[34] Parloff, in reviewing nearly 500 rigorously controlled studies, concluded that, "the research evidence... has not met the needs of the policy makers and does not greatly enhance the credibility of the field of psychotherapy."[35]

What remains for us then, after analyzing all these studies and their summaries? Shall we switch off the light and leave? Of course not! After all, research constantly comes across the placebo effect! Despite everything, placebo therapy (and hence most other therapies) does produce some results as compared with no therapy. Perhaps this merits attention. Let us consider for a while the understanding of psychotherapy proposed by Hans Eysenck:

> Many of those who go to the psychoanalyst are not in fact neurotically ill at all. For the majority of them, psychoanalysis constitutes what one critic once termed the 'prostitution of friendship'. In other words, unable because of defects of personality and character to make and keep friends to whom they can confide, they pay the psychoanalyst to serve this function, just as men buy sex from prostitutes because they are unable or unwilling to pay the necessary price of affection, love and tenderness which is needed to achieve a sexual relation on a non-commercial basis.[36]

Perhaps shedding illusions about the efficacy of a given therapy and the correctness of its explanatory apparatus might improve this 'prostitution of friendship' to the benefit of patients? Perhaps it is in this direction that research and therapy should be heading. Eysenck's views, controversial as they may seem at first glance, are not isolated. The psychiatrist Fuller Torrey once described psychotherapy as: "the world's second oldest profession, remarkably similar to the first. Both involve a contract (implicit or explicit) between a specialist and a client for a service, and for this service a fee is paid."[37] This view is shared by existentially minded thinkers, who claim that the so-called professionalization of privacy, which manifests itself in seeking out the help of paid professionals in solving the most intimate problems, is an inherent feature of our contemporary culture.[38]

The elements that are emphasized as decisive for particular therapeutic approaches are nothing else but the realization of Eysenck's suggestions. Lomas refers to the relationship between patient and therapist as a second parenthood,[39] while Sayers speaks of mothering a patient.[40] Most professional prostitutes know rather well—intuitively and without any training—that they should pretend. The better they do it, the more clients they have. As research results and patients' preferences indicate, pretending friendship in a skillful way is probably the most important factor in client satisfaction. Let's take a look how well our paid friends deal with such pretending.

Sadly, one of the first comparative studies whose results were rather shocking for professional psychotherapists demonstrated that they did not have any special skills. Therapists were not any better than other ordinary people with minimal knowledge, common sense, and goodwill.[41] Similar results were yielded by meta-analytical studies, which did not show any differences between psychotherapeutic outcomes in patients "treated" by trained therapists and unqualified persons.[42] In another meta-analysis, professionals, non-professionals, as well as junior and senior psychology students achieved the same psychotherapeutic outcome.[43] Other similar studies spelled even worse news for professional therapists. Not only it was shown that there was no link between a therapist's experience and training on the one hand and therapeutic outcome on the other, but in several cases, patients perceived amateurish therapists as better than professionals.[44] While earlier studies yielded results which were not favorable to professional psychotherapists, a 1995 review of all the previous studies, comparing the skills of amateurs, paraprofessionals (e.g. trained volunteer therapists), and professionals, might be described as devastating.[45] Amateurs without training or supervision achieved the same results as professionals when it comes to changes in patients' behavior. Many studies found that paraprofessionals achieved better results than professionals. There are tons of similar research and most of them show similar results.

Of course, as might have been expected, the studies quoted above, especially the later ones, raised a storm of controversy in psychotherapeutic circles. As it often happens, critics were more concerned with attacking the methodology of the research rather than focusing on careful analysis of the results or on improving training programs and therapeutic systems. That was a rather natural, though totally unconstructive reaction. Time for conclusions: professional therapies bring about the same effect as placebo therapies; amateurs and paraprofessionals achieve the same or better results than professionals.

So does it not follow that the factors crucial for successful therapy operate outside of therapeutic systems and are not connected with skills developed during the training that therapists receive? Is it really not worth investigating these factors? SHouldn't we give up fighting among schools and therapeutic modalities, and train professional "paid friends" instead? In our opinion, it would not only be more effective but also a more honest approach to therapy than deluding patients with insight, repression, resistance and other psychotherapeutic bullshit. We are afraid, however, that this will never happen because limiting the role of psychotherapy to what it really is would strike at the fundamental needs of most therapists: recognition, respect and prestige.

As we conclude this chapter, we would like to return to our question about recommending therapy to someone who is truly dear to us. Shall we recommend psychotherapy at all? Is it worth it? The answer is not quite easy; even if the recommended therapy came down to hiring a "paid friend," let's hold off for a moment. Before making a decision, we should consider other pros and cons.

---

[1] R. A. Harper, *Psychoanalysis and Psychotherapy: Thirty-six Systems* (Englewood Cliffs NJ: Prentice-Hall, 1959).

[2] R. Herink, (Ed.), *The Psychotherapy Handbook. The A to Z Guide to More than 250 Different Therapies in Use Today* (New York: Meridian Books, 1981).

[3] G. Goleman, "Psychiatry: Guide to Therapy is Fiercely Opposed," *New York Times,* (September 23, 1986).

[4] D. A. Eisner, *The Death of Psychotherapy: From Freud to Alien Abductions* (Westport, CT: Praeger, 2000).

[5] T. Maruszewski, *Pamięć autobiograficzna* (Gdańsk, PL: Gdańskie Wydawnictwo Psychologiczne, 2005): 174.

[6] N. Symington, *"The Analytic Experience: Lectures from the Tavistock* (London: Free Association Books, 1986): 9.

[7] H. Strotzka, "The Psychotherapist's Fear of Empirical Research," *Psychotherapy and Psychosomatics, 40(1-4),* (1983): 228-231.

[8] S. Begley, "Why Psychologists Reject Science" *Newsweek,* (October 1, 2009): http://www.newsweek.com/why-psychologists-reject-science-begley-81063

[9] D. G. Myers, *Intuition. Its powers and Perils* (New Haven and London: Yale University Press, 2002), 235.

[10] W. M. Epstein, *Psychotherapy as Religion. The Civil Divine in America* (Reno and Las Vegas: University of Nevada Press, 2006), xi.

[11] British Psychology Society, "Statement on the Statutory Registration of Psychotherapists," *Bulletin of the BPS", 33,* (1980): 353-356. As cited in R. Persaud, *Stay Sane: How to Make Your Mind Work for You* (London: Metro Books, 1977): 64.

[12] E.g., K. Keniston, S. Boltax, and R. Almond, "Multiple Criteria of Treatment Outcome," *Journal of Psychiatry, 8,* (1971): 107-119; D. Horenstein, B. K. Houston, and D. S. Holmes, "Clients', Therapists', and Judges' Evaluations of Psychotherapy," *Counseling Psychology, 20,* (1973): 149-158; S. L. Garfield, R. A. Prager, and A. E. Bergin, "Evaluation of Outcome in Psychotherapy," *Journal of Consulting and Clinical Psychology, 37,* (1971): 307-313.

[13] M. Harty and L. Horowitz, "Therapeutic Outcome as Rated by Patients, Therapists and Judges," *Archives of General Psychiatry, 33,* (1976): 957-961.

[14] D. Rowe, "Introduction*"In* Masson, *Against Therapy.*

[15] A. Bandura, D. H. Lipsher, and P. E. Miller, "Psychotherapists' Approach-Avoidance Reactions to Patient's Expressions of Hostility," *Journal of Counsulting Psychology, 24* (1960): 1-8.

[16] E. J. Murray, "A Content Analysis Method of Studying Psychotherapy," *Psychological Monographs, 70,* (1956): 420.

[17] M. B. Parloff, "Some Factors Affecting the Quality of Therapeutic Relationships,' *Journal of Abnormal Social Psychology, 52,* (1956): 5-10.

[18] J. Welkowitz, J. Cohen, and D. Ortmeyer, "Value System Similarity: Investigation of Patient—Therapist Dyads," *Journal of Consulting Psychology, 31,* (1967): 48-55.

[19] D. Rosenthal, "Changes in Some Moral Values following Psychotherapy," *Journal of Consulting Psychology, 19,* (1955): 431-36.

[20] A. Myerson, "The Attitude of Neurologists, Psychiatrists, and Psychologists toward Psychoanalysis," *American Journal of Psychiatry,* 96, s. 623-641, 1939 (1955): 641.

[21] Rowe, "Introduction."

[22] W. J. Curtis and D. Cicchetti, "Moving Research on Resilience into the 21st Century: Theoretical and Methodological Considerations in Examining the Biological Contributors to Resilience," *Development and Psychopathology, 3,* (2003): 773-810.

[23] Symington, *The Analytic Experience.* As cited in Persaud, *Stay Sane,* 66.

[24] Feltham, C. *What is Counselling? The Promise and Problem of the Talking Therapies* (London: Sage Publications, 1995). As cited in Persaud, *Stay Sane,* 66.

[25] Persaud, *Stay Sane,* 24-25.

[26] H. J. Eysenck, *Decline and Fall of the Freudian Empire* (Piscataway, NJ: Transaction Publishers, 2004), 78.

[27] Authors of this book would like to point out that in a classic situation, when you meet with friends in a pub, you could actually split the bill!

[28] Scientists call a therapy or treatment that inhibits the patient's ability to regain physical or emotional balance a counter-effective therapy.

[29] Eysenck, *Decline and Fall,* 80.

[30] G. Andrews, "The Essential Psychotherapies," *British Journal of Psychiatry, 162,* (1993): 447-451.

[31] R. B. Stuart, *Trick or Treatment: How and Why Psychotherapy Fails* (Champaign: Research Press, 1970), 50.

[32] Dawes, *House of Cards,* 58.

[33] Epstein, *Psychotherapy as Religion,* xi.

[34] Dineen, *Manufacturing Victims,* 122.

[35] M. B. Parloff, "Psychotherapy Research Evidence and Reimbursement Decisions: Bambi meets Godzilla," *American Journal of Psychiatry, 139(6),* (1982): 721.

[36] Eysenck, *Decline and Fall,* 71.

[37] F. F. Torrey, *Witchdoctors and Psychiatrists* (New York: Harper & Row, 1986). p. 1.

[38] P. Lomas, *True and False Experience* (London: Allen Lane, (1973): 3-4, 15-17.

[39] See: P. Lomas, *The Limits of Interpretation* (London: Penguin, 1981).

[40] See: J. Sayers, *Mothering Psychoanalysis* (London: Penguin, 1991).

[41] R. B. Sloane, E. R. Slapes, A. H. Cristol, N. J. Yorkston, and K. Whipple, *Psychotherapy vs. Behavior Therapy* (Cambridge: Harvard University Press, 1975).

[42] D. M. Stein, and M. J. Lambert, "On the Relationship between Therapist Experience and Psychotherapy Outcome," *Clinical Psychology Review, 4,* (1985): 1-16; J. S. Berman, and N. C. Norton, "Does Professional Training Make a Therapist More Effective?" *Psychological Bulletin, 97,* (1985): 401-406.

[43] J. R. Weisz, B. Weiss, M. D. Alicke, and M. L. Klotz, "Effectiveness of Psychotherapy with Children and Adolescents: A Meta-analysis for Clinicians," *Journal of Consulting and Clinical Psychology, 55,* (1987): 542-549.

[44] N. R. Simonson, and S. Bahr, "Self-disclosure by the Professional and Paraprofessional Therapist," *Journal of Counseling and Clinical Psychology, 42,* (1974): 359-363.

[45] D. Faust, and C. Zlotnick, "Another Dodo Bird Verdict? Revisiting the Comparative Effectiveness of Professional and Paraprofessional Therapists," *Clinical Psychology and Psychotherapy*, *2(3)*, (1995): 157-167.

CHAPTER 12

LIGHT AT THE END OF THE TUNNEL: THE EFFECTIVENESS OF PSYCHOTHERAPY TODAY

It was one of those rare books, procurable only if together with other people all queuing up to read it; you waited long enough for your turn to come. When your time eventually came, you had the book for just a fleeting moment. There it was, tattered with its pages dog-eared from constant reading and stained from being held by so many hands. You could only enjoy it for one night, so if you decided to choose a good night's sleep or some other pleasure, you had to pass the book unread to the next person in line. No wonder that rather than missing a unique opportunity for intellectual discovery, we preferred to give up sleeping. At that time, libraries were swelling with boring tomes by orthodox thinkers and scientific writers; those rare literary gems that were circulated by the underground publishers and distributors, smuggled from abroad or miraculously overlooked by the censorship and published in Poland in a limited edition, traveled constantly from hand to hand. That book, though released in a paperback version with an uninviting cover and low-quality, presumably recycled paper, kindled the minds of psychology students in the 1980s. Had it not been for its wide criticism of every single manifestation of Western intellectual output at the time, it would almost certainly have never been made available from the official publisher. What a limited mind must the censor have had to take on such a perspective! That almost pocket-sized collection of texts on psychology and psychiatry contained much more explosive and highly flammable content than all libraries altogether, to us. We discovered Zimbardo's prison study, Milgram's research on obedience to authority, Tomasz Szasz's critique of the mental illness myth, a number of Ronald David Laing's works, to name just a few writings. The book was powerful enough to blow the minds of young people who studied psychology squeezed into the straitjacket of primitive, dialectical materialism.[1]

There was also one more article that affected us particularly deeply at the time: David Rosenhan's "Sane in Insane Places." It was a recount of the author's experiment conducted with the intent to verify whether psychiatrists had been capable of distinguishing mental illness from mental health. Rosenhan conducted his study in a number of mental hospitals across the United States. The study involved the participation of seven additional volunteers who had never struggled with any health mental issues before. They presented themselves in different mental hospitals and each volunteer, when interviewed by a hospital admission doctor, reported that he or she had been hearing voices. Each of the eight volunteers ("pseudopatients") was diagnosed as having psychiatric disorders and was subsequently admitted to hospital, with seven being diagnosed further to suffer from schizophrenia. Each pseudopatient received treatment with strong psychoactive medication. Being one of the pseudopatients himself, Rosenhan was discharged from hospital after two months and reported subsequently the entire experiment to the media. He was straightaway accused of using trickery and spreading slanders. One of hospitals challenged Rosenhan to send more pseudopatients to their facility, confident that their medical staff would be competent to identify them this time. Rosenhan accepted the challenge and in the following month, the hospital was proud to declare that 41 pseudopatients were detected. As it turned out, Rosenhan did not send even a single patient to the hospital.[2] The outcome of his hoax came as a shock to American psychiatry. It unequivocally showed that psychiatrists were devoid of reliable diagnostic tools that would support them in discriminating between mentally healthy people and those suffering from mental disorders.

Rosenhan's study was undoubtedly one of the major landmarks in the history of clinical psychiatry and psychology. It threw our minds into turmoil and stirred up a myriad of questions that we flooded our teachers with. What has changed over the past several decades? Have we made any visible progress in the field of diagnosis and therapy?

For these questions to be answered in an exhaustive manner, one needs to freshen up the knowledge on the present condition of the research into the effectiveness of psychotherapy. Numerous research studies that we have quoted in the previous chapter date from before the 1980s. Although major challenges to be handled by psychotherapy today remain unchanged, the scientific research into its effectiveness has made strides since then and brought new findings that will justify a more optimistic view of this diverse domain. A glance at the history of research allows identification of three phases of its development.

The first one, stretching until the end of the 1960s, attempted to determine whether psychotherapy was capable of bringing about a genuine, positive personality change. As can be predicted, questions of such generality remained unanswered and no conclusive observations were made. The second research phase was initiated in the 1960s by investigators with backgrounds in behavioral therapy. They started to compare the effectiveness of diverse treatments of phobia and other issues with the application of experimental methods, that is, by assigning patients randomly to different groups. Their novel approach caused substantial progress in the field of research methodology. At the same time, owing to them or even by their fault, all those drawbacks we have looked at in the previous chapters were discovered.

The launch of a collaborative research program on depression that took place in the United States in 1980 marked the beginning of the third phase.[3] The said project prompted a methodological turn to a medical research model, as analogous to the one behind the process of drug efficacy evaluation. This phase saw the considerable strengthening of methodological rigor and, consequently, of the internal validity of research.

Rosenhan's study certainly played an important role in those trends. The hoax itself and its outcome sparked a breakthrough in American psychiatry. At first, they provoked impassioned and wide-ranging debate over the value of diagnostic methods and contributed to the subsequent commencement of an extensive research program. Furthermore, they paved the way for far-reaching changes in psychiatric diagnosis, changes that even today many authors unhesitatingly term as revolutionary.[4] The research program in question started in 1974 and was conducted under the supervision of Robert Spitzer, a psychiatrists and a fervent opponent of Rosenhan's confrontational method. The project was a forthwith response to a crisis enveloping the entire profession of American psychiatry, the exposure of which had been possible due to the unorthodox experiment. The program yielded the third edition of *Diagnostic and Statistical Manual of Mental Disorders* (DSM), developed and published under the auspices of the American Psychiatric Association.

If truth be told, DSM had already been available since 1952□ however, its first publication and the subsequent updated edition published in 1968 had been commonly ignored and rejected as a standard for mental health diagnosis.[5] It was only the third revision that succeeded in establishing a common language recognized and spoken by psychiatrists all across the Unites States and in some European countries. This time, the manual edition was based on statistical research and was compliant with *The International Statistical Classification of Diseases and Related Health Problems* (ICD), published by the World Health Organization, and as such, it eventually standardized diagnostic methods and tools. Over the years, DSM-IV and DSM-V were also released.

Unfortunately, nowadays it seems that the wheel of history has come full circle. It was already in the 1990s when critical comments dismissing those tools were first voiced.[6] Emerging as a panacea for solving problems exposed by Rosenhan back in the 1970s, today the manual has proven to be a highly subjective and impermanent concept, additionally cracked by corruption.[7] The expanded DSM-V edition, rejected by the National Institute of Mental Health (NIMH),[8] is a marketing tool by means of which almost anyone can be qualified for treatment rather than a reliable attempt to set up a diagnostic system. Why are such powerful statements made?

In early February 2012, the German weekly *Der Spiegel* ran an extensive article on the issues of mental illnesses and disorders. It quoted the American psychiatrist Leon Eisenberg, known as one of the members of the team who invented Attention Deficit Hyperactivity Disorder (ADHD), and who bore joint responsibility for the disorder being added to the DSM inventory.

"ADHD is a prime example of a fabricated disease," Eisenberg said. "The genetic predisposition for ADHD is completely overrated."[9]

Eisenberg's few statements stirred up uproar, not only among psychiatrists, but also among public opinion. Not surprisingly, the disorder was on everyone's lips following the publication of the article. With the number of ADHD-diagnosed patients constantly on the rise nowadays, (4-8% of all preschool age children), almost everyone either knows someone with recognized ADHD or has been diagnosed with it. Unfortunately, Leon Eisenberg died a few months after the publication of this statement, without ever explaining in more detail what actually he had had in mind. Some concluded that he must have referred to ADHD over-diagnosis, whereas his critics decided that this deathbed confession had only been meant to attract the attention of the media and the profession.

It is not our intention to decide who was right. Instead, let us show additional facts to cast more light on how the DSM diagnostic categories are formed:

> Nearly 70 percent of the DSM-V task-force members report having ties to the pharmaceutical industry. This represents a relative increase of 20 percent over the proportion of *DSM-IV* task-force members with such ties just a decade ago. But it is not only task-force members who have financial relationships with Big Pharma. Of the 137 *DSM-V* panel members (that is, workgroup members) who have posted disclosure statements, 77 (56 percent) reported industry ties, such as holding stock in pharmaceutical companies, serving as consultants to industry, or serving on company boards.[10]

The list of allegations against the DSM authors is far from complete. Lisa Cosgrove faults them on almost utter removal of the description of side effects of psychotropic medications in the DSM-V re-write. She also claims that there are significant gaps in the disclosure policy:

> Unrestricted research grants were excluded from the APA's disclosure requirements, even though the monies from such grants can total hundreds of thousands of dollars or more. There also are no policies for managing indirect financial ties, such as industry funds that are pooled and given to academic departments, hospitals, and medical schools. Moreover, *DSM* panel members are allowed to resume their financial relationships with industry as soon as their tenure on the DSM panels is over.[11]

We leave you to reflect on how the flaws identified by Cosgrove may affect the quality of subsequent psychiatric treatment.

An accurate and correct diagnosis is a precondition for research into the effectiveness of any therapy system. For what can we measure upon the completion of any therapeutic process if we lack confidence as to what we have measured at the outset? Nonetheless, we need to be very careful not to throw out the baby with the bathwater by claiming that due to problems with the diagnosis, it is in general unfeasible to measure a therapy's effectiveness. Let us therefore assume that some cases do allow quite an accurate and, what is most important, reproducible diagnosis. Having done so, we may move back to the analysis of research findings, maintaining the scientific rigor and sticking to our goal, which is finding a dependable psychotherapy that can be recommended to a person in need.

Recent years have brought many reviews of studies about the effectiveness of therapies.[12] It is beyond the scope of this discussion to analyze them all in detail; therefore, only some general conclusions will be presented based on these reviews. By no means should they be treated as definitive or, particularly, as guidelines when choosing a specific therapy system. It is very probable that between the time these words are written and the publication of this book, new research findings become available and are likely to challenge some of our statements here. Our purpose is to provide you with an overview of psychotherapy's effectiveness.

The majority of research explicitly supports the efficacy of behavioral, cognitive or cognitive-behavioral methods in treating anxiety, mood and eating disorders, in large part, research focused on the treatment of such disorders as general anxiety, social phobia, agoraphobia, panic, post-traumatic stress, bulimia nervosa and depression. A review of studies into the treatment of addictions has pointed to behavioral methods again. This refers specifically to the exposure to alcohol-related stimuli applied in the phase of elimination of drinking behaviors. Other treatments were labeled "probably efficacious." These were the methods that gave positive results in only one study that was not repeated (or repeated by the same research team). The "probably efficacious" category also encompasses cases where a treatment demonstrated efficacy for at least three study participants and the study itself was conducted by three independent research teams. With regard to addictions, such methods as social skills training, spouse assistance in medication taking and marital therapy were labeled "probably efficacious." In drug abuse, supportive-expressive therapies focused on solving emotional problems along with the behavioral therapy were listed as "probably efficacious."

Treatment of personality disorders proved efficacious when based on the psychodynamic and cognitive behavioral methods. In both cases, however, the number and frequency of sessions were critical: the higher number of therapeutic sessions attended, the more distinct the outcome became. As regards to schizophrenia symptom relapse prevention, behavioral and supportive family interventions were found to be efficacious. A systemic family therapy was classified as "probably efficacious" and so were the social skills training and the application of reinforcements. As with the previous therapies, also here the timeframe and frequency were of significance. Long-term therapies (with the duration of more than 9 months) were close to reaching the "efficacious" status. In handling marital discord, behavioral therapy was determined as "efficacious," whereas cognitive therapy along with psychodynamic therapy was "probably efficacious." Sexual dysfunctions were treated with sets of combined exercises supported by behavioral therapy. Interventions that involved monitoring of the mental aspects of somatic disorders are of significance. Behavioral therapy was "efficacious" in relieving pain and increasing the activity of patients with rheumatism. It demonstrated similar efficacy for other conditions related to chronic pain, such as migraine, irritable bowel syndrome (IBS), cancer or Tourette's syndrome. It was only "probably efficacious" in AIDS treatment.

This brief overview of the research gives us the following picture: behavioral therapy augmented by cognitive therapy components is taking the lead as it has most consistently been judged efficacious in research studies. Such a conclusion is nothing new or revolutionary. Eysenck had already appreciated the value of behavioral therapy. "There is evidence for systematic desensitization, and indeed in their assessment, Smith and her colleagues find that the behavior therapies are significantly superior to talking therapies in general."[13] This view is echoed by contemporary experts. Steven D. Hollon, professor of psychology at Vanderbilt University, said, "Evidence-based therapies work a little faster, a little better, and for more problematic situations, more powerfully."[14]

When interviewed by TIME Magazine and asked about types of individual therapy that are supported by evidence, Alan Kazdin, professor of psychology and child psychiatry at Yale University and president of the American Psychological Association, in large part indicated approaches with cognitive behavioral origins.

> There are now many forms of individual therapy with strong evidence. Two prominent examples for adults are graduated exposure for the treatment of anxiety, and cognitive therapy for the treatment of depression. There are internet and self-administered versions of these that also are effective. Two prominent examples for children are behavior analysis for children with autism spectrum disorders, and parent management training for the treatment of children with severe aggressive and antisocial behavior. These and other evidence-based treatments can help a variety of clinical disorders.[15]

Findings similar to those presented above can be found in many other research studies, for example in a report issued by APA Division 12, whose area of interest and professional activity is clinical psychology.[16] The report was drawn up as an antidote to the market of pseudo-therapy services as well as to its turbulent and uncontrolled expansion taking place across the United States. It marked a new chapter in the history of therapy research, with therapy efficacy being monitored by scientists, experts in the field of clinical psychology. With time, it led to the publication of other reports and papers designed as its extensions and updates.[17] All these efforts have invariably brought results that validate behavioral, cognitive and cognitive-behavioral therapies as being far more efficacious as compared to other approaches. They have also indicated moderate efficacy of interpersonal therapy. In fact, all investigated types of therapies are occasionally reported to be efficacious.[18]

However, these reports do not meet the criteria set forth by the Task Force on Promotion and Dissemination of Psychological Interventions, Division of Clinical Psychology within the American Psychological Association. For a treatment to be considered empirically validated, these criteria include, among other things, two well-constructed group experiments demonstrating differences between specific groups with either control or experimental groups receiving medication, placebo therapy or other types of therapy. A therapy method applied in the experiments must be conducted based on therapy manuals, so only manualized therapies can be considered for scientific investigation. Obtained results must come from at least two independent investigators or investigating teams.[19]

The most recent trends in psychotherapy research have taken the features of the evidence-based medicine (EBM) tradition. Termed also as fact-based medicine, it was initiated and developed in 1980s at McMaster University in Canada as a response to the basic science paradigm (pathophysiology, biochemistry) being dominant within the field of medicine. The tradition has its origins in clinical epidemiology, which is a scientific discipline that focuses on medical experimentation issues and the assessment of clinical research credibility. In psychotherapy, the term "evidence-based treatment" (EBT) is rather used. The recent years have seen the publication of a number of extensive monographs whose authors offer detailed description of the rudiments of efficacy measurements for psychotherapy, along with principles for the monitoring of effects based on solid theoretical grounding. Needless to say, due to their rigor and highly structured design, it is cognitive-behavioral therapies that represent major EBT trends.[20]

Thus, it appears that we have found true leaders in the tangle of therapies: behavioral, cognitive and cognitive-behavioral approaches. Their success likely to stems from the fact that they have been created and structured by scientists who were oriented towards specific and measurable goals when conducting their treatments. Moreover, these are among a few therapies in general guided by detailed manuals. This means that every professional who has chosen a particular method will follow the same order and the same set of steps during the entire treatment. Such standardization of methods is a tool used not only to facilitate measurement but also to educate future therapists. Treatment algorithms are intended to manage the therapist's decision as to what should be done in a particular situation. How different this approach is from the one we have discussed previously, that is, one that sees therapy rather as an art.

We have repeatedly sought to understand how it is possible to teach therapy that represents an art. Most likely, as in every art form, a special talent is required and a few available methods include the observation of the "master" at work, as well as an effort to gain an insight into the master's wisdom and skills. Many a time, masters of "artistic therapies" cannot clarify the mechanisms underpinning their treatment methods. This is how Bert Hellinger, a German psychotherapist associated with a therapeutic method best known as Family Constellations and Systemic Constellations, commented on an attempt to explain how his own therapy system worked: "Actually, the theories aren't important to me. I can see *that* these things happen, and explanations after the fact don't add anything to the practical work. Many people would be interested in an explanation of exactly what happens and how it's possible, but I don't need an explanation in order to work with the phenomenon."[21]

It would be interesting to know whether Hellinger's students and followers also have this gift of 'seeing' and thus do not need any explanations to understand what the master is doing. I wonder how they pass on this knowledge. Theory has to this day been the only known source of knowledge and description of the world. Another one, relevant to religious cognition exclusively, is revelation. Can it be that in psychotherapy, we have already reached this boundary?

Eclectic psychotherapy, which currently represents one of the most popular approaches, is also absent from the quoted reports and research studies. The eclectic approach allows the therapist to choose methods and measures that, in their view, will work for a particular patient. On what grounds? This must undoubtedly be that unerring intuition and experience that in no way can be explicated to an amateur. Imagine a novice artist-painter striving to emulate his master's style or to develop his own. Before he reaches perfection, many canvases are just thrown away or reworked with a bit of luck. It takes a lot of tools and supplies in order to achieve a satisfactory outcome. However, the cost of artistic experimentation is limited to the cost of canvas, stretcher bars, paint, as well as the time spent on using them all together. How do psychotherapy-artists develop their professional competence? Is psychotherapy by any chance the only art form that can be mastered without making mistakes? And if not, then which works of this art find their way to the waste heap of psychotherapy?

Many years ago, while talking to a practicing therapist who wishes to remain anonymous, one of us learned that "a good therapist is one who has had at least several patients bite the dust." Let me remind the reader, who may be displeased with our cynicism, that we wander around in a maze of 500 different therapies to choose from for your loved one. Which one to choose? Which criteria should guide our decision? Notwithstanding the rare, well-grounded studies cited above, scientists do not put much interest in investigating the matter. As William O'Donohue and Kyle E. Ferguson straightforwardly reveal, there is "a lack of will to boldly face this problem. Scientifically oriented practice seems to… ignore bad practice rather than to fight it in some way."[22]

At the same time, they suggest that any therapy that has not been empirically validated should be termed "experimental" and patients who receive such treatment "human subjects." They do so drawing analogies to test and acceptance procedures for new medications. In the United States, no drug can be brought to the market if its developer has failed to submit it to relevant laboratory testing and clinical trials. How different is the practice of introducing new therapies into use?

Just when we are writing this, in the United States, Great Britain and Poland, almost anyone can turn into a therapist and start their own private practice based on self-made, even weird theories. In 1993, Jeff Blyskal wrote in a popular magazine:

> Last February, I decided to become a psychotherapist. I found a comfortable office in the East Fifties for a mere $875 a month, IS Furniture Rental was willing to outfit the place in traditional style, with plenty of rich burgundy tones - cherry desks, medical-file cabinets, couch, even oil paintings - for only $335 a month. The cost of business/appointment cards would come to $70; the phone, installed, would cost $621.81; a month long radio-ad campaign (60-second spots, four times a day) would reach a quarter million listeners for only $2,000. So for just $4,000, I could have become a professional healer with absolutely no training, credentials, or license.[23]

Tana Dine summarized Blyskal's hoax as follows: "Blyskal didn't hang up his shingle but he could have because, contrary to public belief in most North American jurisdictions, there is nothing to stop such a scam."[24] If one is additionally clever enough to deck out their offices in diplomas and certificates from pseudo-scientific institution, they will surely entrap vulnerable people ready to entrust them with their money and mental health. Things are no different in Great Britain:

> In the UK, therapists and counsellors aren't regulated. Anyone can set themselves up as one with no training or proper oversight. While bad doctors can be struck off and barred from practicing in the UK, there's no way of stopping conversion therapists or any other practitioner who is preying on vulnerable people and damaging rather than helping them. They can be kicked out of their professional bodies and lose their accreditation, but nothing stops them going on to humiliate and abuse more vulnerable patients in future.[25]

In Poland, anyone who has officially registered their business can become a therapist without any certification required. All you need to do is to fill in an on-line form and start paying your National Insurance contributions. To support this claim empirically, one of us took actions in order to set himself up as a therapist, "specializing" in the radical forgiveness approach. He shared doubts with trainers of this system, whose responsibility was to certify future practitioners, whether he could become a therapist without any background in psychological education. As he was solemnly reassured, the course was designed to train not therapists but "personal improvement trainers" and therefore not even an academic degree in psychology was needed. In our view, the said trainers displayed a great deal of tact and excessive caution. The market of psychology-related services is not subject to any regulation from any regulatory authority of any type. It is truly hard to imagine a more fertile soil for charlatanry to thrive on.

In order to make an informed choice, shall we limit the spectrum to cognitive-behavioral approaches and those that are recognized as empirically grounded? Is it safe to decide on one of the well-established, empirically supported therapies? Again, considerable caution is needed also in this case. Conclusions regarding therapy efficacy based on experimental research may not apply to clinical practice because certain conditions of research studies are in many ways inconsistent with the said practice. Treatment studies essentially concentrate on homogenous types of disorders, whereas in practice, therapy is offered also to patients who struggle with more complex issues. In controlled studies, the treatment is highly organized and based on a number of specifically defined procedures. This is intended to structure the therapist's activity and contributes to a positive outcome. In practice, a more individualized and flexible approach is preferred.

In addition, it should be pointed out that the treatment of homogeneous disorders regularly proves to give better results than treatment of complex mental health issues and, simultaneously, the structured approach is more efficacious than the flexible one. All this permits the expectation that the effectiveness of clinical practice will be lower than results obtained in research, with the latter not being shocking when it comes to the value of achieved ratios.

At this point, we must mention what scientific methodology terms internal and external validity. If we are able to isolate a phenomenon well and examine it with no extraneous distortions, we think that we are measuring well what we intend to measure. If this is the case, one can speak of high internal validity in a study. External validity evaluates to what extent results of our study can be applied outside laboratory and generalized to an array of other situations. In this sense, research into therapies shows a high degree of internal validity but regrettably low external validity. Clinical practice should therefore be expected to bring different, that is, most likely worse results than laboratory studies.

These concerns are frequently raised by critics of the EBT approach, who indicate that its advantage is only of a marketing nature. "Few outside the mental health professions realize the term "evidence-based therapy" has become a form of marketing. ... It refers to therapies conducted by following instruction manuals, originally developed to create standardized treatments for research trials. These 'manualized' therapies are typically brief, highly structured, and almost exclusively identified with cognitive behavioral therapy or CBT."[26] Another much-discussed argument is also a rift between what therapists claim they do and what they actually do with patients. "Only half of the clinicians claiming to use CBT use an approach that even approximates to CBT."[27]

Given these points, inferences derived from research only speak to the efficacy of a therapy that is thoroughly professional, well-controlled, highly structured and reproducible. This is an ideal, and as such, it is indeed a far cry from today's clinical practice. Though imperfect, therapies with high efficacy rates obtained in treatment studies are most likely closer to this ideal and they should be recommended in the first place. However, we propose to refrain from the final decision yet. A slight change in topic, dear reader, is something that you should know about …

---

[1] K. Jankowski, ed., *Przełom w psychologii* (Warszawa: Czytelnik, 1978).

[2] D. L. Rosenhan, "On Being Sane in Insane Places," *Science, 179 (70)*, (1973): 250-258.

[3] L. Elkin, M.B. Parloff, S.W. Hadley, and J.H. Autry, "NIHM Treatment of Depression Collaborative Research Program," *Archives of General Psychiatry, 42*, (1985): 305-316.

[4] R. Mayes, and A. V. Horwitz, "DSM-III and the Revolution in the Classification of Mental Illness," *Journal of the History of the Behavioral Sciences, 41(3)*, (2005): 249–267; A. Speigel, "The Dictionary of Disorder: How One Man Revolutionized Psychiatry," *The New Yorker, 80(41)*, (2005): 56-63; M. Wilson, "DSM-III and the Transformation of American Psychiatry: A History," *American Journal of Psychiatry", 150(3)*, (1993): 399–410.

[5] G. N. Grob, "Origins of DSM-I: A Study in Appearance and Reality," *American Journal of Psychiatry, 148(4)*, (1991): 421–31.

[6] See: S. A. Kirk and H. Kutchins, *The Selling of DSM: The Rhetoric of Science in Psychiatry* (Hawthorne, NY: Aldine de Gruyter, 1992); Kirk and Kutchins, *Making Us Crazy: DSM: The Psychiatric Bible and the Creation of Mental Disorders* (New York; Free Press, 1997).

[7] Detailed criticism of developement and implementation of the DSM system can be found here: J. Davies, *Cracked. Why psychiatry is Doing More Harm than Good* (London: Icon Books, 2013).

[8] C. Lane, " The NIMH Withdraws Support for DSM-5," (2013): http://www.psychologytoday.com/blog/side-effects/201305/the-nimh-withdraws-support-dsm-5

[9] V. B. Jörg, "Schwermut ohne Scham," *Der Spiegel*, (February 6, 2012): http://www.spiegel.de/spiegel/print/d-83865282.html

[10] L. Cosgrove, "Diagnosing Conflict-of-Interest Disorder," *Academe*, (November-December, 2010): http://www.aaup.org/article/diagnosing-conflict-interest-disorder#.UtuaetKtaWh

[11] Ibid.

[12] M. J. Lambert and A. E. Bergin, "The Effectiveness of Psychotherapy," In *Handbook of Psychotherapy and Behavior Change* (4th ed.), eds., A. E. Bergin and S. L. Garfield, (Oxford, England: John Wiley & Sons, 1994): 143-189; S. McPherson, P. Richardson and P. Leroux, Eds., *Clinical Effectiveness in Psychotherapy and Mental Health: Strategies and Resources for the Effective Clinical Governance* (London: Karnac, 2003); L. Seligman and L.

W. Reichenberg, *Selecting Effective Treatments: A Comprehensive, Systematic Guide to Treating Mental Disorders*, 4th ed., (Hoboken, NJ: Wiley, 2011).

[13] Eysenck, *Decline and Fall*, 78. Desensitization is one behavioral method based on diminishing emotional anxiety by gradually increased, repeated exposure to anxiety-inducing stimuli until adequate adaptation is achieved.

[14] E. Jaffe, "Debate Over Cognitive, Traditional Mental Health Therapy," *Los Angeles Times*, (January 11, 2010): http://articles.latimes.com/2010/jan/11/health/la-he-psychotherapy11-2010jan11

[15] M. Szalavitz, "Q&A: A Yale Psychologist Calls for Radical Change in Therapy," *Time*, (September 13, 2011): http://healthland.time.com/2011/09/13/qa-a-yale-psychologist-calls-for-the-end-of-individual-psychotherapy/#ixzz2qISGEzAx

[16] Task Force on Promotion and Dissemination of Psychological Interventions, Division of Clinical Psychology, American Psychological Association, "Training in and Dissemination of Empirically Validated Psychological Treatments: Report and Recommendations," *The Clinical Psychologist*, *48(1)*, (1995): 3-23.

[17] D. L. *Chambless* et al., "*Update* on *Empirically Validated* Therapies II," *Clinical Psychologist*, *51 (1)*, (1998): 3–16; A. D. Reisner, "The Common Factors, Empirically Validated Treatments, and Recovery Models of Therapeutic Change," *Psychological Record*, *55 (3)*, (2005): 377-399; W. C. *Sanderson*, and S. *Woody, Manuals for Empirically Validated Treatments: A Project of the Task Force on Psychological Interventions (Oklahoma City: American Psychological Association, Division of Clinical Psychology, 1995)*.

[18] E.g., M. E. P. Seligman, "The Effectiveness of Psychotherapy: The *Consumer Reports* Study," *American Psychologist*, *50(12)*, (1995): 965-974.

[19] More in: Chambless et al., *Update on Empirically*, 3–16.

[20] See C. D. Goodheart, A. E. Kazdin, and R. J. Sternberg, eds., *Evidence-based Psychotherapy: Where Practice and Research Meet* (Washington, DC: American Psychological Association, 2006); A. E. Kazdin, "Evidence-based Treatment and Practice: New Opportunities to Bridge Clinical Research and Practice, Enhance the Knowledge Base, and Improve Patient Care," *American Psychologist*, *63,(3)*, (2008): 146-159; L. Luborsky and E. Luborsky, *Research and Psychotherapy: The Vital Link* (Lanham, MD: Jason Aronson, 2006).

[21] B. Hellingerand and G. ten Hövel, *Acknowledging What Is: Conversations with Bert Hellinger* (Phoenix, AZ: Zeig Tucker & Theisen, 1999), 66. More about Hellinger: http://www.hellinger.com/

[22] W. O'Donohue and K. E. Ferguson, "Evidence-based Practice in Psychology and Behaviour Analysis," *The Behaviour Analyst Today*, *7(3)*, (2006): 335-349.

[23] J. Blyskal, "Head Hunt: How to Find the Right Psychotherapist - for the Right Price," *New York*, (January 11, 1993): 29.

[24] Dineen, *Manufacturing Victims*, 81.

[25] G. Davies, "The Psychotherapy Scandal," *Progressonline*, (October 28, 2013): http://www.progressonline.org.uk/2013/10/28/time-to-regulate-psychotherapists-and-prevent-abuse/

[26] J. Shedler, "Bamboozled by Bad Science," (October 31, 2013): http://www.psychologytoday.com/blog/psychologically-minded/201310/bamboozled-bad-science

[27] G. Waller, H. Stringer and C. Meyer, "What Cognitive Behavioral Techniques Do Therapists Report Using when Delivering Cognitive Behavioral Therapy for the Eating Disorders? *Journal of Consulting and Clinical Psychology*, 80, (2012): 171-175.

CHAPTER 13

FIRST, DO NO HARM

"Everybody should know, then, that to step into the office of a psychotherapist … is to enter a world where great harm is possible." This is the warning given by Jeffrey Masson in his book *Against Therapy*, based on years of observations and analyses of various studies.[1] Unfortunately, the warning was unknown to a reader of the Polish edition of our book who described her situation in a letter to one of us:

> For some time I was a patient at a neurosis clinic, where I was counseled by a doctor and a psychotherapist (psychodynamic approach). The psychotherapist forbade me to take my hypothyroidism medications during the therapy and the doctor agreed. I was seriously depressed, anxious, could not concentrate and had suicidal thoughts, and could not function normally in my work and family life. As a result, I had to give up work just as unemployment was on the rise, and had problems getting back [into] the labor market. After a while, I stopped that ineffective psychotherapy, started taking medications and my health returned to normal. However, I still haven't regained my professional position or my lost earnings.

We know dozens of similar examples, which is why, despite the note of optimism from the previous chapter dealing with the latest research results, here we will continue, together with Masson, as well as other researchers and practitioners, to drive you away, dear readers, from psychotherapists' offices. Perhaps you will manage to find a good cognitive-behavioral therapist who works according to precisely defined standards, is goal-oriented and very much interested in achieving that goal. Perhaps this therapist will be able to repeat the therapeutic process that brought excellent results in the past. If this happens, then you have our congratulations. Why this may not necessarily happen as described in the previous sentence; we will tell you more later, but before we do that, we will say openly: if you can, give therapists a wide berth. There are many reasons for doing so. Owing to the uncontrolled spread of the phenomenon, it is difficult to say which of them is the most serious. To us, they all seem sufficient to scare people off therapy.

The first reason for our distrust lies in the very definition of therapy and its nature. According to one of commonly accepted definitions:

> Psychotherapy is taken to mean the informed and planful application of techniques derived from established psychological principles, by persons qualified through training and experience to understand these principles and to apply these techniques with the intention of assisting individuals to modify such personal characteristics as feelings, values, attitudes, and behaviors which are judged by the therapist to be maladaptive or maladjustive.[2]

Let us leave aside the vagueness of the term "established psychological principles," which in no way refers to empirical knowledge, and according to which, any psychoanalytical invention can now become a commonly established principle, (as can in the future, a number of increasingly accepted idiocies). What arouses the strongest opposition and doubt on our part in this definition is the word *values*. We have to admit that we have no idea what this modification of personal values should look like. Who should decide whether specific values are maladaptive or maladjustive? The patient? The therapist?

Let us assume that it should be the former. The patient concludes that some of his values are maladaptive, that they bother him. Let him be a man who believes in being an individual of excessive integrity. Even such prosaic events such as copying a newspaper article using the company photocopier, irritate him greatly. In his opinion, this is simply not right. He is equally intolerant when someone uses a company car or office hours to deal with their private matters. He himself never does that. On many occasions, he reprimands others for such behavior, which usually leads to conflict, change of position or even change of job. Once the owner of a company to whom he informed of such abuse laughed at him. He himself does not know how to behave in such situations. Turn a blind eye and say nothing? Notify the employer? All this generates considerable emotional costs for him. The man has come to the conclusion that his idea of integrity is maladaptive and has decided to ask a therapist to help him cope with himself so that he can treat similar situations as normal without incurring any additional emotional cost.

Should the therapist undertake such a task? Can he or she "treat" honesty or integrity? If so, then on what basis does he or she decide what is still moral and what is not? If the decision is left to the patient, then will the therapist also undertake to "treat" remorse after committing a theft? Or perhaps after a murder? Who and on what basis authorizes therapists to make such judgments? Is it enough for them to have an international NLP certificate or a certificate issued by a psychoanalytical society? Perhaps the therapists themselves discover a calling to fulfill such a role? As we keep wondering about it, we can sense shivers running down our spines.

What suggests itself at this point is a rather obvious conclusion shared by many therapists—that the patient's value system cannot be interfered with. However, this is by no means simple. Let us imagine the following situation. We have a family in which the father appreciates what can be described as, shall we say, Spartan values. They include self-discipline, self-control, perseverance, responsibility and so forth. On the other hand, we have a hedonistically oriented mother. This leads to a bitter conflict over the raising of their children. Concerned about their future, the father tries to instill in them perseverance, responsibility, self-discipline. The mother, who would do anything for them, takes over from them duties imposed by the father, fills their lives with pleasure and makes sure their childhood is carefree. Like water, which always flows from top to bottom, choosing the easiest route, so too the children choose the mother's lifestyle. A serious conflict arises. The family reports for therapy. What should the therapist do? Leave them to themselves, giving up therapy? Undertake the therapy, trying to persuade both sides to modify their value systems to meet somewhere half-way? Conduct the therapy in such a way that the two sides come to an agreement without the therapist's interference? Or propose his own system of values?

Won't any of the solutions, to a greater or lesser extent, more or less consciously, conform to the therapist's system of values? Is there any possibility that it could be otherwise? Studies mentioned in Chapter 11 suggest that this is in fact impossible. There are many more such studies. The impact of therapy on changes in value systems and moral judgments of patients has been demonstrated by, for example, Rosenthal[3] and Parloff.[4]

Paradoxically, in these circumstances, any therapy with a clear ideological or even religious orientation seems to be more honest. If a patient comes to a therapist who is also a Catholic priest, he can rest assured that any modification of his value system will be in accordance with the Catholic value system. If a woman goes to a therapist with a feminist orientation, she, too, knows what to expect. It is not so straightforward with therapists who are "neutral" when it comes to their worldview and religion. After all, they all have their own value systems to which they will bend their patients. So perhaps therapists should consciously define their worldviews when advertising their services? Perhaps a solution can be for the therapists to inform their patients about their value systems and for the patients to consider this before deciding on a therapy? To put it as briefly and as simply as possible, it seems highly desirable for the therapists' system of values to be spelled out on the doors to their offices and to be included in the so-called therapeutic contract.

Having been lost in daydreaming, let us assume that we have found a therapist whose value system corresponds to ours and that we have decided to start therapy. Is there any other threat to us? Unfortunately, there is, and this is quite a serious threat. We may become dependent on therapy. At first, this might seem a little flippant, a little ridiculous. Indeed, it is difficult to speak in this case of physiological dependence. In most cases, psychotherapy does not cause big enough changes in our brains nor in our physiology to allow us to compare it to addictions associated with smoking, alcohol, medication or drug abuse. We will not write about those few instances when it might happen because the scale of this phenomenon is modest. Far more dangerous is the threat that cannot be measured but is, in our opinion, very common. Raj Persaud illustrates this very well: "One patient of mine indicated that she had finally got over her dependence on therapists, having, over several decades, jumped from one counselor to another. I enquired how she had achieved this. 'Well' she said brightly, 'I have found this new person who isn't therapist. He is a healer who I see once a week and that has finally helped'."[5] In his book, Persaud repeatedly warns against the threat of dependence, and other authors agree with him in this respect.[6] Dependence on therapy naturally diminishes our immunity to life problems. People used to counseling in important matters in their lives can lose their independence and in some cases, can even become helpless. Repeated attempts at psychotherapy that bring no results, as it was the case with Persaud's patient, can also lead simply to the emergence of the learned helplessness syndrome, which denotes a conviction that there is no link between our actions and their outcomes. This is just one step away from depression. Canadian clinical psychologist, author of "Manufacturing Victims" wrote: "While people have become accustomed to hearing about all sorts of victims, from those of sexual harassment and verbal abuse, to those of 'dysfunctional families,' divorce, academic discrimination, even vacation cancellation and home renovation, they have not yet paid attention to the psychological techniques which are being used to create and cater to these 'victims.' Nor have they noticed how it is the psychologists who are benefiting in the end from this victim-making, as the industry creates 'users' dependent on their services."[7]

We are probably equally convinced that therapy often leads to dependence, and further, that a risk of dependence and the distant prospect of depression will not prevent patients from throwing themselves into the arms of one therapist after another. That is why we will write about some other threats awaiting patients deciding to try therapy. One of them can even have disastrous consequences, as it stems from a virtually complete lack of medical training among therapists. We will discuss it using a real-life example, the details of which have been changed: Let us imagine a young girl who starts feeling unwell and has problems with, among others, walking. The girl goes to a psychologist who, after examining her, concludes that the reason why she feels unwell and experiences slight movement disorders lies in her psyche. The girl simply seeks to draw attention to herself. The psychologist, certain of his diagnosis, convinces the girl's parents that he is right. As a result, the girl finds herself in a psychiatric hospital where she experiences a nightmare, unsuccessfully trying to persuade those around her that her symptoms are not made up. When the girl's condition begins to deteriorate, she undergoes some neurological tests, which reveal a brain tumor. Unfortunately, by that time it has become too big for her to be saved.

Is it one of those stories with which tabloids like to shock their readers? Unfortunately, people are commonly suspected of having psychosomatic symptoms, though the finale is not always so tragic. Someone who suffers from headaches and who is labeled as having a "psychosomatic disorder" often ceases to be credible in their physical suffering to those around them. We have met very few psychologists or psychotherapists who, before embarking on a course of therapy, consider any possible biological causes of their patient's condition, and yet we still remember from our diagnosis classes at the university that the first and foremost task of a diagnostician is to make sure that the disorder *is of no organic origin*. The fact that this recommendation is so easily disregarded today is probably nothing more than a result of the vogue for the so-called holistic approach to the patient's problems. For years, doctors were attacked for not taking the psyche into account in their diagnoses, treating their patients only as biological organisms. In their worship of treating a human being as a psychosomatic whole, psychologists sought the cause of almost all diseases in the psyche – even to the absurd extent of radical forgiveness therapy where every illness, including cancer, is attributed to problems caused by a lack of forgiveness in the past. Today we demand that every doctor should take into account the patient's psyche, whereas virtually no one demands that psychologists apply a similar approach, which would mean treating the somatic sphere just as seriously as the mental sphere.

Unfortunately, these are not all the threats posed by therapy. Quite often therapy brings negative results and a considerable deterioration of the patient's condition. Successive studies have demonstrated that as a result of therapy, no fewer than 10% of patients feel worse and even display psychotic reactions. Many couples who underwent therapy went along with their therapists' suggestions to such an extent that their family lives disintegrated as a result. What is more, those recommendations and suggestions were formulated solely on the information presented by only one of the partners, that is, the one who decided to undergo the therapy, completely ignoring information from the other party involved.[8]

Interesting conclusions can be drawn from the results of the Cambridge-Somerville Youth Study:

> In evaluating the effectiveness of a project designed to prevent delinquency in underprivileged children, 650 boys of six to ten years old were randomly divided into two groups with equal chances of delinquency. One group received individual therapy, tutoring and social services; the other received no services. The treated boys rated the project as "helpful" and the counselors rated two-thirds of the group as having benefited. However, the researcher, Joan McCord, followed the boys over time looking at effects on criminal behavior. The results showed little difference in terms of the number of crimes, but the counseled group committed significantly *more* serious crimes. A thirty-year follow-up showed the same pattern and revealed that, in terms of alcoholism, mental illness, job satisfaction and stress-related diseases, *the treatment group was worse*. McCord summarizes the results as "'More' was 'worse': the objective evidence presents a disturbing picture. The program seems not only to have failed to prevent its clients from committing crimes… but also to have produced negative side effects…"[9]

When interpreting the results of her research, McCord identified three factors that might contribute to the harmful effects: *encouraged dependency, false optimism* and *externalized responsibility*. She suggested that, 1) "Through therapy, the psychologists might have fostered a dependency among the boys, rendering them less able or inclined to cope with life's problems on their own;" 2) "the supportive attitudes of the counselors may have filtered reality for the boys, leading them to expect more from life than they could receive;" and 3) "counseling may have taught the boys that they were not responsible for their behavior because it was a consequence of their underprivileged childhood experiences - an external cause to blame."[10]

This was by no means the only such study. Keith Ditman and associates studied three groups of alcoholics who had been arrested and charged with alcohol-related offences.

> The court had assigned these individuals to AA, an alcoholism clinic, or a non-treatment control group. A follow-up found that 44 per cent of the control group were not re-arrested, compared to 31 per cent of the AA group and 32 per cent of those treated in a clinic; 47 of those that received treatment did worse than the untreated. "Not one study," Peele asserts, "has ever found AA or its derivatives to be superior to any other approach, or even to be better than not receiving any help at all. Every comparative study of standard treatment programs versus legal proceedings for drunk drivers finds that those who received ordinary judicial sanctions had fewer subsequent accidents and were arrested less."[11]

Other documented cases of adverse effects of therapy include decompensation, that is, the failure of the adaptive mechanisms of an individual overburdened with difficult situations, problems and tasks; deterioration of depressive states, including attempted suicide; lower self-esteem coupled with feelings of shame and humiliation; weakened self-control manifested in aggressive behavior or uncontrolled sexual behavior; the already mentioned dependence on therapy and/or therapists, and loss of a sense of responsibility for one's own life.[12] There have also been cases of incomprehensible impulsive behavior of patients undergoing therapy in their own social environment. The behavior in question did not occur before the start of therapy.[13] In another article, Handley and Strupp demonstrated that most psychotherapists agree that the problem of harm done during therapy is real and that it often leads to suicide.[14]

Moreover, the research results we have referred to do not really worry the therapists who make no effort to eliminate them.[15] Robert Spitzer of the New York Psychiatric Institute once said with disarming honesty that, "negative effects in long-term outpatient treatment are extremely common."[16] Carkhuff was of a similar opinion: "the evidence now available suggests that, on the average, psychotherapy may be harmful as often as helpful, with an average effect comparable to receiving no help."[17] Is it possible that all these cases of adverse effects were a result of improvements in therapy treated as an art? Perhaps these are the discarded canvases, damaged stretchers, wasted lumps of precious metals that help artists to improve?

If all this were not enough, a patient about to undergo therapy has a considerable chance (of at least 1:10) of coming across a therapist who sexually harassed his patients in the past. This number comes from studies carried out many times in the United States in order to determine the scale of the phenomenon. The studies began only in 1973, when 460 doctors, including psychiatrists, were asked about sexual relations with their patients. It turned out that between five to 13% of them had engaged in some kind of erotic contact with their patients.[18] Of course, these figures concern only those respondents who admitted to such contacts. Unfortunately, as offenders are usually not very keen to reveal themselves in such surveys, we need to accept that the figures this time are likely to be underestimated. By how much? It is difficult to say, but there is no doubt that even when a survey is fully anonymous, many respondents prefer not to reveal such relations. An overestimation of the figures is rather unlikely.

However, psychotherapists made up only a percentage of the questioned mental health physicians. It is, therefore, worth adding some more data to these figures. In 1977, the American Psychological Association commissioned a survey which involved only practicing psychologists. Among the 703 respondents, 10.9% admitted to some sort of erotic contact with their patients.[19] The figures aroused the interest of women, who quite rightly noted that victims of harassment were usually women. A study carried out in 1983 by women revealed that no fewer than 15% of therapists had had sexual contact with their patients; in this case, too, the figures were based on voluntary disclosure. Another survey carried out among psychiatrists again set the figure at 10%.[20]

In order to realize the scale of the phenomenon, let us note that these figures are similar to those we can obtain when surveying defendants who are asked in court whether they plead guilty. What comes to mind is an analogy to studies into the integrity of scientists. Here, too, most results are based on voluntary declarations. The real scale of the problem could be determined only if we carried out a well-designed survey with patients being the respondents. As far as we know, no one has carried out such a survey yet. However, we do have at our disposal a large collection of testimonials from former or present patients. Especially reliable are those in which these patients were not victims but... witnesses. Yes, you have read correctly; some psychotherapists have no reservations even when it comes to harassing their patients in the presence of others. "The strangest thing I saw during therapy was an attempt to seduce a patient during hypnosis, of course without the consent of the hypnotized individual. No restraint whatsoever, a loss of control, hedonism, not to mention a lack of a general and professional code of ethics. How come that people with such personality disorders, with no control over their desires and a compulsion to seduce and exploit women can be psychotherapists and supervisors?"[21] Usually, such incidents take place in the semi-darkness of psychotherapists' offices and rarely do they come to light, unless their scale is so huge that the conspiracy of silence is broken. This is what happened in the case described below:

> One recent high-profile scandal has become totemic for the psychotherapy industry. Psychotherapist and art therapist Derek Gale was accused of multiple cases of inappropriate sexual contact, falling asleep in sessions and offering his patients illegal drugs. A hearing of the Health Professions Council (HPC), the government-appointed body that regulates most health professionals, also heard allegations that Gale regularly went on holiday with patients, asked them to act out scenes of sexual abuse and on one occasion advised a patient to take advantage of "unlimited sex."[22]

As we write these words, the deadline passes for Lechoslaw Gapik, a well-known Polish psychology professor, sexologist and therapist, to report to prison. Gapik has been sentenced to four years in prison and has been banned from medical practice for six years for molesting six of his patients. Sexual harassment was revealed after capturing a therapeutic session conducted by him with a hidden camera. Soon other molested patients contacted the prosecutor. Gapik defended himself, claiming that what he had done was his original method of treating frigidity. However, the judge was in no doubt, the sentence is final and binding, and the therapist is now trying to avoid prison by hiding in a hospital.

A similar story in the USA led to the founding of an organization to help victims exploited by therapists.

> In the spring of 1989, *The Boston Globe* publicized the issue of therapy exploitation by highlighting cases involving two prominent psychiatrists in the Boston area. Three women who had been exploited by other psychotherapists, and who had met one another through a mutually known psychotherapist, contacted the women whose stories had been publicized. The five women got together that July to explore the possibility of starting a networking and support group. One of these founders, a former advertising copywriter, suggested the acronym TELL to stand for Therapy Exploitation Link Line, a name that was quickly accepted by all.[23]

The organization's website features an embarrassingly long list of books devoted to therapists' abuses[24] as well as articles[25] and stories told by the victims.[26]

Another type of data shedding some light on the analyzed problem comes from people who have their feet firmly on the ground, that is, representatives of insurance companies. For some time, US insurance companies have been refusing insurance against charges of sexual abuse to psychotherapists. Prejudice? Stereotypes? We don't think so. For an insurance company, a psychotherapist is a client like anybody else, but here again figures speak for themselves. In terms of the amounts of claims, the sexual harassment charge is the main type of charges in successful malpractice suits brought against American therapists in 1976–1986. It covered 44.8% of the total amount of compensations awarded. This is why therapists are the only professional group in the United States to which insurance companies refuse insurance contracts on the grounds of too frequent sexual harassment-related malpractice suits.[27] It is worth bearing in mind that many members of this group took the Hippocratic Oath. Let us once again refer to its fragment used as a motto for the present part of the book:

> Whatever houses I may visit, I will come for the benefit of the sick, remaining free of all intentional injustice, of all mischief and in particular of sexual relations with both female and male persons, be they free or slaves.

Dear readers, having read all this, would you be willing to entrust a loved one to a therapist? If you are determined, you should also be aware of the fact that analyses of the personalities of therapists-to-be have revealed that by practicing this profession, they want to satisfy very specific needs characteristic of people unsure of their identity, alienated or having social inhibitions, people who are dependent, like to dominate, have sadistic inclinations, cannot express their own feelings and sometimes are masochists.[28] We can get some idea of these needs from what one of the most famous therapists in the world, Milton H. Ericsson, once said to a female patient:

> These terms are absolute, full, and complete obedience in relation to every instruction I give you regardless of what I order or demand. ... You will be told what to do, and you will do it. That's it. If I tell you to resign your position, you will resign. If I tell you to eat fresh garlic cloves for breakfast, you will eat them. ... I want action and response—not words, ideas, theories, concepts. Once you come, you are committed to therapy, and your bank account belongs to me as does the registration certificate for your car. I will tell you what to do and how to do it, and you are to be a most obedient patient.[29]

Our grim portrait of psychotherapists is complemented by opinions of representatives of other professions.

> Jay Schadler, a journalist, after watching hypnotists and recovered-memory specialists at work, thinks that "some... may be as sick as their patients," perhaps drawn into the business because of their own problems. Alan Gold, a lawyer who has cross-examined psychologists in court for twenty years, thinks that some may lack the intelligence necessary to succeed in other fields. According to him, "the 'softer' the area of alleged expertise, the easier it is for dumbness to survive."

Yet, if in spite of our efforts so far, you have decided to take part in atherapy, ask your chosen therapist what happens when during the therapy the patient is no longer able to pay for it. Will the therapy be continued or discontinued in accordance with market principles? Think how such a situation can affect the patient. In addition, it is worth bearing in mind that factors which definitely hinder a return to mental health include focusing on oneself and on the problems one is experiencing, that is, on what comprises the essence of most therapies. Still, even if you do decide to enter a therapist's office, bear in mind the warning with which we began this chapter: "Everybody should know, then, that to step into the office of a psychotherapist ... is to enter a world where great harm is possible."

---

[1] Masson, *Against Therapy*.

[2] J. Meltzoff and M. Kornreich, *Research in Psychotherapy* (Atherton, NY: Transaction Publishers, 2008), 4.

[3] R. Rosenthal, "Changes in Some Moral Values following Psychotherapy," *Journal of Counseling Psychology, 19*, R., (1956): 431-434.

[4]Parloff, "Some Factors Affecting," 5-10.

[5] Persaud, *Stay Sane*, 77-78.

[6] W. Dryden, *Indyvidual Therapy in Britain* (London: Harper and Row, 1984).

[7] Dineen, *Manufacturing Victims*, 8.

[8] M. J. Lambert, A. E. Bergin and J. L. Collins, "Therapist Induced Deterioration in Psychotherapy," In *The Therapist's Contributions to Effective Treatment*, eds., A. S. Gurman and A. M. Rogers (New York: Pergamon, 1977), 452-481.

[9] Dineen, *Manufacturing Victims*, 54.

[10] Ibid, 55.

[11] Ibid, 55.

[12] See: M. Lakin, *Ethical Issues in the Psychotherapies* (New York: Oxford University Press, 1988).

[13] See: H. H. Strupp, S. W. Hadley and B. Gomes-Schwartz, *Psychotherapy for Better or Worse: The Problem of Negative Effects* (New York: Jason Aronson, 1977).

[14] S. W. Hadley and H. Stupp, "Contemporary Views of Negative Effects in Psychotherapy," *Archives of General Psychiatry, 33*, (1976): 1291-1302.

[15] D. T. Mays and C. M. Franks, *Negative Outcome in Psychotherapy and What to Do about It.* (New York: Springer, 1985).

[16] R. Spitzer, as cited in Sarason, *Psychology Misdirected*, 42.

[17] R. R. Carkhuff, *Helping and Human Relations: A Primer for Lay and Professional Helpers*, (New York: Holt, Rinehart & Winston, 1969): 122.

[18] S. H. Kardener, M. Fuller and I. N. Mensh, "A Survey of Physicians' Attitudes and Practices Regarding Erotic and Non-Erotic Contact with Patients." *American Journal of Psychiatry, 130*, (1973): 1058-1064.

[19] J. C. Holroyd and A. M. Brodsky, Psychologists' attitudes and practices regarding erotic and non-erotic contact with patients. *American Psychologist, 32*, 843-849.

[20] N. Gartrell, et al., "Psychiatrist-patient Sexual Contact: Results of National Survey, 1: Prevalence. *American Journal of Psychiatry, 143*, (1986): 1126-1131.

[21] Private correspondence.

[22] J. Doward and C. Flyn, "Sex Scandals, Rows and Mavericks: Is it Time to Regulate Psychotherapy?" *The Observer*, (May 9, 2010): http://www.theguardian.com/ life-andstyle/2010/may/09/rogue-psychotherapy-regulation-row

[23] http://www.therapyabuse.org/about_us.htm

[24] http://www.therapyabuse.org/RS_sugreadings.htm

[25] http://www.therapyabuse.org/papers.htm

[26] http://www.therapyabuse.org/topics.htm

[27] R. Z. Folman, "Therapist-patient Sex: Attraction and Boundary Problems," *Psychotherapy, 28*, (1991): 168-173.

[28] D. I. Templer, "Analyzing the Psychotherapists," *Mental Hygiene, 55*, (1971): 234-236.

[29] E. L. Rossi, *Innovative Hypnotherapy: The Collected Papers of Milton H. Erickson on Hypnosis* vol. 4, (New York: Irvington Publishers, 1980), 482-490. As cited in Masson, *Against Therapy.*

CHAPTER 14

WITH WHAT CAN WE REPLACE PSYCHOTHERAPY?

Probably every critic of psychotherapy faces the question stated in the title of this chapter. Therapists, psychologists, psychiatrists, patients and even psychology students keep asking this particular question, awaiting critical or constructive answers. Their reactions are quite easy to predict, as if we were facing the brain-washed members of a religious cult: "What would you like to see in place of psychotherapy?" Well, the answer is simple and brutal – nothing. We will explain why by quoting Jeffrey Masson, one of the most famous critics of psychotherapy:

> In reply, I would note that, as one feminist friend put it, nobody thinks of asking: What would you replace misogyny with? If something is bad, or flawed, or dangerous, it is enough if we expose it for what it is. It is almost as if once it has been determined that something exists, we decide it must be there for a reason (undoubtedly true) and then slide into the false position that it must be there for a good reason, which is undoubtedly not true. Or it is as if we believed that if we finally rid ourselves of something heinous (like apartheid), then we must replace it with something similar in nature. The truth is we do not know all the wonderful things that could happen once something hateful is abolished. Anyone who has ever oppressed another human being invariably asks what will happen once the oppression is over. What will happen to children once we stop beating them in schools? What will happen to slaves when they are freed from the plantations? What will happen to animals when we stop slaughtering them for food? What will happen to women when we stop subordinating them? What will happen to nonconformists when we do not incarcerate them in psychiatric institutions? What will happen to the wife when her husband no longer beats her?[1]

Perhaps the single biggest mistake lies in the way this question is asked. The major obstacle in accepting our very simple answer is a common belief that psychotherapy is a good of some sort, and that its lack will leave an empty space for bad, evil things. Thousands of therapists worked for decades to ensure that this misconception is deeply rooted in our brains, and it will probably stay there for many years yet to come. However Masson is not alone in his criticism. Raj Persaud expressed similar, but perhaps less categorical views: "It is only because so many people with mental health problems are suffering in silence that senior members of the psychology and psychiatry professions have refrained from voicing their concerns about the lack of regulation within counselling and therapy. They feel that any form of coun-seling is better than suffering in silence. Given my belief in the negative effect of counseling, I differ from my colleagues on this point. In many instances, I believe therapy could be damaging.[2]

Masson and Persaud are not alone in their views, nor are their criticisms based on cherry-picked observations. Their fundaments are quite robust, based on reviews of numerous solid research papers, on analyses of hundreds of cases and on their own practices. Before Masson reached his conclusions, he studied and practiced as a psychoanalyst for over 12 years. Persaud is a practicing psychiatrist in London, UK and at the famous John Hopkins Hospital in Baltimore, USA. In order to say the before mentioned "nothing" that we both strongly agree with, they have thoroughly analyzed and thought over hundreds of different factors, and they have reached very simple, yet not necessarily apparent or evident conclusions.

The first one deals with problems that people often face in their lives. They start at the day of our birth and are later defined by a series of more or less fortunate events, common teenage problems, broken friendships and broken hearts, death of our loved ones, sometimes due to an incurable illness that causes significant suffering, and our own ailments, disappointments or accidents. We will most likely experience pain after breaking up, often more than once, with our life partners. Perhaps someone will betray us or our val-ues. We might become addicted to one of numerous psychoactive substanc-es. We are likely to personally experience some form of aggression or vio-lence and we will often feel overwhelmed and without any control of the world around us. These events will be often mixed with positive events such as pride in our children, falling in love, personal satisfaction with our own achievements, great times with friends or families, and so forth. At the end, we will die, and most of us will not be fortunate enough to die in our sleep. Most of us, just before our lives end, will experience physical or emotional suffering to some degree. This is essentially what all our will lives look like.

Now let's take a peek at what patients present with in therapists' offices. Are those problems anything else than those little, expected particles of our normal, everyday lives? Or perhaps they were normal before therapy was invented and have now been transformed into "therapeutic problems" that can be "resolved" in exchange for your money? Isn't it typical today to bring any slightly more complex situation that life throws at us into a therapist's office? Don't you agree that humans have had to deal with any of those less fortunate events simply because they were a part of everyday existence, ver-sus today, when they have become "traumatic events" that need to be "worked up" in someone's office?

In our relentless pursuit of joy, convenience and carefreeness, we have erased the dead, sick and old from our horizon. Nowadays, a 30 minute visit in a funeral home deals with everything that used to happen over the course of many days in the homes of those who passed away. The sick are kept separated in hospitals, the elderly and infirm are locked in dedicated "homes" and those dying stay in hospices. In the very same way, we have removed from our streets mentally impaired and mentally ill people. For over a hundred years, we have been trying to sell our normal, everyday problems to therapists. And they are very happy to buy them from us, appropriating new areas of human misfortune. Some might believe that such a transaction is beneficial to both sides. But is it really?

Before you start paying your therapist for not winning a lottery, let us remind you of the research that we have presented in the previous chapters of this book. There is no doubt that to avoid confronting our own problems results in progressively diminished psychological and mental strength. Escaping our problems means that smaller and smaller problems become more and more overwhelming, eventually leading us into nervous or mental breakdown. Seeing a therapist often leads to therapy addiction, decreased immunity and other negative outcomes. But those are not the only reasons for our very critical views about the dishonorable dealing called psychotherapy. This is exactly the place where ethical concerns arise. Let's try to analyze our doubts by working on a representative example.

A young woman, let's call her Aggie, is not satisfied with her looks, which in the era of digitally enhanced and mass-produced images of non-existent bodies, haircuts, complexions, eyes, lips and other body parts, is a rather a common problem among many young people. There are a few ways in which Aggie can approach her problem, but here we will focus only on two of them. Firstly, she can do what most other girls choose to do – get an appointment with her hair stylist, signup for yoga classes, visit a beautician, make-up artist, aesthetician, stylist or a designer's clothing outlet. She can transform the way she looks. The effects can be astounding, but are primarily related to the skills of the artisans she chooses. She might suddenly find that others start to notice her at work. Her self-esteem might improve as she transforms from the previous ugly duckling into a swan. She might want to apply for a job that she didn't previously consider, or she might just agree to finally go out on a date with her friend. The hard work of the skilled craftsmen might actually have a very real impact on her life.

However, scenario number two is becoming increasingly popular: Aggie decides to go to therapy. She spends exactly the same amount of time and money to pay a therapist. She wants to learn to accept herself just the way she is. Surprisingly, after a few sessions, she actually starts to like herself. She looks in the mirror without previous disgust and aversion and she starts to believe that her value does not necessarily depend only on the way she looks. Of course we might argue as to which option would give Aggie better outcomes and why, but it is not important in this example. We have used this case to ask a question, what makes a therapist any different from a hairdresser or a make-up artist? They all solve the same problem in different ways. What makes any solution any better than the other?

Let's now look at the following analogy: when somebody has a problem and feels emotionally down, he can simply drop into a local pub for a few drinks. It might make someone feel better (or at least his subjective perception will improve) in exactly the same way some patients subjectively feel better after leaving their therapist's office. Yet for some reason, we do not think of hairdressers or barmen/barmaids as the saviors of humanity, as we tend to label psychotherapists. Are we going to systematically encourage people to visit local pubs, renowned hairstylists or plastic surgeons, just as we encourage visiting therapists? Should we defend workers of the local bar or fashion designers from criticism in the same way we protect therapists?

Now a more important issue: does therapy actually make any difference by relieving people from their suffering, or does it only provide subjective pleasure or satisfaction? The first person to focus on the ethical side of psychotherapy was Karl Popper, followed by Raj Persaud.[3] The problem is not in the answering whether we should provide people with pleasures. Those are always in demand and always will be. The real issue here is whether it is ethical to sell pleasure under the label of relieving suffering. Nobody who spends an evening in a movie theater and nobody who visits a prostitute actually tries to claim a refund from his health insurance, despite the fact that both those activities can help to resolve some emotional issues.

We don't have a shadow of a doubt that reimbursing and supporting people with mental illness is necessary. However the doubts begin to pile up when we start to include certain areas into the therapy that until very recently, had nothing to do with mental health and wellbeing. Let's take for example, helping people with a substance abuse problem, for example alcoholics. Isn't the creation of Adult Children of Alcoholics (ACOA) Syndrome nothing more than an attempt to include completely healthy people into therapy by trying to convince them and those around them that a childhood spent in a family where at least one person had problem with alcohol results in social maladjustment for their entire life?[4] We also protest against calling people treated for alcoholism alcoholics. The studies clearly show that alcoholism can be considered cured at a certain stage. The claim that we have recently read about: "I am a non-drinking alcoholic, I haven't put alcohol in my mouth for 56 years" is a result of indoctrination by therapists trying to enter another area, where in reality, there is nothing for them to do. Judith Herman said, "The reconstruction of the trauma is never entirely completed; new conflicts and challenges at each new stage of the lifecycle will inevitably reawaken the trauma and bring some new aspect of the experience to light."[5] Such clear statement always makes us suspicious of a purposeful attempt to permanently bind a customer to his therapist by inducing emotional dependence on therapy.

This kind of behavior, apart from being morally ambiguous, has well-defined, negative outcomes on patients, as we have presented earlier. Accepting the views of our therapist or convincing ourselves that some events in our lives determine all future behaviors, is simply imprinting your life pass-port with a stamp that reads: "I am entirely OK. I am not responsible for my problems. I blame my drinking father or on the fact that I was molested." Stan Anderski believes that such interpretation can sometimes be a source of success:

> The case of the British psychiatrist Ronald Laing offers a good example of the common phenomenon of how one can become the center of a cult by holding views which correspond to what a large number of people want to hear. ... His popularity among the trendy crowds is due to the message (which is the only conclusion that clearly emerges from his tenebrous disquisitions) to all the youthful (and not so youthful) freaks that no matter what, they are entitled to blame it on other people, especially their mothers.[6]

On the other hand, when those therapists lend a helping hand by provid-ing their customers with such explanations, they imprint them with a belief that it is impossible to leave all those experiences behind. Therapists fre-quently provide such interpretations in order to bilaterally bind people to themselves or to the therapy. Where is the border between discomfort and suffering? Do the examples described above qualify as suffering? Is the fact that our life partner leaves us (and this happens at least once to large part of human population), so devastating that we are unable to pick ourselves up? Do we need a therapist? Perhaps only some of us suffer? Who is to decide? Are problems related to low self-esteem a suffering or a discomfort that can be "treated" with a small dose of pleasure? If a student does not cope with the pressure from his chosen university, but his colleagues are doing rather well, does he suffer? We cannot provide answers to those questions in this book. Those dilemmas are less common among doctors because they often have clearly defined borders between health and disease: the diagnosis of many medical conditions can be based on measured parameters or on the presence/absence of biological markers. Perhaps we will never be able to answer those questions, but since psychological help in many countries is included in reimbursable health insurance schemes, people who provide ther-apies should be able to provide those answers.

Even if we give away the right to decide who is actually suffering to pa-tients, the question remains whether his choice is the best possible solution? Shall we all, as potential clients of therapists, be buying such services. Perhaps it would be best to ignore those never-ending dilemmas, reformulate the problem and try to think in a completely different way? Why do certain events cause suffering in some individuals, while those very same triggers motivate others to action? Why do some of us give up when diagnosed with an incurable disease, while others are stimulated to get the best out of the life that they have left? Maybe those are congenital predispositions of our nerv-ous system? Some research results similar to those previously described by us, focus on analysis of past events and show that for example, in children exposed to severe trauma (such as sexual molesting, physical and emotional abuse), at least half of them do not experience any permanent symptoms.[7] Actually, some psychologists are convinced that our mental strength is defined by our genes or, in the best case scenario, develops in the first few days of life. An extreme example of such beliefs was presented by Judith Herman: "The most powerful determinant of psychological harm is the character of the traumatic event itself. Individual personality characteristics count for little in the face of overwhelming events."[8] Many therapists equate with such views.

Psychologists who do not rely only on their hunches, experiences or beliefs, reformulated the problem a few decades ago. This is when an epic journey into exploring the limits of human capabilities started. Researchers asked scientific questions analogous to those in the previous paragraph, and formulated the so-called salutogenic paradigm, as opposed to the dominating pathogenic paradigm. What are those paradigms and where was the revolutionary change in handling problems related to therapy, mental health and psychological wellbeing? Medicine and clinical psychology are dominated by something called the pathogenic paradigm. Simply speaking, it means focusing on the disease as the subject of research and interest. This approach tries to deal with disease and measures the effectiveness of various therapeutic approaches. The Salutogenic paradigm focuses on health instead of disease. According to this paradigm, science should primarily research the conditions of staying in good health, answer questions on how to remain healthy and how to care for our wellbeing, rather than on fixing problems that have already appeared. Supporters of the new paradigm believe that focusing on the disease does not allow removal of its sources. A good example of the old paradigm from the world of medicine would be treatment methods that do not lead to eradication of the disease, such as new medicines for arterial hypertension or diabetes. The Salutogenic approach would focus on developing new vaccines that protect against getting sick in the first place, avoiding disease and thus are far more effective than any form of therapy. This is how this new paradigm can answer the question asked at the beginning of this chapter. Understanding what pushes us into psychotherapists' offices and the conscious control of our own lives is far more important than never-ending therapies.

Researchers who asked the right questions, independently, were Aaron Antonovsky and Suzanne C. Kobasa, followed by Salvatore Maddi. The result of their work almost completely fill the empty space left after getting rid of psychotherapy. Antonovsky focused on former World War II concentration camp prisoners. He wondered why some of them quickly adapted to the new reality after regaining their freedom, while others were completely broken down and depressed, developed psychiatric disorders or even died. Kobasa focused on people who found themselves in less extreme circumstances but more characteristic for our current reality: middle-grade managers from Illinois Bell Telephone during their restructuring, involved in redundancies, requalifications and so forth. Those changes were stressful enough that some research subjects experienced significant health problems. The rest, on the other hand, (similar to Antonovsky's research), coped quite well and never suffered any health-related issues. Based on his work, Antonovsky announced his theory of coherence in 1979.[9] This concept is usually known by psychologists, but remains unpopular among therapists, psychiatrists and among professionals dealing with mental prophylaxis.

Suzanne C. Kobasa independently announced her theory of hardiness in the same year.[10] Hardiness, understood as mental strength or invulnerability, is composed of three elements. The first one is commitment – a tendency to consciously get involved in everyday personal and professional life activities and involves curiosity about the surrounding world, regardless of what kind of involvement or activity it is or of what problems or difficulties are associated with it. Committed individuals have a general feeling of purposefulness that allows them to perceive their own life as important and to identify themselves with events and people in their surroundings. They are hard to put pressure on, as they have significant impact on their environment. Committed individuals prefer activity over avoidance. They usually have a clear path to follow, often even a sense of life duty or a mission. They plan to realize their concepts, therefore they focus on the present moment (as it allows them to fully engage in current activity) and on the future (as this is where the goals are). They rarely think about the past, almost never re-analyze defeats, but they reinforce past successes. On the other side of commitment is alienation, apathy and withdrawal from active life.

The second disposition is control, defined as a tendency to believe and act as if one can influence the events taking place around oneself through one's active efforts. It is not a naïve belief in an ability to fully control external events, rather an ability to identify and understand the real opportunities for leverage. A sense of control strengthens resistance to stress because events are perceived and evaluated as natural consequences of someone's activities, not as random, external, uncontrollable, sudden or overwhelming events. A sense of control allows the inclusion of all difficult events into an individual's life plan. Treating them from this perspective lessens their importance. It also provides individuals with broader and more varied array of tools at their disposal to deal with and control stress, especially when facing particularly difficult situations.

The last element is challenge disposition, defined as the belief and perception that change, not stability, is a natural and obvious element of everyday reality. These people follow the motto: The only sure thing in life is change. People with high challenge disposition try to predict the change and get motivated by the opportunities that changes present for personal growth. They do not consider a change as a threat to their security. Those are individuals with "appetite for life." People with low readiness to accept challenges see every change as a threat to their own existence and are looking for ways to secure themselves from it. Individuals with high challenge disposition are persistently trying to understand what has happened to them and why, despite potential failures. In contrast are those who believe that they do not deserve their fate and are haunted by bad luck.

Hardiness, seen through the prism of commitment, control and challenge, influences the retention of a good mental condition and prevents psychological or psychiatric disorders from developing, despite confronting events generally considered stressful or traumatic. Such individuals are coherent, have high levels of integrity and are convinced that their lives are meaningful (are understandable), can be controlled (can be shaped by an individual), and are not a burden (but rather something pleasant). Many research studies point to the connection between hardiness and better physical and mental health, as well as to a higher resistance to life's adversities.[11]

Did the hardy ones receive their gift from mother nature? Many clues point out to the fact that at least some inherited temperamental traits can improve the way we handle difficult situations. Researchers of resilience also confirm those findings and they claim it depends on the neuro-hormonal balance.[12] But there is also a lot of evidence that hardiness can be developed during our lives. In the 1950s, it was demonstrated that rats exposed to electric shocks at a young age were much more resistant to stress at later stages of their lives.[13] Newer studies on humans confirmed that exposure to stress can build up resistance to stressful situations. For example, hospitalized children handled separation from their parents significantly better if they experience a similar situation in the past, for example during stayovers at their friend's house.[14] Similarly, research conducted on Australian teenagers leaving the country for the first time as exchange students also confirmed the above findings. When compared to the control group that stayed in the country, in their home environment, the long-distance long-term journey changed the way they handled stress. Teenagers who spent time abroad were far less likely to suffer from neurosis than their peers who stayed at home.[15]

Exposure to stress results in increased immunity to stress if the initial stressor was resolved and handled by the individual themselves. If people who experienced a stressful situation were not able to handle those events on their own, the end result of a series of stressful experiences would result in dejection or nervous breakdown. Australian teenagers participating in the student exchange programs could expect full support from people who cared for them. Their guardians would also make sure that the stress levels do not cross the line, where it would be too difficult to handle it.

The importance of immunity to stress usually surfaces in extreme situations. Raj Persaud wrote that hardiness is a trait that cannot be seen in everyday life, but emerges only under severe circumstances. We think that accepting such point of view should be followed by investigating current day military conflicts, as they provide us with incredible opportunity to access knowledge otherwise inaccessible in psychological laboratories. Invariably, properly trained, equipped with refined armaments, appropriately supplied and well fed armies lose battles and wars against much weaker, starving insurgents. On one side, we have well-educated, super-fit soldiers who are often deprived of any stressful experiences, who never experienced a situation of shortage of food, who are put under specialist care every time they participated in live-fire exchange with enemy forces to "treat" their Post-Traumatic Stress Disorders (PTSD), where dozens of therapists, nurses and doctors provide world-class care, rehabilitation and support. On the other side of the barricade, children are often found among the insurgents. These are children who experience famine and misery daily, for whom death always was and forever will be an accepted part of their lives; children who regularly face freezing-cold nights, lack of water and any kind of comfort. They probably don't even know what a psychotherapist is. Of course, we understand rather well that psychological hardiness is not the only reason that gives insurgents advantages in combat. Excellent knowledge of their terrain, local support, unconventional tactics and other similar factors are also important. We do believe however that the mentioned conditioning and resistance to stress are underestimated; even the most refined technology will not help when the morale of troops who could make use of this technology falters.

Research results and observations allowed the authors of the conception of hardiness to create entire stress conditioning programs. Systematic trainings are conducted today primarily at The Hardiness Institute, Inc. in California, but other institutions use them as well. Hardy Girls Healthy Women from Waterville is one example. The description of hardiness programs can be found in several books,[16] and their effectiveness is systematically evaluated and researched.[17]

The concept of hardiness and specialized training programs sound like an echo of the Victorian era, where in accepted values systems, self-control was usually among the most important virtues. The propaganda of spontaneity and sensitivity, still carried on the waves of humanist psychology, rejected those values and made them not only redundant, but also considered them as factors that limited self-realization or 'being oneself.' As the result, modern western civilizations raised "sensitive" citizens, who, unable to cope with their own emotions, unload them by emptying the clip of an automatic weapon at random people in a local school or supermarket. Lack of self-control is also a reason why we struggle with the realization of a very simple goal: living a happy life. At least that's what Mihaly Csikszentmihalyi says, creator of a concept of flow and a renowned psychologist in the developing trend of so-called positive psychology. "In other historical periods, such as the one in which we are now living, the ability to control oneself is not held in high esteem. People who attempt it are thought to be faintly ridiculous, 'uptight,' or not quite 'with it.' But whatever the dictates of fashion, it seems that those who take the trouble to gain mastery over what happens in consciousness do live a happier life.[18]

Why are the concepts of resilience and coherence, (hardiness in particular), largely ignored and do not get deserved attention? The answer to this question is quite brutal. Let's quote Masson one more time: "The very mainspring of psychotherapy is profit from another person's suffering."[19] What benefit would therapists receive from propagating any ideas that would help to build and maintain psychological or emotional strength? There is not much income in it. Propagating the idea of self-improvement and mastering self-control are also not very popular. Patients prefer approaches that take all the responsibility for their own lives off their shoulders; after all, it is so much easier and so much more pleasant. It seems that a vast majority of clinical psychologists and psychotherapists are still rooted in the pathogenic paradigm.

Others, when mentioning resilience, joyfully indulge in pseudo-scientific delusions. An excellent and rather grotesque example is "A Seminar for Leaders of the Future," marketed as *Beyond Illusions* that one of us recently got invited to. The primary goal of the training was already formulated in a particularly disturbing way: "The seminar aims at boosting psychological resilience and at retaining [the] ability to think rationally in the upcoming crises. Another goal is to focus attention on the impact of climate changes on current instability on global markets." We could spend decades trying to understand how resilience is connected with climate change and the global market. Our inability to make appropriate connections was luckily clarified in the last sentence: "This program is such a novelty that we decided only to invite people with wide cognitive horizons, open minds and focused not only on personal success, but also on spiritual development and devoted to pursuing the answer to the most important questions in life: WHO AM I – WHERE AM I GOING AND HOW MUCH IS MY LIFE WORTH?" [Emphasis theirs]

It is clear now that our inability to make those connections was simply due to the lack of open-mindedness and because of our limited "cognitive horizons." Aware of our cognitive limitations, we decided not to participate in the seminar, although the program seemed to be particularly attractive. The participants had a chance to listen to lectures on *Planetary Fields of Consciousness*, or *Beyond Illusions – Realization of Visions of Timeless Success*. A seminar that was in fact an introduction to a coaching event, labeled as *Managing the Rest of Your Life*, was organized by the European Association of Research and Development of Information Engineering, PsychoCybernectics and Neuro-Linguistic Programming in cooperation with "The Soul Clinic." Bear in mind that this is all perfectly legal…

As one of us is a member of National Chamber of Training Companies in Poland, he immediately reacted by sending a letter to the chamber's board protesting against the pseudoscientific content and propagation of quackery, calling to retain a rational approach in spreading any novel ideas. Unfortunately, we have never received any answers.

This chapter was titled, "With What Can We Replace Psychotherapy?" Perhaps, apart from answers presented by us above, which all readers can accept or reject, it would be worthwhile to suggest yet another, very simple answer proposed by Jeffrey Masson: "What we need are more kindly friends and fewer professionals."[20]

---

[1] Masson, *Against Therapy*.

[2] Persaud, *Stay Sane*, 61.

[3] K. Popper, *"The Open Society and Its Enemies* (London: Routledge & Kegan Paul, 1966). As cited in Persaud, *Stay Sane*, 61.

[4] See: M. Windle, J. Searles, eds., *Children of Alcoholics: Critical Perspective* (New York: Guilford, 1990).

[5] Herman, *Trauma and Recovery*, 195.

[6] S. Andreski, *Social Sciences as Sorcery*, (London: Andre Deutsch, 1972), 139.

[7] M. Rutter, "Resilience in the Face of Adversity: Protective Factors and Resistance to Psychiatric Disorder," *British Journal of Psychiatry, 147,* (1985): 598-611.

[8] Herman, *Trauma and Recovery*, 57.

[9] A. Antonovsky, *Health, Stress and Coping*, (San Francisco: Jossey-Bass, 1979).

[10] S. C. Kobasa, "Stressful Life Events, Personality, and Health: An Inquiry into Hardiness," *Journal of Personality and Social Psychology, 37,* (1979): 1-11.

[11] V. Florian, M. Mikulincer, and O. Taubaum, "Does Hardiness Contribute to Mental Health during Stressful Real-life Situation? The Roles of Appraisal and Coping," *Journal of Personality and Social Psychology, 68,* (1995): 687-695; S. C. Kobasa, S. R. Madi, and S. Kahn, "Hardiness and Health: A Prospective Study," *Journal of Personality and Social Psychology, 42,* (1982): 168-177; S. R. Madi and D. M. Khoshaba, "Hardiness and Mental Health," *Journal of Personality Assessment, 63,* (1994): 265-274; S. Kobasa and C. Puccetti, "Personality and Social Resources in Stress Resistance," *Journal of Personality and Social Psychology, 45,* (1983): 839-850.

[12] Curtis and Cicchetti, "Moving Research."

[13] S. Levine, J. A. Chevalier and S. J. Korchin, "The Effects of Early Shock and Handling on Later Avoidance Learning," *Journal of Personality, 24,* (1956): 475-493.

[14] M. Rutter, "Psychological Resilience and Protective Mechanisms," *American Journal of Orthopsychiatry, 57,* (1987): 316-331.

[15] G. Andrews, A. Page, M. Neilson, "Sending Your Teenagers Away." *Archives of General Psychiatry, 50,* (1993): 585-589.

[16] See D. M. Khoshaba and S. R. Maddi, *Hardy Training: Managing Stressful Change* (Newport Beach, CA: Hardiness Institute, 2005); S. R. Madi and D. M. Koshaba, *Resilience at Work: How to Succeed No Matter What Life Throws at You* (AMACOM American Management Association, 2005).

[17] See J. Sudkins, L. Furlow, and T. Kendricks, "Developing Hardiness in Nurse Managers," *Nursing Management, 14,* (2007): 19-23; S. R. Maddi, "Hardiness Training at Illinois Bell Telephone," In *Health Promotion Evaluation*, ed., J. P. Opatz (Stevens Point, WI: Wellness Institute, 1987): 101-115.

[18] M. Csikszentmihalyi, *Flow: The Psychology of Optimal Experience* (New York: Harper Parennial, 1991), 23.

[19] Masson, *Against Therapy.*
[20] Ibid.

CHAPTER 15

SHIP OF FOOLS:
THE REASONS FOR THE PRESENT STATE OF
PSYCHOTHERAPY

Illuminated in characteristic museum-like fashion, an unusual picture, *The Ship of Fools*, painted on a small board by Hieronymus Bosch, hangs on a wall in the Louvre. The deep green background contrasts with the brightly lit group of people in a boat depicted in the foreground. It is not so much people, however, as their vices that are the main theme of the painting. Debauchery, drunkenness, gluttony, greed and other lusts, which drive people to madness, were loaded onto a ship by the Netherlandish master and sent out to sea. Such endeavors were familiar to the contemporary audience to whom Bosch addressed his message. In the Middle Ages, the insane and mentally handicapped were often "taken care of" by being sent on a sea voyage. Those people who did not fit into established social order were frequently dispatched to sea without any particular destination. But it was not mentally ill men and women that Bosch put on board his ship. Instead he depicted representatives of his present-day society: a monk, a nun, peasants, burghers, and a jester. The painting is an allegory of madness brought on by those desires.

Today, as we look at psychotherapy, it rather reminds us of the macabre visions from Hieronymus Bosch's paintings, and like him, we feel an irresistible urge to load all that decay on board a ship and send it to sea without any definite destination. We might thus symbolically sum up the history of the madness of our civilization: a civilization which was unable to deal with its own madness, initially simply disposed of it, and later developed therapies which evolved into such forms and magnitudes that the only sensible solution might be to set most of them on the ships of fools and let the follow all of those who forever disappeared behind the horizon.

How did we get to this point? Why do we have to hark back to the gloomy visions of the Flemish master without being able to see a bright and glowing vision of the future? We have already touched upon many causes for the miserable shape of psychology in the previous chapters, but it is worth sorting them out now. This general picture of psychotherapy is due to various groups of causes. Some of them stem from the therapists' attitudes towards therapy. Those could be classified as internal factors. We have wrote about them primarily in the previous chapters. These include:

- Therapists' reluctance to research into the process of therapy and its effectiveness;
- Evasion in specifying therapy's objectives;
- Proponents of various therapies taking advantage of the placebo effect, time variable, regression to the mean, and similar natural factors as arguments in favor of therapies' effectiveness;
- Considering therapy as an art;
- Lack of studies with high external validity;
- Common acceptance of subjective criteria for evaluation of therapeutic outcomes;
- Therapists' perception of the therapeutic process as an opportunity for satisfying their twisted or morbid needs;
- Lack of a clear distinction between giving patients pleasure and satisfaction on the one hand, and relieving their pain on the other.

The above factors contribute to the disintegration of psychotherapy from within. Another one, which we have not mentioned so far, is related to the systems of supervision. These are very diverse and depend on the country's legal system, which imposes particular forms of controlling psychotherapy, controlling the therapeutic modality (which does or does not involve a system of supervision), may involve external control by observers from outside a particular therapeutic school or control only by internal audits and so forth. The most restrictive control system that we have personally encountered is that in force in Germany. Great Britain has a very liberal one, while Poland is a country where literally everything is possible.

Apart from the internal factors, there exists a number of external ones that contribute to the preservation of current status quo. These can be further divided into factors related to patient behavior and into external circumstances that contribute to perpetuating and deepening negative phenomena in psychotherapy. Naturally, the purpose of this classification is to sort out our thinking about their causes. We do not claim that our classification is in any way authoritative or exhaustive.

Let us first take a look at factors connected to patient behavior that contribute to perpetuating and deepening negative phenomena in psychotherapy. Two of them will be repeated from the above list. The first one is the previously mentioned fact of the common acceptance of subjective criteria when evaluating therapeutic outcomes. Obviously, this is possible only with patients' approval. Moreover, we think it was patients who imposed this way of perceiving the effectiveness of therapies on therapists. Naturally, therapists quickly realized that it was easier to manipulate subjective impressions than measuring real effects, so they have willingly agreed to these expectations.

It also seems that some scientists grossly contributed to cementing the use of subjective criteria in the evaluation of therapeutic outcomes. One of the chief contributions in this respect was made by Martin Seligman, who in 1995 published a paper in *American Psychologist* about the results of a survey conducted by *Consumer Reports* among the clients of psychotherapists.[1] The first pages of the paper feature a sort of confession by Seligman:

> But my belief has changed about what counts as a "gold standard." And it was a study by *Consumer Reports* (1995, November) that singlehandedly shook my belief. I came to see that deciding whether one treatment, under highly controlled conditions, works better than another treatment or a control group is a different question from deciding what works in the field (Muñoz, Hollon, McGrath, Rehm, and VandenBos, 1994). I no longer believe that efficacy studies are the only, or even the best way of finding out what treatments actually work in the field. I have come to believe that the "effectiveness" study of how patients fare under the actual conditions of treatment in the field can yield useful and credible "empirical validation" of psychotherapy and medication. This is the method that *Consumer Reports* pioneered.[2]

What was so groundbreaking in this research that it shook the belief of the future APA president? Well, while carrying out a survey about people's satisfaction with cars they owned, *Consumer Reports* asked them also about their satisfaction with psychotherapy. There were no control groups, no randomization (the sample, which consisted mostly of middle-class, well-educated and predominantly female individuals with a median age of forty-six, was not at all representative of the general population), no objective indicators – only subjective evaluations. It was like they were measuring customers' satisfaction with a vacuum cleaner or a toaster. Very few respondents replied to the survey (only 3.9 percent). Yet, despite so many methodological flaws (we only pointed out some of them), *Consumer Reports* concluded that: "our groundbreaking survey shows that psychotherapy usually works. ... Longer psychotherapy was associated with better outcomes."[3]

Why did a recognized psychologist and researcher publish a paean of praise for some marketing analysts from *Consumer Reports*? Was it only because he was the main consultant of the team conducting the poll? We can only guess what the correct answer is. Even though the source article and its discussion by Seligman have been subjected to devastating criticism,[4] it seems that the disgraceful aim of introducing subjective indices into research on the effectiveness of psychotherapy was achieved. Unfortunately, today not only therapists and patients, but also many researchers commonly use this approach.

Many therapists willingly propagate another myth, even though some probably do not even believe it, that psychotherapy treats causes and not symptoms. This assertion also implies contempt for symptomatic treatment. Thus, if a patient suffering from fear comes to a psychotherapist's office, then the use of simple therapeutic methods to reduce anxiety (including pharmacological ones) is considered merely symptomatic treatment. A therapy aimed at treating causes, for example psychoanalysis, assumes to identify the root cause first, gain insight into them and then eliminate the sources of fear. It does not matter that it persists for years or that the symptoms do not disappear. Therapists argue that the 'real' process of arriving at causes must be long and difficult, and we must not be discouraged by failures. In fact, it is often the other way around. Here is how Raj Persaud writes about this:

> In fact, the treatment which has the best scientific evidence of its effectiveness is a behavioral approach, where the therapist is not at all interested in why a behavior started, and doesn't even need to know this to bring about a change in the client. ...Spending much therapeutic time endeavoring to determine the reasons behind a behavior is always going to be highly speculative and there is little scientific evidence that this helps the process of change in any way. And let us not forget that what clients want from treatment is change. But the worthwhile agenda of change usually gets set aside in counseling and talking therapies and is replaced by an agenda of exploring causes, so change gets neglected, which explains why it is only rarely achieved.[5]

The other recurrent factor is the lack of a clear distinction between giving patients pleasure or satisfaction, and relieving their pain, anxiety or other symptoms. This phenomenon, too, is contributed to by the fact that patients and therapists share similar views here. It is patients, rather than therapists, who create demand for ineffective psychotherapeutic services, which are far from delivering real, life-changing results.

Why do people seek support when dejected, when taking major decisions, when trying to deal with issues such as losing a job, a partner's infidelity and so forth? Perhaps, as we identify another factor, we touch upon one of the most important and very primal human characteristics: the tendency to seek out someone much stronger and/or smarter, someone in whom we could totally trust and who would be able to take our hand and lead us? We are always on the lookout for an authority figure. How grateful we are when someone helps us to make a difficult decision; how very uplifting is the conviction that the person who supports us understands the nature of our problems better than we do?

This is exactly what most often drives us to psychotherapy. We pay only for our own illusions, not for the work of therapists nor for the questionable results of their work. On what grounds can we say that they are really stronger, wiser, or that they better understand other's problems? Who made psychotherapists such persons? A training course in a therapy without scientific foundation? It seems so, because as Judith Herman claims: "Therapists who work with survivors report appreciating life more fully, taking life more seriously, having a greater scope of understanding of others and themselves."[6] And they discovered it themselves, one might add, paraphrasing the aphorism of Stan Lec, "I am beautiful, I am wise, I am good. And I discovered it myself!"[224]

But is arriving at such a conclusion and acquiring self-marketing skills the result of psychologists' training, which consists of showing interest and compassion to patients on the one hand, but also behaving in a way that earns customers' trust on the other? The basic skills of any psychologist include self-confidence or expressing one's own opinions in an authoritative manner. Isn't this simply a marketing technique…and a rather clever one? Here is what Charles Sykes writes about it: "The therapists transformed age-old human dilemmas into psychological problems and claimed that they (and they alone) had the treatment… The result was an explosion of inadequacy."[7]

Meanwhile, those therapists who are not afraid to own up to their feelings while encountering patients who seek their support say, as Sandor Ferenczi: "We greet the patient in a friendly manner, make sure the transference will take, and while the patient lies there in misery, we sit comfortably in our armchair, quietly smoking a cigar. We make conventional and formulaic interpretations in a bored tone and occasionally we fall asleep. In the best of cases, the analyst makes a colossal effort to overcome his yawning boredom and behave in a friendly and compassionate manner."[8] Such attitudes are not unusual. Jeffrey Masson shares similar views on the matter:

Many times I sat behind a patient in analysis and became acutely and painfully aware of my inability to help. Many times, indeed, I did feel compassion. But at times I also felt bored, uninterested, irritated, helpless, confused, ignorant, and lost. At times I could offer no genuine assistance, yet rarely did I acknowledge this to the patient. My life was in no better shape than that of my patients. Any advice I might have had to offer would be no better than that of a well-informed friend (and considerably more expensive). I must assume that none of this was unique to me. Everything I experienced in the situation must have been felt by other therapists as well.[9]

And this is often the only thing we pay for. No wonder that patients, who like moths seeking the light of an authority figure, fail to realize that this is merely reflected light. The absurdity of this search is well illustrated by a joke about a patient who comes to a cardiologist and tells him he thinks he is a moth. The doctor, somewhat uneasy, explains to him that this case is rather outside of his area of expertise, and that a psychiatrist, whose office is on the next floor, might be better qualified to help him. "I know, doctor," replies the patient, "the thing is your office was the only room with the light on!"

Jeffrey Masson points out that trying to find an authority figure in a psychotherapist is a complete dead-end for a person seeking help.

When I began my psychoanalytic training, I was a Sanskrit scholar who had become disillusioned with the notion that life could ever provide a guru, a person with unique insights into the internal life of another person. I thought this claim was unique to Indian culture, one that had caused people a great deal of unhappiness, though no doubt many would claim that it had also brought them great happiness, even joy and bliss (just as some people who have had electroshock claim that it did them a great deal of good). ... And yet, here I was, eight years later, coming to the same unhappy conclusion about psychotherapy: There are no gurus. Maybe I was touching on one of the characteristics of the human animal, the need to seek somebody apparently stronger, wiser, better, happier, from whom guidance could be sought.[10]

This blind search for an authority figure is often accompanied by tremendous reluctance to accept responsibility for one's own life. We have already mentioned this phenomenon several times. It is patients who create demand for psychotherapeutic services. They buy "explanations" of their problems, which allows individuals to shake responsibility off their shoulders. It feels amazing! Now we are able to decline any responsibility for what our life looks like. We have all been training this ability since childhood. As early as elementary school, we graduate convinced that we have no aptitude for mathematics; none of us ever admits that we simply did not work hard enough. Having flunked our French, Spanish or German paper, we "know" we have never had a gift for languages; we miss the fact that we did not set aside enough time to study our vocabulary or grammar. When a patient whose relationship is breaking up comes to a psychotherapist, he or she will accept every interpretation except that he did not try hard enough to save it. In the chapter titled "The Nightmare of Recovered Memories," we quoted a person whose therapist helped her to 'understand' that her adult life was miserable and hopeless only because she had been sexually abused as a child (even though she did not remember this at all). Such therapeutic solutions are offered by all sorts of institutes, especially those based on pseudo-scientific assumptions, from psychoanalysis to radical forgiveness. Probably the common correlation is that the more responsibility a given therapeutic modality places on the patient, the less popularity it enjoys.

Another factor that depends on patients is called 'craving for miracles.' When we, or our children, fall ill, and especially if the diagnosis comes to us like a bolt from the blue, we find ourselves at the end of our tether. The first reaction in such situation is rebellion and a denial of reality, many times described in the literature. In such circumstances, our rational thinking is stifled by fear and protest. We lose the ability to properly take stock of the situation. We are ready to believe in every promise, even such as those offered by the so-called Colin Tipping's approach, also known under the name 'radical self-forgiveness':

> Yet science, particularly psychoneuroimmunology, is showing that forgiveness and other forms of emotional/spiritual healing, is *extremely powerful*, [emphasis ours] and that anyone with cancer who doesn't include forgiveness in their treatment protocol is ignoring a huge part of the problem and, by extension, a large part of the solution. ...Bringing that trauma, whatever it is, to the light and processing it relatively painlessly through Radical Forgiveness is the best way to

heal this. The beauty of Radical Forgiveness is that it is not a therapy in the normal sense of the word. It does not require you to go digging up the past and rehashing the pain. That's exactly what cancer patients have spent a lifetime avoiding! …The reason that I suggest Radical Forgiveness over other forms of forgiveness is that it works at the energy level, so it happens very quickly. Ordinary forgiveness takes its own time. You can't force it. Time will heal it, we say. But when you have cancer, that's not good enough. You want results NOW! Radical Forgiveness is quick, it is easy to do and is therapeutically non-invasive.[11]

These immediate effects are of course produced by the Colin Tipping's approach.

There is even a more treacherous way of exploiting the fears and concerns of parents whose children suffer mental illness or neurological disorders. Glenn Doman, when founding his Institute for the Achievement of Human Potential in Philadelphia, announced the following theses (among others):

5. Our individual genetic potential is that of the human race.
6. Our individual genetic potential is that of Leonardo, Shakespeare, Mozart, Michelangelo, Edison, and Einstein.
7. Our individual genetic potential is not that of our parents or grandparents.
9. All intelligence is a product of the environment.[12]

The claims above, like most of those formulated by Doman, clearly contradict the results of scientific research. Today, even the greatest skeptics do not deny the significance of genetic factors in the development of intelligence. It is commonly accepted that genes account for at least 35-50% of variation in our intelligence. The views of Doman, however, offer great hope to everyone struggling with even the most hopeless case. Glenn Doman is attracting many followers. They promise to cure every brain disease, including mechanical damage. In Poland for example, a variant of his method is sold under the name of "Neuro-Re-Education." Examples of similar false promises would easily fill an entire book; add to it all the false promises and faked claims of alternative medicine and several volumes would hardly be sufficient.

But a craving for miracles occurs not only in the face of an unexpected disease or a difficult psychological situation. It also manifests itself in chronic situations and is sometimes disguised behind more rational explanations. A case in point is a friend of ours who has suffered from incurable chronic pain for many years. Conventional medicine helps him to address the symptoms only, even when using measures such as morphine. As a skeptical and rather rational person, he rejected the promises of various miracle-workers for years. But now and then, he sometimes lets himself be talked into a new fantastic method of controlling his pain. When asked why he does so, he answers quite rationally: "And how can it harm me?" So far it has not. Perhaps some side effects or the placebo effect do sometimes give temporary results or a subjective sensation of improvement. It is not infrequent that a patient gives up conventional methods of therapy for unconventional ones. Professor Mark Pawlicki from the Institute of Oncology in Cracow estimates that at least five thousand Poles die every year as a result of replacing conventional treatment with the so-called alternative methods.[13]

The problems with psychotherapy are further perpetuated by laziness and ignorance. Laziness drives us to choose a therapy where the therapist takes responsibility off our shoulders and does not force us to expend much effort in working on ourselves. Thus, sloth will lead us to the offices of those therapists who can help us deal with our problems with their hocus pocus and, conversely, will discourage us from therapists who suggest we need to work hard during therapy. But we are also too lazy to make an informed choice of psychotherapeutic approach. Very often, we miss a great deal of information, we make many errors, and we tend to be easily deluded with false promises.[14] Typically, we rely on information from one source, for example from a friend who has 'heard' a lot about the subject. Rarely do we look for additional information about therapists and the modalities they use; we do not show much initiative in this respect. As a result, even the worst therapists will still have a supply of patients.

The laziness is compounded by ignorance and vice versa. People without psychological training derive their knowledge about psychology and therapies from popular magazines, often not even from those dealing with popular science. This information, in turn, is filtered by journalists who try to market attractive material while often being just as ignorant as their readers. This, as a matter of fact, explains why nonsense such as psychoanalysis has persisted for decades.

The following example was taken from a web forum run by one of us and illustrates rather well the way of thinking of potential consumers of psychotherapeutic services:

> Let me speak from the viewpoint of a parent whose child, instead of being given from the start an accurate medical diagnosis and proper treatment, was sent to psychologists instead. This triggered an avalanche of problems that are now very difficult to resolve; there's no denying that my ignorance was largely to blame, and that's why I'm determined to somehow fix the situation...
>
> As a non-specialist, what guarantee do I have that the state-sponsored psychologist (and hence, it might seem, a reliable one) who is to administer "therapy" to my child is not a sort of legitimized Renata A? None. After all, I have no knowledge about the subject. I have no idea which therapy is right and which is not (if I had, I would administer it myself without anyone's help); I can hardly be expected to give up my scientific work in a completely different field which I find fascinating, in order to search for information (on the Net? in books? and what qualifications do I have to verify these sources?) on whether a therapy is good or bad, honest or quackish. So what should I do, if I can't verify that, and my child definitely needs some therapy (I don't know which, it's not my depart-

ment; there must be some progress and that's that; this is what specialists are for); he should be referred for some treatment where something will be "done" to him.[15]

Regrettably, nothing excuses us from critical thinking. When choosing a therapy, we must realize that having a degree in psychology, being employed in a state-sponsored psychotherapy clinic and so forth, is no guarantee of reliability. We also believe that the educational, legal and scientific systems should allow a patient to make a decision on the basis of such considerations and not worry if it was right. Unfortunately, as we wrote earlier, those do not work, or at least, not as they should.

From a certain vantage point, the author's line of thought is correct. If we wanted to gain knowledge about the difference between diesel and gas engines before going to a car mechanic, find out about the chemical composition and properties of various fillings before a visit to a dentist, or get acquainted with the recipes for the dishes we want to eat before we set off to a local restaurant, our life would certainly turn into a nightmare. We would not be able to function normally. But the point is that the decisions we make are not equally momentous. A badly replaced air filter in your car will most likely not ruin your life, and we can always make a complaint. Similarly, we can always go to another dentist to fix the botched job of the first one; we can also return a poorly cooked meal to the kitchen and go to another restaurant. However, the decision to entrust someone with your life savings, to lie down on the operation table, or to put your psychological well-being into someone's hands is often irreversible. Should we not, perhaps, gather some more information before making potentially life-changing decisions?

There are circumstances which most fraudsters find very advantageous, for example, the hope of obtaining something extraordinary coupled with ignorance. The media regularly informs us about groups of people who gave away their life savings and went to work abroad and were left on their own. We hear about tourists who, after arriving at a very attractive destination, spent a week in ghastly conditions or had to return home, furious, at their own expense. We learn about gullible victims of yet another financial pyramid scheme and so forth. As we hear those pieces of news, are we inclined to put the entire blame on the fraudsters? Don't common thoughts revolving around the term "stupidity" come to our minds as well? Stupidity seems just a stronger description of ignorance.

The paragraph above brought us rather smoothly to another group of factors; we called them *external circumstances favoring the perseverance of poor practices in psychotherapy*. The most prominent among them are the market conditions in a given niche. These are quite specific for psychotherapy and have been accurately pinned down by Stanislav Andreski:

Once an activity becomes a profession – that is, a way of making a living – dedicated amateurs tend to fall into second place, greatly outnumbered by the practitioners guided primarily, (if not solely), by the normal motives of the market place – which commonly boil down the desire to get the most at the least cost to themselves. In other words, as soon as it becomes apparent that there is money in it, the salability of goods rather than their intrinsic excellence becomes the domination criterion. Hence the quality of the goods is attended to only in so far as the buyers are interested in it, able to judge and willing to pay for it. Only under such circumstances is honesty the best policy. … Though easily deceived on finer points, people will not go on buying soap which does not remove dirt at all, or knives and forks which break as soon as they are used; whereas with products [that] do not serve as instruments for a clear and obvious purpose, there is no natural limit to shoddiness, particularly when the canons of taste can be manipulated by vested interests.[16]

It is hard to disagree with this line of thinking. The interpenetration of the market and psychological services will be described in detail in the next part of this book. To add to Andreski's comment, we can say there are some specific conditions which foster the development of psychotherapy against the logic of the market. Psychotherapy has an invisible bodyguard that will protect it even if individual tastes and preferences are not quite satisfied. This role is fulfilled by a specific taboo that accompanies psychotherapy and which is yet another factor on our list.

Until recently—and in some circles even now—openly admitting that one went to a psychologist, psychiatrist or psychotherapist amounted to admitting to being mentally ill. There still is a social stigma associated with mental disorders. Even if someone decided to go into therapy, he would typically travel to another town, and it was only his nearest and dearest that knew anything about it, but sometimes even they were kept in the dark. In such circumstances, market information about a poor therapist has no chance of surfacing and being circulated. While word spreads freely about negligent cobblers, mechanics or craftsmen which weeds them out of the market, quack psychotherapists might operate with impunity in the market for many years.

Recently a lot has changed in this respect. In certain circles, it is even necessary to have one's own therapist; however, these are actually only apparent changes. When one of us was involved in the making of the TV program *I can't Live without Therapy*,[17] the producers had a hard time finding three people willing to speak in front of the cameras about their positive experiences with psychotherapy. When we were helping to do other TV programs about problematic areas in psychotherapy, we always encountered the same problems... there were no victims of therapy! Out of dozens of people who had approached us as victims of unethical or even extremely destructive therapists, only one agreed to publically speak about it. Some of the victims cited fear of their therapists' revenge as the main reason for refusal! And it was not a groundless fear but rather based on information from other patients who had learnt the hard way. Some of them did not want to reveal their identity, despite assurances that their faces might be hidden and their voices altered. Only one refusal was due to a poor mental state. Given such an attitude, is it even possible for any negative information about psychotherapy to spread?

It is worth adding that reporters who tried to elicit comments from therapists, representatives of the Polish Psychological Association, and even psychiatrists, would almost invariably meet with resistance, evasive answers, and refusals to appear in front of the cameras. They interpreted this resistance as an unambiguous case of conspiracy of silence and an attempt to conceal irregularities they had previously heard about in casual conversations. Lack of statements from victims, experts or therapists, make such tv programs come across as bland, as if they raised non-existent problems. As a result, some of them were never aired despite the huge effort that had gone into making them.

Moreover, everything that takes place in a psychotherapist's office is also shielded by a taboo. Some therapists additionally reinforce this taboo by entering into a so-called contract with the patient whereby they swear him or her to secrecy about everything that is said during therapy. Such secrecy does bind all therapists, just like the seal of confession binds all confessors, but not patients! By the same token, the seal of confession does not extend to penitents! Treating this secrecy as symmetrical merely makes for covering up the wrongdoings.

Even if we assume that there is a minimal chance of word being spread about bad or dishonest therapists, it is further minimized by a cognitive dissonance that contributes to diminishing patient dissatisfaction with therapy. This phenomenon is best explained by an experiment conducted in 1959 by Leon Festinger and Merrill Carlsmith. It was very simple. The researchers asked students to do boring and monotonous tasks for an hour. Afterward, they explained to them that they wanted to check if people worked better after being told that the job was interesting. At this stage, they told them a made-up story that an assistant, who usually imparted this information to the participants, was absent from work that day. For this reason they asked the students to tell subsequent participants that they were in for some interesting and engaging tasks. According to the experimenter, this information was supposed to be more credible if conveyed by students rather than himself. When they had done their job, the students were paid for participating (lying). Half of the students were paid 20 dollars, and the other half only one dollar. At the very end they were asked to fill out a questionnaire where they rated, among others, the tasks they had carried out. What were the results? Those who received 20 dollars considered the tasks boring and monotonous, while those who only got one dollar found the same tasks to be more interesting and engaging! Why was that?

The most important aspects of the whole experiment were three elements: belief about the real nature of the task, the fact of lying and the reward. If I think that the task is tedious, but I am paid 20 dollars for lying, I have a sufficient justification for telling lies, i.e., I lied for money. If, by contrast, I believe that the task is boring and am paid only one dollar for lying, then the question emerges: Why did I lie? Leon Festinger, originator of cognitive dissonance theory, explained this phenomenon in the following way. Our belief and the fact of lying are two cognitive elements that appear incompatible with each other. This discrepancy produces some discomfort and motivates us to try to reduce it, just like hunger or thirst causes us to look for food or drink. Since the students' behavior was inconsistent with their beliefs, they had to change one of the cognitions in order to reduce the ensuing dissonance. It would have been difficult for them to modify their behavior, which had already become the fact. They could only change their attitude and belief. Thus, the students persuaded themselves that the task had been interesting.

Now let us imagine the thought process that might be taking place in the head of a patient who, after long wavering, has finally made the difficult decision to go into therapy. During the sessions, the therapist is performing convoluted analyses and asking him to carry out the most bizarre tasks; between them they ponder connections between the patient's childhood and his problem without approaching the latter at all. Additionally, the client pays for the sessions, and the sum is by no means small. Will he conclude at some point that all this was one big mistake? Or will he rather rationalize the whole situation in line with the decision he took, the choice he made, and the money he invested? The cognitive dissonance concept predicts that most people will choose the second alternative, provided that the therapist does not go beyond the point of absurdity (which is also the tolerance level). This border varies from person to person.

This combination of taboo, secrecy, cognitive dissonance and fashion for psychotherapy is grist to the mill of all sorts of charlatans. The fact that they are not threatened with any consequences fosters the constant development of the psychotherapeutic market. This rapid process is contributed to by the media and often by eminent authorities too. Journalists crave stories on sensational and revolutionary therapies. Adding a few question marks and explaining that the research is still underway saves their journalistic conscience, but people who are desperate and/or ill perceive such news as tidings of salvation. With poor regulation of psychotherapeutic services, psychobusiness, as we like to call it, thrives. It looks like a giant fair of miracles. The next part of this book will be about what happens at those fairs.

---

[1] *Consumer Reports*, "Mental Health: Does Therapy Help?" (Novemeber, 1995), 734-739.

[2] Seligman, "The Effectiveness of," 966.

[3] *Consumer Reports,* "Mental Health," 734.

[4] E.g., Dineen, *Manufacturing Victims,* 62-64.

[5] Persaud, *Stay Sane,* 58.

[6] Herman, *Trauma and Recovery,* 153.

[7] C. J. Sykes, *A Nation of Victims: The Decay of the American Character* (New York: St Martin's Press, 1992), 37.

[8] Ferenczi, *Journal Clinique,* 246.

[9] Masson, *Against Therapy.*

[10] Ibid.

[11] C. C. Tipping, *Radical Forgiveness — A Complementary Treatment For Cancer,* (n. d.): http://www.radicalforgiveness.com/member/free-stuff/

[12] G. Doman, *The 89 Cardinal Facts For Making Any Baby Into A Superb Human Being* (n.d.): http://www.iahp.org

[13] I. Cieślińska, "Homeopatia, czyli uzdrawianie wodą z... mózgu. *Gazeta Wyborcza,*" (March 31, 2009): http://wyborcza.pl/1,75476,6448687,Homeopatia__czyli_ uzdrawianie_woda____z_mozgu.html

[14] See more on this subject in Myers, *Intuition.*

[15] Private correspondence. Renata A. is a fictional character who will reapper in the last part of this book.

[16] Andreski, *Social Sciences as,* 43.

[17] P. Młynarska, Executive Producer, *Nie mogę żyć bez terapii* [Television broadcast]. (Warszawa: TVN Style, March 31, 2008).

PART III

BEYOND CONTROL: PSYCHOBUSINESS

CHAPTER 16

FAIR OF ILLUSIONS, MIRACLES AND HOPES

It must have been a truly extraordinary therapy, if its authors deserved such an enthusiastic opinion by the indisputable therapeutic guru, Carl Rogers: "A group of very honest young therapists tell, with great candor and openness, about the new kind of therapy they are developing and the mutuality of relationship it involves."[1] Not only did experts enthuse over the therapy, the media fell over itself to interview the therapy's creators. Enthusiasm peaked just as the book was published.

> "The Dream Makers," a press release from Crowell dutifully relayed, "contains ideas so powerful that readers across the country are reporting that merely reading the book has changed their lives. "Newspaper stories were no less knee-jerk: Joe and Riggs, gushed [in] a Honolulu paper, "may be the biggest thing to hit psychotherapy since Freud kicked cocaine." In 1977 and 1978 alone, founders -- usually Joe and Riggs -- were guests on 134 radio and 104 television shows, including back-to-back appearances on Merv Griffin and *The Tonight Show* and four appearances on *Good Morning America*. Geraldo Rivera reported from the Gardner Street compound that "all of them coexist in what apparently is one big, happy family."[2]

The patients, too, were convinced that its authors were modern day successors of Freud and that they would soon be nominated for the Nobel Prize. The newly founded treatment center was growing rapidly, as were the numbers of patients. At its peak, it had 350 resident patients and 2,000 members, including various branches. It seemed that nothing could stop this psychotherapeutic revolution, and that finally, a therapy that could solve most of people's problems had been found. Yet, one day towards the end of the 1970s, the bubble burst. This is what we can read about the therapy today that was so beautifully presented beforehand:

> I suspected the worst. But the worst I could dream up was nowhere close to the reality that I would discover two years later, after it all ended. I would litigate against the Center therapists for 5 years for what they did and stand over administration proceedings to revoke their licenses, taking part in over 225 days of depositions, countless motions, reading a room full of diaries and note books of over 40 patients covering the mean time of five years each, many 9 years.

What came out was what some have called the most horrific and brutal school of psychotherapy that probably ever existed and ultimately licensing revocation hearings that were among the longest in California History. ...The most outrageous psychological malpractice in the history of the profession.[3]

So what happened at the Center for Feeling Therapy, as the place was called, that it deserved such scathing words?

Relationships were controlled, and children were forbidden. Thus time would all be spent on the Center projects, which created businesses that it controlled and from which patients worked... and they could better serve the therapists personal needs and recruit. Like at Synanon, pregnant women were told to have abortions and one who resisted was made to carry a doll around with her until she finally surrendered and aborted. She surrendered when weights were attached to the doll. Another woman, today a psychologist and cult expert, Donnie Whitsitt, was told to give up her children to her ex-husband as she was not fit to be a mother.

For patients to see their sins, assignments were given designed to humiliate. One woman had to moo like a cow; another accused of being a "suck on the group" had to bring a black dildo to group and suck on it; a former beauty queen was to wear her beauty pageant gown for a week sans makeup [and] shower and sleep in it; a man had to wear diapers, suck on a bottle and sleep in a crib. A patient who failed to lose her directed amount of weight was ordered to gain 20 pounds in order to experience being fat. A woman said to be too promiscuous was told in her group to get ready to be laid by every man in the group. This person claimed that when she went to Riggs, he physically attacked her for saying she was leaving. Another claimed during her intensive [treatment] a naked man grabbed her by the hair, pulled her down the hallway and dumped her in a room in front of Riggs where she was told she was too crazy and then locked in a room to think about it.

One Mexican-American patient who had criticized Riggs was assigned by Steve Gold for 2 weeks to be a housemaid for others while another was given laughing gas and then told to discuss sexual fantasies. Another was to insert her fingers into her vagina during group therapy and another was ordered to bang her head against the wall until allowed to stop. A therapist wrote that one patient was to become so humble she would not feel her "herpes hurting her cunt." A document confirms that one patient was ordered to court a girl he did not like, i.e. have a "surrogate girlfriend," which not only indicates the cruelty to the patient, but to disregard for even 3rd persons, i.e. the non-member, targeted girl. Another patient was busted for failure to "get laid."

Activity was regulated. Schedules were made for various people from the moment they woke up to the moment they went to sleep, including times for eating, seeing friends and sex. Even the type of sex was sometimes directed.

Center patients had to have permission to date and permission to break up. Permission [about] who they could live with was also needed. They were told what clothes to wear, what their weight should be and what television program should be watched or not watched.[4]

Is this not reminiscent one of Hieronymus Bosch's paintings? We will not analyze the motives that prompted people to undertake such therapy, nor the mechanisms that made some of them live in such conditions for up to nine years. We should, however, ask ourselves what motivated the people who created and preserved this horrifying reality? Were those people monsters? Paul Morantz, a lawyer who represented the victims of the founders of the Center for Feelings Therapy in court, is not in much doubt when talking about their motives: "the goal appeared to be primarily money, a lot of it."[5]

Welcome to the world of psychobusiness, a world dominated by money, a world where human suffering is a commodity, often merchandized with total impunity. This is what happened here. The founders of the Center were banned from practicing in California because of lawsuits initiated by patients against them. Victims accused the therapists of rape and numerous other mistreatment. The most interesting part however, is that no criminal charges were ever brought against them. Moreover, the victims never received any apology from people who caused them such significant harm.

> In September 1987 the longest, costliest and most complex psychotherapy malpractice case in California history came to an end when the Psychology Examining Committee of the California Board of Medical Quality Assurance revoked the licenses of Joseph Hart and Richard Corriere, former heads of the Center for Feeling Therapy. For more than two years, the state had been trying its case against thirteen members of the Center's former professional staff, and now all those accused of incompetence, gross negligence, fraud, patient abuse or aiding and abetting the unlicensed practice of psychology had either lost, surrendered, or, as in two cases, had severe restrictions placed upon their professional licenses.[6]

That was all. No additional consequences for the therapists except for those related to their professional practice.

In the previous parts of this book, we looked at psychology as a science and psychotherapy as its applied form – born in academia, but abused by charlatans of all kinds who have nothing in common with proper science. However, this perspective is not sufficient to fully define the significance and role of psychology and psychotherapy in the world today. To change our point of view, we need to leave the walls of academia and join a fair of miracles, illusions and hopes. Vendors offer their goods and services, food and cosmetics, shoes and clothes, medications and snake oil. Among these sellers, we will also find psychotherapists and counselors, self-improvement courses and lots of other products and services offered by the psychological industry, because apart from all, or perhaps even above all, psychology is a business – just like any other commerce on the market today.

If we look at psychology from this perspective, it will be easier for us to understand all those numerous distortions and the reluctance to the falsify claims on which psycho-services can be sold. We might just find out that all those outrageous things that we have described earlier are actually very clear and understandable from the marketing perspective. Looking at psychology as at any other business, we will see that it actively tries to attract customers, and that any attempt to find the boundary between scientific psychology and psychobusiness is necessarily doomed to failure.

The market already begins to operate when scientists decide to choose a research area based on the possibility of obtaining a research grant. Then investigators carry out their work and write articles in a way that would allow for their publication in higher ranked journals. They are reluctant to have their research replicated, just like manufacturers, who are usually not very eager to submit their products to consumer tests. Members of academia often have second jobs outside their universities – at the funfair that we have just visited. That's where they are forced to fight for their customers. And they fight, using advertising and marketing tools and skills, using tried and tested methods of exaggerating, overlooking and distorting. As a result, the walls separating academia from the market cease to play any protective role, (even symbolically).

We are even more inclined to treat psychology and psychotherapy as a psychobusiness after we have analyzed the increasing incidence rate of mental illnesses in recent years. The World Health Organization estimated that today over 350 million people worldwide suffer from depression alone. At its worst, depression can lead to suicide. Suicide results in an estimated 1 million deaths every year.[7] In most countries, the number of people who would suffer from depression during their lives falls within an 8–12% range.[8] Worldwide, more than one in three people fulfill all diagnostic criteria for at least one mental disorder at some point in their life.[9] If we take those numbers into account, we can draw only two logical conclusions. First, we are dealing with a pandemic of mental illness on an unprecedented scale in the history of the human race. By comparison, the Spanish flu, which may have been the biggest pandemic ever, claimed merely 50-100 million lives. If we decide to treat these incidence and growth rates seriously, we would have to conclude that the entire human race is going mad. The second logical conclusion is that we are dealing with... incredibly efficient marketing. As scientists, we believe that the most probable scenario is the most likely to be true. Let us, therefore, take a closer look at this second possibility and analyze the mechanisms of psychobusiness.

Today, people who have problems with their own life or health have virtually unlimited access to therapeutic services. In bookshops we can find entire walls of shelves with books promising all kinds of assistance, offering solutions to every imaginable misfortune or proposing complete life-transforming instructions. Magazines devoted to health promotion and to the popularization of psychology are another source of information about various therapies. When we get online, we are dazed by the abundance of all kinds of therapeutic bargains. What has recently become particularly popular is counseling via the Internet, without personal contact with the therapist. Actually, some of these consultations are done by chatterbots, a simple software that emulates conversations with a real human being.

An average patient-to-be, if they are actually able to analyze their problems rationally, can ask themselves a number of questions. How do I choose the right therapy? Should I listen to temptations, which sound like promises? Should I ask friends, even though my problem is rather embarrassing? Or should I be guided by more rational considerations, like the scientific foundations of the various therapeutic systems? Or, should I analyze the context in which information about a course of therapy appears? The stakes are very high indeed. At best, a bad choice may mean a waste of time and money, but it may also mean a more serious crisis, such as a nervous breakdown, or lead to more serious life problems. The worst-case scenario, but still likely, is suicide or death as a result of psychotherapy.[10] What should we do then, in order to make the right choice?

As we have written in the previous chapters, the decision is not easy because the ordeal of choosing means selecting the right therapy from hundreds of different modalities available. The list is by no means closed either, with new, "revelatory" and "revolutionary" therapies being added to it daily. We have called the phenomenon of creating new money-making trends in psychotherapy and in self-development psychobusiness. This is a rather mild expression, as some authors suggest that such practices should be simply regarded as criminal activity and that any therapy without empirical foundations should be called experimental.[11]

Psychobusiness emerges mainly on the periphery of science, usually taken over by pseudoscience. History is full the most bizarre pseudoscientific constructs resembling intricate sand castles built by children on the beach. Some of those ideas have become nothing but historic curiosities or have even been consigned to the dust bin of science. They include Edmond Halley's concept of hollow earth, Wilhelm Reich's orgonomy, the theory of heterogenesis, or the ideas of the brilliant (in his own opinion) Alfred W. Lawson, such as the zig-zag-and-swirl concept that was supposed to replace entire physics.[12] Other concepts have been functioning on the borders of science for decades or even centuries. This group includes psychoanalysis, astrology, parapsychology or even Lamarckism, which still gets revived from time to time.

The commercial success and longevity of a pseudoscientific concept depends to a large extent on its ability to imitate a scientific system, usually based on emulating the structures functioning in the academia, or sometimes by simply mimicking scientific terminology. Today, the dominant strand in pseudo-psychology is called neurolinguistic programming (NLP). Many companies selling these services call themselves "institutes". Unfortunately, under the fancy name of "international institute" we often find sole traders. A simple Google search for "institute" and "NLP" shows how common this practice is. Below we have listed institutes found on just the first two pages of search results:

- International Association of NLP Institutes;
- NLP Training and Coaching Institute;
- The NLP Training Institute of Chicago;
- NLP and Coaching Institute;
- Training Institute for NLP;
- International Institute for NLP;
- The Performance Institute of NLP;
- NLP Institute Arizona;
- Irish Institute of NLP;
- International Institute for NLP & MORE

These "institutes" offer seminars and lectures, organize conferences, congresses and symposia. Their members create international associations and publish their own periodicals like the *Journal of NLP*. Some of them use academic titles they have never obtained (for example, Richard Bandler, who told people to refer to him as "doctor"). Others just buy their degrees for a few hundred dollars from institutions like Belford University. The procedure is very simple.

> Diane Cerulli's colleagues have always sought her out for health advice, and as a certified medical assistant, she did a "pretty good job" of diagnosing their aches and pains.
> But she couldn't see the scam coming -- an online offer to get a medical degree by taking a test based on life experience for a mere $1,400.
> "I always wanted to be a doctor, and I thought this was a dream come true," said the 59-year-old from Matawan, N.J.
> After taking the multiple-choice test, Cerulli received a letter from Belford University, one of many online schools that purport to be accredited. It read: "You are now a doctor. Diplomas and paperwork will be mailed to you after you pay $1,400 for the degree."
> Cerulli said, "Foolishly, I did that. I was told I could see patients and prescribe medicine." ...
> Online-based classes have become increasingly popular for students of all ages.
> According to a 2008 survey from the Sloan Consortium and Babson Survey Research Group, 3.9 million students were enrolled in at least one online course in 2007, a 12 percent increase over the previous year.
> Many of these online institutions are reputable, but many are being described as "diploma mills" that dupe those looking to advance their education. ...
> Belford's Web site touts: "Get a degree for what you already know!" It also offers "easy access" to degrees in all fields for "as low as" $249!"[13]

Fortunately, that institution was eventually brought to justice in 2011 and after the court ruling in 2012☐ has discontinued its operations. However, new entities keep springing up in its place. The "List of Scam Schools - Beware the Scams", published by the *Consumer Fraud Reporting*, contains dozens of fake universities just from the English-speaking countries alone. Plenty of similar and still unidentified institutions keep emerging in Europe, Asia and other parts of the globe. One of such examples is the Jupiter European Academy of Integrative Psychology, which must be given the credit for being an absolute leader when it comes to imitation capabilities and the spread of pseudoscience in Poland. Stories about this school read like a science fiction novel:

> The head [of the Academy] is Ramaz Zacharian, a researcher from the Moscow Institute of Aviation, whose research interests also include integration psychology "understood as a synchronization of classical methods of psychology and psychotherapy with metaphysical knowledge". This is followed by a list of Russian scientists with credible sounding recommendations, mixed with specialists in various arcane disciplines (e.g. detection of barely visible objects).
> Polish lecturers at Jupiter include the famous psychotronic man, founder of the church of Svetovid, and [an] expert in paranormal phenomena Lech Emfazy Stefański, (who "also develops his paraphysical capabilities like bending metals"); Leszek Matela, a world-renowned geomancer, who also deals with dowsing, suggestopedia and other phenomena not explained by science; Marek Turski, who studies physical evidence of the presence of the UFO on Earth (author of the hypothesis according to which the interior of the Earth is inhabited by highly developed humanoids dealing with complex genetic engineering, the products of which included Egyptian gods, centaurs, sirens and the Minotaur of Crete); and Helena Wilda-Kowalska, an eminent expert in numerology, who explains the esoteric meaning of letters and numbers contained in them.[14]

And yet it would be very difficult to accuse these organizations of using terminology until recently used only by legitimate scientific institutions. After all, there has never been any official copyright on these names. That is why any organization can easily take advantage of the trust we often put in scientists and academia. Creating a scientific background is a simple way to lend additional credence to any made up concepts. As the pseudoscientific entity expands and becomes firmly established, its supporters claim 'social proof,' by simply stating that so many people, institutions and associations cannot possibly be all wrong.

New pseudoscientific concepts spring up like mushrooms after a warm summer rain. In the time between this book going to press and appearing in the bookstores, probably several new concepts will emerge that no one has ever dreamed of. A good recent example is so-called quantum psychology,[15] which is supposed to be a synthesis of psychology and quantum physics, and which has experienced quite a level of success in recent years. In fact, it is nothing but the discredited NLP, but sold in new packaging. As this book was being written, quantum psychology caused a scandal in Poland, which may be regarded as an indication how common this pseudoscience has become in our country. One of our Polish universities published a book entitled *Consciousness and Its Unknown Face in the Light of Quantum Psychology*,[16] and soon an application was submitted for *habilitation*, a post-doctoral degree, to be conferred. The applicant, Adam Adamski, in his study of quantum psychology invokes Akimov's torsion field theory. This is what he writes about his major discipline: "Quantum psychology is able to accept the thesis that consciousness in its process of operation may behave as a wave or a particle, and it can also include linear and nonlinear processes."[17]

When presenting his contribution to the development of science, he stated that, "Drawing on years of research, I have demonstrated that house, heavy metal and techno music has an adverse effect on the development of the reproductive system in girls, the menstrual cycle of biological rhythms, and causes infertility, problems with carrying a pregnancy to term, somatic diseases and a tendency to developmental dysfunctions.[18]

Fortunately, the qualification commission was not persuaded and did not grant the degree in question, but after analyzing the candidate's "accomplishments," it was found that he had been teaching students and pupils the whole time, and had even supervised several doctoral students. In addition, he had promoted his achievements in schools and therapeutic centers, teaching infertility prevention methods. As if that were not enough, he had also worked as an expert witness for the courts. It is true that one of the reviewers demanded that the ministry send inspectors to the institution that employed Adamski, but as we are writing these words, he is still teaching young psychologists and continues to promote his "accomplishments."

The pseudosciences love using the prefix "neuro" these days. Thus, we have witnessed the resuscitation of buried theories, now referred to as neuro-psychoanalysis, neurotechnology, (which is nothing more than the utilization of biofeedback), neurokinesiology, neuro-re-education and the like. Why "neuro"? The neuron – the basic nerve cell, which constitutes the building block of the nervous system – has recently become a sort of a symbol, like the DNA helix before. The extraordinarily dynamic development of the brain and nervous system research we have seen since the introduction of ground-breaking methods using neuroimaging has led to a situation in which neuro-physiology, alongside genetics, has become a science that brings new discoveries again and again. Journalists do not hesitate to raise hopes associated with these discoveries and have no qualms either about creating catastrophic visions of the applications of these sciences, with various charlatans eager to take advantage of all this splendor.

Do patients looking for help have a chance of finding a genuine answer to the questions that bother them in the academic world? This is highly unlikely, because the odds of various pseudo-therapies surviving and being added to respectable lists are determined by the tacit or active approval of academic society. Active criticism on the part of the representatives of science, or at least their interest in pseudoscientific concepts, is rare these days. When a therapy receives such tacit support and enters into circulation, it is only a matter of time before it will seduce hope-seeking patients. The indifference of representatives of scientific psychology with regard to psychobusiness creates the perfect conditions for it to flourish—either at universities or under the aegis of science. How does pseudoscience force its way into academia? Let us carefully analyze one such case, which is rather symptomatic of our times.

---

[1] C. Rogers, A note on cover of paperback edition of Hart, J., Corriere, R. & and Binder, J. *Going Sane.* (New York: Aronson, 1975).

[2] C. L. Mithers, "When Therapists Drive their Patients Crazy," *California,* (August, 1988): http://web.archive.org/web/20090413103617/http://templeofdreams.com/center1.html

[3] P. Morantz, "Escape from the Center for Feeling Therapy (The Cult of Cruelty)" (October, 2010): http://www.paulmorantz.com/cult/escape-from-the-center-for-feeling-therapythe-cult-of-cruelty/

[4] Ibid.

[5] Ibid.

[6] Mithers, "When Therapists Drive"

[7] World Health Organization, "Depression. Fact sheet N°369," (October 2012): http://www.who.int/mediacentre/factsheets/fs369/en/

[8] L. Andrade and A. Caraveo, "Epidemiology of major depressive episodes: Results from the International Consortium of Psychiatric Epidemiology (ICPE) Surveys," *Internation Journal of Methods in Psychiatric Research, 12* (2003): 3–21.

[9] "Cross-national comparisons of the prevalences and correlates of mental disorders. WHO International Consortium in Psychiatric Epidemiology," *Bulletin of the World Health Organization, 78* (2000): 413–26.

[10] See P. Lowe, *"Plea Bargain in "Rebirthing" Death,"* Denver Rocky Mountain News, 3B. P. (2001).

[11] O'Donohue, & Ferguson, "Evidence-based Practice"

[12] An interesting overview of these concepts can be found in M. Gardner, *Fads and Fallacies in the Name of Science,* (Mineola, NJ: Dover, 1957).

[13] S. D. James, "Online Degrees: Schools Scam Aspiring Students," *ABC News,* (August 17, 2009): http://abcnews.go.com/Business/PersonalFinance/online-schools-scam-students-fake-degrees/story?id=8322412&singlePage=true

[14] J. Podgórska, "Lewitujacy uniwersytet," *Polityka, 4,* (January 29, 2005): 28-30.

[15] R. A. Wilson, *Quantum Psychology: How Brain Software Programs You and Your World,* (Tempe, AZ: New Falcon, 1990).

[16] A. Adamski and J. Sławiński *Consciousness and its Unknown Face in the Light of Quantum Psychology,* (Oświęcim: Napoleon V, 2011).

[17] Ibid., 30.

[18] A. Adamski, *Autoreferat.* Bielsko-Biala: Unpublished manuscript, (2012), 10. http://www.ck.gov.pl/images/PDF/Awanse/AdamskiAdam/autoreferat.pdf

THE ART OF HAMMERING NAILS OR THE THEORY AND PRACTICE OF NLP IN THE EYES OF A SOCIAL PSYCHOLOGIST

And those things do best please me
That befall preposterously.
—Shakespeare

Have you ever happened, dear reader, to hammer nails? Did you find it difficult? How many nails did you bend before you properly drove your first one into a solid piece of wood? Did you use any theory to do that? No? Are you sure? Let us then show you that without it, you could never had accomplished your goal.

Let us recall the first time we were driving in nails. The face of the hammer hits the nail slightly slantwise and the whole of it or its head slightly bends. Instead of driving it deeper into the wood, the next blow, a more precise one, bends it even more. Now it makes no sense to go on hammering the nail. You pluck it out then, and you strike every next one with less force but more accuracy. You drive it in with more numerous, but lighter and more precise blows. You know that the hammer must not hit the nail head at an angle. If you continue to learn the difficult art of driving nails, you will certainly come across an exceptionally tough hardwood. Suddenly, it turns out that you have to start learning the skill you thought you had already mastered almost from scratch again. But now you know that you have to hit the head of the nail even more precisely, and that the nail must not be too thin, especially as compared to its length. You are also gaining knowledge about hammers. A too light one is useless, a too heavy one tends to bend nails, and so forth.

There you are! In this way you became the discoverer and owner of a theory; a theory to the effect that the angle between the vector of the hammering force and the vector of the nail's resistance force must equal 180 degrees. The force of the blow must be inversely proportional to the length of the shank that has not been driven into the wood yet. A clever physicist would no doubt come up with a neat theoretical description of your experiences. Actually, you must have done the same because otherwise none of the nails you were hammering would have gone into the wood up to the head. Most probably, you did not use the term "vector," did not describe angles, resistances, vector sums, and so forth. You think you achieved it intuitively.

You could not be further from the truth. You know full well how you did it, because if it were necessary to teach this skill to someone, you would surely come up with a lot of hints that would help him or her to reach your level of proficiency much faster than by trial and error.

Social scientists know very well that most of our activities often have some sort of underlying theoretical model. That is why they study people in their natural environments and try to discover what makes a successful salesman, or how a politician influences others to gain support. While fathoming the secrets of influencing others, researchers often use the method suggested by the recognized social psychologist Robert Cialdini, called "full-cycle social psychology."[1] According to its tenets, the researcher first observes people in their natural environment. He or she looks at the way they try to influence others, make their audience change their views or convince their clients to buy some goods, and so forth. Especially suitable material for observation is provided by those professional groups that exert influence upon others by dint of their work: salesmen, politicians, insurance agents, authors of advertisements and promotional campaigns, peddlers, fundraisers, negotiators, and lawyers. The first stage of the "full cycle" procedure yields a thorough description of the observed methods of exerting influence. The second step is an empirical verification of the methods' efficiency. At this stage, the researchers check whether their observations and conjectures are corroborated in experimental scenarios they invent. The experiments are often carried out in natural conditions, even on the street. The most interesting is the third stage, when the researchers attempt to identify a psychological mechanism that accounts for the efficacy of a given method of exerting influence. They argue and discuss with one another in search of the mechanism's most precise description. This stage is very important since without a correct description of the mechanism, it is not possible to return to everyday life and thus reach the last phase of the "full cycle." At this point, the researchers try to influence people in a controlled way and manipulate them without their knowledge, thus verifying how accurate their description of the mechanism was. This stage concluded, the technique of exerting influence becomes a ready-made, rational and precise way of impacting other people. Here ends the role of a psychologist-researcher. In just the same way, a theory might be developed of hammering nails or of any other practical human activity. What does all that have to do with the concept of neuro-linguistic programming (NLP)?

In the 1970s, Richard W. Bandler and John Grinder came up with a brilliant idea to create a practical therapy model. They established that outstanding psychotherapists acted on the basis of implicit theories which ensured their effectiveness and great rapport with patients. Furthermore, they concluded that observation of their most skillful contemporaries at work should result in the discovery of patterns, which could be then generalized, verified on an empirical basis and put into therapeutic practice. For several years, they observed such therapists as Fritz Perls, Milton H. Erickson and Virginia Satir at their work. The gathered material that enabled them to formulate the tenets and hypotheses of NLP.

The central philosophy of NLP is summed up in the sentence "The Map is not the Territory."[2] That means, each of us operates on the basis of our internal representation of the world (the map) and not the world itself (the territory). The maps that we create are mostly limited and distorted. The therapist's task is to understand and operate on the basis of the client's map of the territory.

The maps that people make of their world are represented by the five senses: visual, kinesthetic (tactical and visceral sensations), auditory (noises and sounds), olfactory (smell), and gustatory (taste). Each experience in the world is composed of information received through these said systems of senses, different in terms of quality, and which are termed representational systems by the original proponents of NLP.[3] As they suggested, each of us processes the majority of information using one primary representational system (PRS). Following the example of the most outstanding therapists, to work effectively with a patient, one should necessarily match the patient's PRS so as to be able to use their "map".

Another discovery that the NLP originators were particularly proud of was the idea that access to representational systems is possible through so-called accessing cues: precisely specified eye movements. Careful observation of these movements should enable the NLP therapist to unequivocally identify the PRS of the patient, engage the patient by way of it, and in consequence, facilitate matching their PRS. All other hypotheses of the NLP system relates to assumptions about how mental disorders arise; the type of therapy, communication and so forth stem from these assumptions.

When analyzing how the NLP concept was formulated, it is worth making analogies between the manner in which it developed and the research methodology called the full-cycle approach in social psychology. Bandler and Grinder followed the full-cycle method, but regrettably, they somehow omitted the stage of empirically verifying their claims. They found that verification of their hypotheses was inessential and instead moved straight to the creation of an entire system, which they then put into practice. Bandler, known for his open contempt of scientific testing of NLP hypotheses, claimed that his system represented an art, not a science; hence testing its assertions was pointless or even impossible. NLP founders distorted the full-cycle approach, creating a quasi-process which included only three stages. Despite the contempt expressed by Bandler, due to its widespread use, many researchers tested its theoretical underpinnings based on an empirical approach.

In order to obtain a coherent empirical image of NLP that was free of the emotions of patients undergoing therapy, independent of the beliefs of therapists, and stripped from the superficial opinions of unknown academic psychologists, one of us[4] conducted a systematic analysis of scientific articles devoted to NLP.[5] A most extensive register of such studies (Neuro-Linguistic Programming Research Data Base) can be found on the web pages of the NLP Community.[6] At the time of the analysis, it contained abstracts and bibliographic information, with reference to 315 articles, written by 287 authors and published between 1974 and 2009. The database was created at the University of Bielefeld in Germany in 1992 and moved to Berlin in later years. It was designed to gather and organize available empirical studies on NLP from all over the world. Its creators refer to the database as a "state-of-the-art resource." It is regularly updated and recommended on an ongoing basis by numerous institutions worldwide. In Poland, this database is recommended by The Polish Institute for NLP, whose founder and chairman, Benedykt Peczko, personally suggested that it is the "all-embracing," global source of scientific studies on NLP. Out of several bases of articles developed by NLP's proponents, this one offers the highest number of entries.

There were three major arguments in favor of my choice of this particular base. First, we came to the conclusion that the 18 years of work performed by people on the base who are committed to showing the empirical underpinnings of the concept must give better results than those we could have achieved if searching through other available bases in a short time, such as PsychLit, PsychInfo or MEDLINE. Second, the fact of using the base established by followers of the concept meets their possible accusations that we were biased and partial in preparing this review. Thirdly, analysis of the base contents, of the manner in which it is updated and of a selection of articles, might disclose additional information on how the image of NLP as a science with empirical foundations is created.

In order to obtain the very essence of the empirical material available in the database, I performed a number of operations on the base. The first was to select the most reliable studies for further analysis. To this end, I evaluated them based on the criterion of whether the magazine in which the given articles were published is noted on the Master Journal List of the Institute for Scientific Information in Philadelphia. Although there are many doubts raised about this list, magazines from the Master Journal List are much less likely to have published unreliable articles than others. As a result of the initial selection, of the 315 articles I had 63, accounting for 20% of the entire base, left for further analysis.

Of interest are the findings of quantitative analysis of publications in individual years. Scientific activity peaked in the 1980s. It experienced a minor renaissance at the beginning of the present century. NLP as a research issue enjoyed immense popularity in the period directly following the formulation of its empirical underpinnings in the 1970s. In the subsequent years, the research interest in NLP decreased. The activity of researchers having their studies published in renowned magazines was proportional to the entire sample.

The sample of 63 studies selected for further analysis included articles published in 30 different magazines. This high number of magazines may be treated as indirect verification of the reliability of the gathered empirical evidence. It will be difficult to claim that one of the magazines (or a group of them) was biased in favor of NLP or that their activities were aimed at deprecating the concept. It should also be emphasized that the scope of the magazines was very wide indeed.

I put the selected sample of articles through a qualitative analysis, as a result of which, three categories of studies emerged:

1. Thirty-three empirical articles were found that tested the tenets of the concept and/or the tenet-derived hypotheses.
2. Fourteen articles comprising polemics, discussions, case analyses, or empirical works in which NLP represented little significant aspect and so were of no empirical worth from the point of view of my analysis.
3. Sixteen works having nothing in common with the NLP concept were available in the base, most likely by chance or due to other reasons that were unknown to me.

The first category is a subject of a more detailed quantitative and qualitative analysis presented in the subsequent part of this chapter. The second category comprises such studies as, for instance, a phenomenological account of the first author of horse riding lessons,[7] a discussion of the application of lateral concepts in the new millennium,[8] an analysis of a new therapy carried out by Virginia Satir,[9] an analysis of 2 cases of rape victims,[10] an analysis of a case of recovering from clinical depression[11] and many others. Accepting the analyzed sample as 100%, the articles on NLP that proved to be useless for the purpose of empirical analysis, represented 22.2%. I found the third category puzzling. It encompassed for example an essay on changes in Soviet psychology on the path of perestroika,[12] the social and ethical limitations of psychology,[13] intuition,[14] mimicry,[15] the status of the pharmacists,[16] the application of alternative therapies by children with dyslexia[17] and many others. Articles not related to the NLP concept account for as much as 25.4%. What is interesting, articles from this category represented an insignificant share in the 1980s and increased gradually towards contemporary times.

Empirical studies published in magazines of the Master Journal List of the Institute for Scientific Information in Philadelphia constitute the essence of the empirical material with reference to NLP. There were as many as 33 such papers, representing 52.4% of the selected sample. Qualitative analysis allowed me to single out the following subcategories:

1. Nine works supporting NLP tenets and tenets-derived hypotheses (27.3%).
2. Eighteen works non-supportive of NLP tenets and tenets-derived hypotheses (54.5%).
3. Six works with uncertain outcomes (18.2%).

The numbers indicate unequivocally that the NLP concept has not been developed on solid empirical foundations. Less than one-third of the analyzed works shows supportive evidence, more than a half is non-supportive, and the remaining papers uncertain results and doubts. Let's move beyond the mere numbers. *Argumenta ponderantur, non numerantur* - the force of the arguments lies in their weight, not numbers. It is often the case that one empirical study weighs as much as a number of others. Some works verify basic assumptions of a theory, others only a less significant aspect of the problem.

The studies reporting outcomes, which I rated as supportive of the concept, tested its basic assumptions in a very small number. In this respect, the study by Kinsbourne[18] is exceptional as it tested the hypotheses concerning eye movements, as well as Yapko's experiment[19] in which he revealed that matching the primary representational system had a positive influence on the depth of hypnotic relaxation as compared to the control group. Dooley and Farmer's study[20] may also be possibly classified into this category.

A high number of the remaining papers in this category lacked a control group. Usually only the initial and final measurements of the same group of subjects were taken. This was the case with the research of Duncan, Konefal and Spechler.[21] These authors provided a 21-day residential training to a group of subjects and then compared the pre-training and post-training status based on self-actualization measures of the Personal Orientation Inventory. Two years later, Konefal, Duncan and Reese[22] carried out almost identical research. The 21-day training was also provided, but what the authors measured this time were changes in trait anxiety and in the locus of control. In subsequent research performed under a similar procedure and without a control group, Konefal and Duncan[23] measured changes in social anxiety. Studies on the application of NLP by employees of southern Indian companies were likewise conducted without any control group,[24] and so were studies on the application of NLP for treating post-traumatic stress disorder.[25] The latter work from this category showed positive effects of neuro-linguistic mirroring in cross-cultural counseling.[26]

It is most likely that any type of intensive 21-day effort undertaken on self-development, based on any concept, would result in similar changes as those measured in the quoted research. The placebo effect is relatively frequent both in therapy as well as in other forms of exerting effect on people. Similarly, it is difficult to state whether the positive effects of mirroring resulted from the application of a specific NLP technique or from the necessity to put more focus on observation of the interlocutor, which in turn, was positively evaluated by them.

With respect to the category of non-supportive articles, the majority of studies concerned the basic NLP tenets. Several works were devoted only to tests of the eye movement hypothesis.[27] They all provided unequivocally negative results. The preferred modality was researched by Gumm, Walker and Day,[28] and also by Coe and Scharcoff.[29] In both cases, the results did not support the neuro-linguistic programming theory.

Other studies tested NLP tenets in a more complex manner, investigating several hypotheses in one study. Fromme and Daniell[30] researched the imagery and sensory mode, as well as communication. They were unable to find any support for the NLP-derived hypothesis that subjects showing differential ability across sensory modes would choose word phrases reflecting their preferred sensory mode. No support was found for the NLP-derived hypothesis that subjects matched for visualization ability would communicate information more accurately than would mismatch subjects. Elich, Thompson and Miller[31] tested claims that eye movement direction and spoken predicates are indicative of the sensory modality of imagery. Again, these tests did not find any support for NLP-derived hypotheses. Graunke and Roberts[32] tested the impact of imagery tasks on sensory predicate usage. The findings proved to be incongruent with R. Bandler and J. Grinder's conceptualization of representational systems.

Particular attention should be given to two reviews of research. In the first, the author carried out a very thorough analysis of 15 other research studies.[33] What is interesting, as many as 11 of these works are not available in the database in question. A few conclusions taken from that review are worth quoting:

> The identification of this PRS (if it is a PRS and not merely current language style) by either eye movements or self-report is not supported by the research data. ... The existence or stability of the PRS is irrelevant to predicate matching as a counseling process, and parsimony argues for the process rather than the yet unverified theory. ... Of most importance, there are no data reported to date to show that NLP can help clients change.[34]

The second review is even more conclusive.[35] It was written as a response to a critical paper by Einspruch and Forman. The authors of the review analyzed 39 studies devoted to NLP which indicated methodological errors and a lack of sufficient knowledge about the theoretical underpinnings of NLP as demonstrated by authors of the studies.[36] Sharpley took into account works analyzed by Einspruch and Forman, expanded that sample with seven additional ones and performed an analysis similar to mine, reviewing 44 empirical studies (of which two are not included in the base either). It turned out that six papers (13.6%) provide evidence supportive of NLP-derived theses, 27 (61.4%) failed to lend support for NLP tenets, and 11 (25%) show only partial support. The author investigated all available works, starting from doctoral dissertations to those published in high scoring magazines. This is how he summed up his review:

> There are conclusive data from research on NLP, and the conclusion is that the principles and procedures suggested by NLP have failed to be supported by

those data. (p.105) Certainly, research data do not support the rather extreme claims that proponents of NLP have made as to the validity of its principles or the novelty of its procedures.[37]

The subsequent three research studies referred to the influence of counselors or therapists predicate matching based on the effectiveness of their actions and quality of rapport.[38] None provided support for NLP-derived predictions. Studies on the effectiveness of specific therapeutic techniques failed to provide data supportive of NLP tenets too. Krugman, Kirsch and Wickless tested Bandler and Grinder's claim for a single-session cure of anxiety.[39] They did not find any support either. Similarly, Matthews, Kirsch and Mosher verified the effectiveness of double hypnotic induction.[40] Comparison of the experimental group against the control group did not yield findings supporting the hypotheses. In addition, the application of pacing and metaphor to overcome client resistance did not prove the reliability of Bandler and Grinder's claims.[41] Additionally, NLP proved to be of little use as a method of enhancing human performance as considered by the US Army. "The conclusion was that little if any evidence exists either to support NLP's assumptions or to indicate that it is effective as a strategy for social influence."[42]

The third category comprised six studies with uncertain results. Mercier and Johnson managed to obtain limited support for NLP theory in their research, with much data contrary to the theory.[43] The same was the case with research by Hammer on matching perceptual predicates.[44] The findings created more doubts than conclusive data as to perceptual predicates. In terms of studying eye movement as an indicator of sensory components in thought, Buckner and Mera found support for the visual and auditory portions of the model, but the kinesthetic portion was not supported.[45] Wertheim, Habib and Cumming found similar partial support for the hypothesis that eye movements relate to processing imagery in the research.[46]

Other research was carried out based on the assumption that the NLP tenets were true and tested. The examination by Durand, Wetzel and Hansen may serve as an example here with the researchers analyzing the content of written statements, telephone communications and electronic mail messages in terms of the occurrence of sensory predicate by means of computer software.[47] Similar procedures were followed in other research.[48]

Among there were studies classified as supportive of NLP tenets, there were none which would indicate in unequivocal terms the existence of different representational systems. Similarly, there is no support found for the idea that subjects were using primarily one predominant representational system in different life situations. Apart from one study,[49] there is no strong evidence that matching the primary representational system brings beneficial effects in communication and therapy. Two studies supporting some claims on eye movements should be replicated several times in order to treat their outcomes as supportive of the hypotheses. Moreover, there are no more extensive and comprehensive research reviews. The only one, which might perhaps be classified into this category, constitutes a criticism of the existing empirical papers and it does not provide any data to support NLP tenets.[50] The analyzed works show numerous methodological errors and shortcomings, such as the lack of a control group, only one research hypothesis being tested or one factor measured.

A much higher methodological level characterizes the studies classified into the non-supportive category. The majority allows for comparison against a control group, provide measurement of a number of variables and use a higher number of indicators. Among the studies are two articles offering extensive and reliable reviews of research. Most results of research from this category were replicated.

Comparison of both categories, both in terms of quantity and quality, unequivocally indicates the predominance of articles that do not lend support for NLP tenets, with the ratio of non-supportive to supportive of 3:1. When evaluating the whole empirical research devoted to NLP, one should also consider the file drawer effect.[51] In view of that, NLP supportive studies should have a greater chance of publication than those showing lack of support. It may be easily assumed that some studies that did not find any support for NLP hypotheses were filed away in researchers' drawers.

A review of the articles issued in sources other than those from the Master Journal List indicate the existence of many review studies showing the lack of support for any NLP underlying principle. Two of them are worth mentioning. Heap analyzed 63 empirical studies and came to the conclusion that the assertions of NLP writers concerning representational systems have been objectively and fairly investigated and found to be lacking.[52] What follows, the hypothesis about the possibility to identify PRS through careful observation of eye movements, was not confirmed either. In his view, these conclusions, and the failure of investigators to convincingly demonstrate the alleged benefits of predicate matching, seriously question the role of such procedure in counseling. Dorn, Brunson, Bradford and Atwater also concluded from their review of the literature that there was no demonstrably reliable method of assessing the hypothesized PRS.[53]

While conducting my analysis of empirical studies, I noted a certain historical aspect of the issue related to performing research supportive of NLP. As I realized, most of the research was carried out in the 1980s and partially in the 1990s. In the subsequent years, the number of such studies has decreased and they were concerned with secondary aspects of the concept, or were performed based on the assumption that the fundamental principles of NLP are true. It looks as if the world of science was losing its interest in the concept put forth by Bandler and Grinder, having already confronted it with the research findings; further, its proponents lack the motivation to undertake any type of research into, for instance, the effectiveness of methods offered by the concept.

Another facet which is worth discussing and which emerged during my analysis, is the matter of investigating how the database is utilized by its administrators as well as by its users. The base is commonly invoked by NLP followers and indicated as evidence for the existence of the solid empirical grounds of their preferred concept. It is most likely that most of them have never looked through the base. Otherwise, they might have come to the conclusion that it provides evidence to the contrary – the lack of any empirical underpinnings. Moreover, they not only fail to browse through the database, dare I say, but they also do not read articles available therein. The reading of two review papers[54] would enable them to first discover that the base lacks as many as 13 entries, and second, to update it appropriately. Luckily, the missing 12 entries are not included on the Master Journal List.

The number of theoretical studies in the base, such as polemics, dissertations and discussions, is so high that referring to it as to the Research Data Base is a considerable misinterpretation as well. What is even stranger is the fact that works completely unrelated to NLP are added to the base. While reading such articles, we strengthened our belief that it was only due to single key words that the NLP-related status of those papers was approved. This gives rise to the suspicion that even the database administrators do not read the articles, not to mention the abstracts.

All this leaves us with an overwhelming impression that the analyzed base of scientific articles is treated just as theater decoration, being the background for the pseudoscientific farce which NLP apparently is. Using "scientific" attributes, which is so characteristic of pseudo-science, is manifested in other aspects of NLP activities also. It is primarily revealed in the language – full of borrowings from science or expressions referring to it, devoid of any meaning whatsoever. It is seen already in the very name—neuro-linguistic programming—which is a cruel deception. At the neural level, it provides no explanation at all and it has nothing in common with academic linguistics or programming. Similarly, impressive sounding but empty expressions are used to formulate the tenets of the concept, for example sub-modalities, pragmagraphics, surface structure, deep structure, accessing cue, non-accessing movement and so forth.

Since my way of analyzing the research studies raised a number of methodological objections, one year later, in November 2010, I took it upon myself to reanalyze them, following a clearly specified methodological path with a view to obtaining a systematic review of research literature.[55] As more than 20 years have passed since the last systematic and comprehensive research reviews were published, I wondered whether the NLP's scientific status has changed since it was last examined. Are there any new research findings in this area? Are the outcomes of previous studies supportive enough to ultimately allow NLP claims as empirically verified, or should we rather pass the final verdict?

To answer this question, I conducted a search of the PsycINFO database to determine if there were any additional studies not included in the aforementioned NLP database. As it turned out, 240 publications were recorded in both bases and 80 articles available in the NLP base were not found in PsycINFO. Furthermore, PsycINFO included 81 publications which were not available in the NLP base. In consequence, the basis for the present analysis has been compiled from all publications from both databases, that is, 401 articles in total.

What was the final result of this analysis? Of this entire sample, 21 empirical studies of the highest methodological value, published after 1986, have been singled out. Out of 22 empirical studies, 9.5% are supportive of the tenets of NLP and the effectiveness of NLP techniques, 19% prove partially supportive, and 71.5% non-supportive. Analysis of these results contradict the claim of the empirical bases of NLP.

Both presented analyses lead undeniably to the statement that NLP represents pseudoscientific rubbish, which should be forever buried deep underground. One may even come to believe that this analysis was a vain effort after all. It yielded the same conclusions as the ones arrived at by Sharpley[56], Heap[57] and others. Without doubt, NLP represents big business, tempting people with amazing changes, personal development and, what is most concerning, therapy. In this respect, our analysis has updated the previous ones with the time elapsed since they were conducted, and provides arguments sufficient to answer an ethical question. Is it even ethical to use and to sell something that is not effective, that does not even exist?

The response will surely be similar to the statement given once by Einspruch and Forman: "the effectiveness of NLP therapy, undertaken in authentic clinical contexts of trained practitioners, has not yet been properly investigated."[58] Additionally, we will certainly be told that NLP works and this should be good enough reason to use it. Proponents of the concept are more than eager to throw such arguments around as if forgetting that the burden of proof with respect to finding evidence of the effectiveness of NLP therapy lies on them. Here we would like to refer to the statement expressed by O'Donohue and Ferguson, who propose that each type of therapy that does not have supportive empirical evidence of its effectiveness should be called experimental.[59] They also put forward a suggestion that each case of performing such therapies without informing the clients about its experimental status and thus earning money should be referred to and treated as criminal activity. We completely agree with this view. Would you like it if your doctor was prescribing you an experimental, unverified therapy of unknown and unconfirmed outcome without mentioning this fact to you? Luckily, in most modern societies this would never happen in a medical doctor's office. Similarly, we cannot even imagine that a pharmaceutical company would be marketing medicines of uncertain or unknown effectiveness and side effects. Yet we allow any psychotherapy to be practiced, in many cases without any relevant research. Why is it allowed with psychotherapy?

If NLP's assertions on the existence of PRS and the possibility to enhance communication just through matching proved to be true, it would revolutionize neuroscience, cognitive psychology and some other disciplines. If NLP's claims on the duration and effectiveness of the proposed therapies proved to be true, the entire area of psychotherapy would experience a sudden shock and research reports with respect to the effectiveness of therapy would have positioned NLP therapy at the top. Regrettably, nothing like this has or is taking place. Instead, we find NLP on the list of discredited therapies. Norcross, Koocher and Garofalo (2006) sought to establish consensus on discredited psychological treatments and assessments using the Delphi methodology.[60] A panel of 101 experts participated in a 2-stage survey reporting familiarity with 59 treatments and 30 assessment techniques, rating them on a continuum from *not at all discredited* (1) to *certainly discredited* (5). Neuro-linguistic Programming for treatment of mental/behavioral disorders was averagely assessed 3.87 (SD=0.92).

The analysis of the Neuro-Linguistic Programming Research Data Base (state of the art) by all measures was like peeling an onion. To reach its core, first one has remove some useless layers, and once you reach the bottom, you will be crying. Today, after 35 years of research devoted to this made-up concept, NLP is just another unstable house built on the sand, rather than an edifice founded on the empirically based pillars. Over 20 years ago, back in 1988, Heap already passed a verdict on NLP. As the title of his article indicated, it was an interim one. In the conclusion he wrote:

> If it turns out to be the case that these therapeutic procedures are indeed as rapid and powerful as is claimed, no one will rejoice more than the present author. If however these claims fare no better than the ones already investigated, then the final verdict on NLP will be a harsh one indeed.[61]

We are fully convinced that we have gathered enough evidence to finally announce this harsh verdict now.

Do NLP proponents offer anything in support of their hypotheses? Yes, but this seems very unimpressive against empirical data. First of all, they consider practice to be more important than theory. They contend that even though few of us know any theory of walking, we have little difficulty in moving about; therefore, it is not important to have a theoretical knowledge of anything as long as the effects are positive. The theory might as well be a burden. Thus, the scanty theoretical basis of NLP is rather its advantage. Its followers claim that the method is effective and that wonderful transformations happen during NLP workshops. Participants emerge purified, confident in their own strength, in people, and with hope for the future. Moreover, they are convinced that NLP therapy is much quicker than traditional methods. As few as several sessions may bring about effects comparable to years-long classical psychotherapy. In their view, classical therapeutic schools make patients focus on their problems and thus perpetuate them. By contrast, NLP strives to turn a person's attention to his or her goals and mental well-being. When someone puts a spoke in their wheel and claims, as research results show, that NLP is a pseudoscience, they... well, threaten to sue.

Such a situation took place when the Daily Motion published the film entitled *Characteristics of pseudoscience*[62] on its website. Since it included materials devoted to scientology and NLP as examples of pseudoscientific strategies, representatives of both movements joined forces and demanded the film to be removed, threatening to bring action for libel. Fortunately, the administrators of the website did not cave in and we can still watch the film, as well as read about the threats and related events in a discussion below it.

Despite all this, for more than twenty-five years, therapies, personal development training courses and other forms of working with people advertised as based within the Neuro-Linguistic Programming (NLP) framework enjoy enormous popularity on the market of psychological services. Unfortunately, NLP practitioners are found among university employees, and advertisements of NLP-related institutions appear in popular science magazines. Students of psychology attend courses where they attain successive degrees of initiation for NLP practitioners. NLP trainings have been provided by companies such as Hewlett-Packard, IBM, McDonald's, NASA, the US Army, and US Olympic teams, and in countless public school systems.[63] It has been suggested that NLP is "being applied widely, if often informally in UK education."[64] We investigated official psychology curricula of the 12 best state universities in Poland. Eight offered content in, and in many cases, even separate courses devoted to NLP. Still, despite that widespread presence of NLP, none of the psychology textbooks that we have heard of (regardless of where they were published) present an in-depth discussion of the concept. What's more, scientific authorities refrain from giving their opinions in this respect. Where does this great popularity and fad for NLP stem from?

So far we have been analyzing the phenomenon from the standpoint of a social psychologist, applying the yardstick of scientific method to it and comparing it with established scientific standards or ethical principles; now, however, when analyzing the phenomenon's popularity, as well as controversies and even pathologies associated with it, we will have to make use of slightly less solid evidence.

Why is NLP so popular despite overwhelming evidence against it? In psychology, like in some humanities, we witness passing trends. This pertains especially to the point where science meets practice. This was the case with psychoanalysis; a few decades ago, positive thinking methods were all the rage; until recently, all sorts of assertiveness workshops have enjoyed enormous popularity; now it is NLP that is very much in demand. In each of these cases, a "fashionable system," apart from its basic tenets which are often limited to narrow fragments of reality, begins to develop philosophical assumptions and borrow methods from other systems. It absorbs and assimilates things that work, that are well received and that guarantee commercial success. Thus, at the peak of their popularity, assertiveness training programs were in fact devoted to broadly understood communication issues. These courses frequently involved the teaching of transactional analysis, the four-side model, and similar elements derived from theoretical or therapeutic models which had been present in psychology for many years. Recently, a participant of an NLP course in picking up girls[65] told us what the "system" taught by the "masters" consists of. It turned out to be an attempt (a blundering one) to use classical conditioning, the credit for which undoubtedly goes to Ivan Pavlov. Most fashionable systems flourish with similar borrowings. NLP also borrowed extensively from other therapeutic systems.

NLP's attractiveness might owe something to its philosophical structure which makes its set of statements interesting and appealing to a wider public. What might seem especially attractive is placing opinions above objective truth. NLP assumes that, by definition, truth is what works in practice. NLP "votaries" believe that reality does not exist objectively and that the only thing that counts is what everyone thinks about it. This conviction may lead to rather serious consequences with regard to the understanding of moral norms and relations with others, more on that in a moment.

Additional elements that make NLP more attractive are its grandiosity, intuitiveness, connection to magic and trance-inducing practices. This attractiveness is also enhanced by the creation of an idiosyncratic set of terms, a move which is quite characteristic of institutions and organizations bearing the hallmarks of sects. Within the framework of NLP, common things such as human relationships are referred to as "dance," the senses are called "sensory modalities," and so forth. The definitions used in NLP are often nothing more than poor metaphors; they might even sound very wise. For example: "NLP is a study of efficiency and model the way."[66]

Opponents of NLP put forward a number of other critical arguments. One of the most serious one is the question of morality. Before we consider it, let us read a story...

> On the morning Corine Christensen last snorted cocaine, she found herself, straw in hand, looking down the barrel of a .357 Magnum revolver. When the gun exploded, momentarily piercing the autumn stillness, it sent a single bullet on a diagonal path through her left nostril and into her brain.
> Christensen slumped over her round oak dining table, bleeding onto [a] glass top, a loose-leaf notebook, and a slip of yellow memo paper on which she had scrawled in red ink, DON'T KILLL US ALL. Choking, she spit blood onto a wine goblet, a tequila bottle, and the shirt of the man who would be accused of her murder, then slid sideways off the chair and fell on her back. Within minutes, she lay still.

As Christensen lay dying, two men left her rented town house in a working-class section of Santa Cruz, California. One was former boyfriend, James Marino, an admitted cocaine dealer and convicted burglar. The other, Richard Bandler, was known internationally as the cofounder of Neuro-Linguistic Programming (NLP), a controversial approach to psychology and communication. About 12 hours later, on the evening of November 3, 1986, Richard Bandler was arrested and charged with the murder.[67]

What prior events led to that tragedy? How did the said men find themselves there? What connected a well-known therapist with a drug dealer? Bandler met James Marino shortly after his divorce in 1980 in a restaurant in Santa Cruz, which was officially known for its beautiful view of the port, and unofficially as the place to find drug dealers and stock up on another packet of cocaine. At the time, Bandler was taking large amounts of cocaine and abusing alcohol. He was increasingly losing control of his life. Bandler made friends with the 18 years older James Marino. Corine Christensen was not only addicted to cocaine, but also deeply in love with Marino, 22 years her senior.

In October 1986, Marino was attacked and badly injured by an unknown perpetrator. Christensen took care of him for some time. When Bandler heard about the incident, he was determined to find out how it had happened. To this end, he met with Christensen and tape-recorded their conversation.

No more than eight hours before her death, he gave an ultimatum: "I'll give you two more questions," he said, "and then I'll blow your brains out."
When James Marino contacted police through an attorney on the afternoon of November 3, 1986, Corine Christensen had been dead for about seven hours. Sheriff's deputies found her lying on the carpeted floor of her dining area, her face caked with blood, one leg draped on a wooden chair. Her home had been ransacked.[68]

From then on, the police, and later also the court, had two versions of this event. What made them fundamentally different from each other was the identity of the killer. It is not hard to guess that Marino maintained that Bandler was the murderer, while the other claimed the opposite. Both testimonies were full of ambiguities and not very coherent, while their details were constantly changed. Both testimonies were only partially confirmed by the evidence gathered by the detectives. Tried for murder, Bandler faced a life sentence.

On the afternoon of January 28, 1988, after five hours of deliberation, a seven-man, five-woman jury voted unanimously to acquit Richard Wayne Bandler of murder in the death of Corine Christensen. Confronted by a paradox – both stories were suspect, yet one man told the truth about the shooting – the jurors voted not guilty. The prosecution, they said later, had not proved beyond a reasonable doubt that Bandler shot her. Although the jurors took pride in upholding the law, more than one remained haunted by the decision. At least two cried after leaving the courthouse. Someone, they knew, had gotten away with murder.[69]

A murder charge against an author of a scientific concept should not rather affect the assessment of his research hypotheses or the value of his theoretical constructs. History has known more than one genius who could not cope with his life. However, when we have an originator of a therapy that enters patients' moral sphere, doubts about the value of the therapeutic system do become fundamental. In the case of NLP, doubts lead to certainty because this system provides a theoretical justification for morally reprehensible attitudes by claiming that you cannot make a mistake, since "there is no failure, only feedback." Hence, under NLP, no matter what you do, you do well. If you harm others, it is not a mistake because you are effective and we have all been enriched by a new experience.

The gun had appeared suddenly, magically, in the palm of Richard Bandler's hand. It was about three inches long, the color of tarnished brass. And Bandler was pointing it toward a psychiatrist who had volunteered for a demonstration.
On this Saturday morning in February 1984, Bandler wanted to illustrate a favorite theory – that anyone can change, with the right stimulus. The psychiatrist was adamant. Nothing could spur him to change a certain aspect of his life. Except, he joked, "perhaps a small-caliber pistol."
For a moment, a smile flickered across Bandler's face. He took the gun out of his pocket. The audience, all advanced NLP students, knew about his confrontational style, and laughed. But, the psychiatrist said, the gun would only work if he knew Bandler was willing to use it.
"I've got news for you," Bandler taunted. "You've got no idea how nuts I am. How many people have one? In their pocket? Waiting for you? And you're going to tell me that I won't do it?" He laughed. "I don't have to kill you, I just have to wound you," he added. "... I've done weird things to clients."
"I know you have," the psychiatrist responded, his voice soothing, compliant, now terrified. As Bandler toyed with the miniature gun, he boasted that he had once so thoroughly cured a man of acrophobia that the man had jumped off a bridge.
"You didn't know it was going to be real, did you?" Bandler mocked. "Now, somehow or other, *you* made it real. That's different than *me* making it real."
Finally the psychiatrist, shaken, surrendered. The change, he said, was possible.
"Absolutely," Bandler responded, his excitement ebbing. "Otherwise, I would have shot you by now."[70]

Not only does the above incident add gloom to the chronologically later happenings described earlier on, but it also helps to understand the careless attitude of NLP proponents toward the problem of manipulation. This is probably the first of the fashionable systems which has no qualms about teaching, propagating, and even encouraging the use of manipulation.

NLP proponents strive to redefine the notion of manipulation, extending its scope in an absurd manner. "Everything is manipulation," they affirm. ... "You cannot help exerting influence.... Every instance of communication, which is ubiquitous, involves exerting influence" (A. Batko). "Every message elicits a response and therefore constitutes manipulation" (G. Halkiew)....Thus, one of the basic principles of neurolinguistic programming, "you cannot not communicate," is overridden by the principle that "you cannot not manipulate" (such an evolution of NLP is observed in many countries). And since we cannot help manipulating, since we are manipulated and do manipulating ourselves, then let us – as NLP proponents argue – do it fully consciously and effectively; let us do it better than others.[71]

And they do. They teach how to permanently get rid of a sense of guilt in five minutes, how to successfully fall out of love and to create magnetic attraction in another person (also in a few minutes). Within NLP, there emerged NLS – neuro-linguistic seduction. This system teaches, among others,

"how to eliminate last-minute inhibitions about having sex" and "how to use subliminal suggestions which create in her mind images and emotions stirring her most secret sexual fantasies"; neuro-hypnotic seduction (NHS) is about how to "captivate her imagination and sexual fantasies," "make her love you to distraction," and "get rid of competitors"; the "Secrets of Seduction" course is a guarantee that "any emotional state in the partner (absolute fascination, desire or infatuation) will be possible to be induced in a few seconds."[72]

It seems that NLP disciples either have no moral qualms or have overcome them (with the help of NLP techniques to be sure).

Apart from this relativist attitude toward moral issues, therapists sometimes criticize NLP for the short-term nature of its effects. To their mind, the system is geared toward bringing about a change in patients' outlook on life and themselves. This might be a valuable outcome and sometimes even sufficient for a permanent change, but usually it is only superficial. A belief that we can solve all our problems does not mean that we can actually do it. This belief certainly makes things easier, but if our problems stem, for example, from our inadequate social skills, a mere belief without practicing these skills will avail us little.

Even if we wanted to examine, in an honest way, the effectiveness of NLP-based therapies, we would run into difficulties once again. On the one hand, there will be reluctance for any research as such, which could settle the question of effectiveness alone. On the other hand, we will face barriers analogous to those that we encounter when trying to take a closer look at the functioning of various cults or sects. The social organization of NLP bears many hallmarks of a sect: access to the closed inner circle is hampered by financial barriers. In order to pass the next stage of initiation, one has to pay a considerable sum of money. A "career" path within the NLP system starts with training courses, which confer the title of the Practitioner in the Art of NLP; sometimes it can be followed up with the certificate of Business Practitioner in the Art of NLP. On finishing the basic course, it is possible to enroll on the next one, Master Practitioner in the Art of NLP. Following on from this, one can pursue his or her career by obtaining the certificate of NLP Trainer or choose the advanced option of Certified NLP Business Trainer.

The certificates are issued only on completion of a paid course. In the world of NLP, trivial things such as academic degrees, scientific achievements or work experience do not matter. Interestingly, "masters" who teach NLP claim that people can become therapists simply by paying for a course, without examinations or psychological studies, not to mention the lack of any supervision of sessions conducted by the fledgling "therapists."

A cursory glance at areas where NLP is used shows that we are dealing with a well-organized and thriving business. It would be difficult to estimate its value, but we may find some clues by looking at Bandler's attempts to monopolize the business.

> It began in July of 1996 when Richard Bandler filed a $90,000,000 lawsuit as a civil action against John Grinder, Carmen Bostic St. Clair, Christina Hall, Steve and Connirae Andreas, and Lara Ewing and 200 John and Jane Does. In that lawsuit, Bandler claimed exclusive ownership of the Society of NLP.[73]

The lawsuit was settled in 2001 with the agreement that Bandler and Grinder would be recognized as co-founders of NLP, so now everybody has unlimited access to its "benefits."

---

[1] R. B. Cialdini, "Full-cycle Social Psychology," *Applied Social Psychology Annual, 1*, (1980): 21-47.

[2] S. Lankton, *Practical Magic*, (Cupertino, CA: Meta, 1980), 7.

[3] J. Grinder and R. Bandler, *The Structure of Magic II*, (Palo Alto, Ca: Science and Behavior Books, 1976); R. Bandler and J. Grinder, *Frogs into Princes*, (Moab, UT: Real People Press, 1979).

[4] All analysis described in this chapter were carried out by Tomasz Witkowski, therefore all narratives are first person singular.

[5] The results of this analysis was originally published as: T. Witkowski, "Thirty-five Years of Research on Neuro-linguistic Programming. NLP Research Data Base: State of the Art or Pseudoscientific Decoration?" *Polish Psychological Bulletin. 41*, (2011): 58-66.

[6] http://www.nlp.de/cgi-bin/research/nlp-rdb.cgi

[7] J. Mathison and P. Tosey, "Riding into Transformative Learning," *Journal of Consciousness Studies, 15*, (2008): 67-88.

[8] M. C. Corballis, "How Laterality Will Survive the Millennium Bug," *Brain and Cognition, 42*, (2000): 160-162.

[9] M. D. Woods and D. Martin, "The Work of Virginia Satir: Understanding Her Theory and Technique," *American Journal of Family Therapy, 12*, (1984): 3-11.

[10] P. Koziey and G. McLeod, "Visual-kinesthetic Dissociation in Treatment of Victims of Rape," *Professional Psychology: Research and Practice, 18,* (1984): 276-282.

[11] A. Hossack and K. Standidge, "Using an Imaginary Scrapbook for Neurolinguistic Programming in the Aftermath of a Clinical Depression: A Case History," *Gerontologist, 33,* (1993): 265-268.

[12] B. Gindis, "Soviet Psychology on the Path of Perestroika," *Professional Psychology: Research and Practice, 23,* (1992): 114-118.

[13] P. J. D. Drenth, "Prometheus Chained: Social and Ethical Constraints on Psychology," *European Psychologist, 4,* (1999): 233-239.

[14] M. D. Lieberman, "Intuition: A Social Cognitive Neuroscience Approach," *Psychological Bulletin, 126,* (2000): 109-137.

[15] J. L. Lakin, V. E. Jefferis, C. M. Cheng, and T. L. Chartrand, "The Chameleon Effect as Social Glue: Evidence for the Evolutionary Significance of Nonconscious Mimicry," *Journal of Nonverbal Behavior, 27,* (2003): 145-162; M. Stel, E. van Dijk, and E. Olivier, "You Want to Know the Truth? Then Don't Mimic!" *Psychological Science, 20,* (2009): 693-699.

[16] S. White, "The Status of the Pharmacist," *SA Pharmaceutical Journal, 76,* (2009): 54-55.

[17] L. Bull, "Survey of Complementary and Alternative Therapies Used By Children with Specific Learning Difficulties (dyslexia)," *International Journal of Language & Communication Disorders, 44,* (2009): 224-235.

[18] M. Kinsbourne, "Direction of Gaze and Distribution of Cerebral Thought Processes," *Neuropsychologia, 12,* (1974): 279-281.

[19] M. D. Yapko, "The Effect of Matching Primary Representational System Predicates on Hypnotic Relaxation," *American Journal of Clinical Hypnosis, 23,* (1981): 169-175.

[20] K. Dooley and A. Farmer, "Comparison for Aphasic and Control Subjects of Eye Movements Hypothesized in Neurolinguistic Programming (NLP)," *Perceptual and Motor Skills, 67,* (1988): 233-234.

[21] R. C. Duncan, J. Konefal, and M. M. Spechler, "Effect of Neurolinguistic Programming Training of Self-actualization as Measured by the Personal Orientation Inventory," *Psychological Reports, 66,* (1990): 1323-1330.

[22] J. Konefal, R. C. Duncan and M. Reese, "Neurolinguistic Programming Training, Trait Anxiety, and Locus of Control," *Psychological Reports, 70,* (1992): 819-832.

[23] J. Konefal and R. C. Duncan, "Social Anxiety and Training in Neurolinguistic Programming," *Psychological Reports, 83,* (1998): 1115-1122.

[24] A. Singh and A. Abraham, "Neuro Linguistic Programming: A Key to Business Excellence," *Total Quality Management & Business Excellence, 19,* (2008): 141-149.

[25] D. C. Muss, "A New Technique for Treating Post-traumatic Stress Disorder," *British Journal of Clinical Psychology, 30,* (1991): 91-92.

[26] D. S. Sandhu, T. G. Reeves and P. R. Portes, "Cross-cultural Counseling and Neurolinguistic Mirroring with Native American Adolescents," *Journal of Multicultural Counseling and Development, 21,* (1993): 106-118.

[27] T. C. Thomason, T. Arbuckle and D. Cady, "Test of the Eye Movement Hypothesis of Neurolinguistic Programming," *Perceptual and Motor Skills, 51,* (1980): 230; A. Farmer, R. Rooney and J. R. Cunningham, "Hypothesized Eye Movements of Neurolinguistic Programming: A Statistical Artifact," *Perceptual and Motor Skills, 61,* (1984):

717-718; S. A. Poffel and H. J. Cross, "Neurolinguistic Programming: A Test of the Eye Movement Hypothesis," *Perceptual and Motor Skills, 61,* (1985): 1262; D. T. Burke, A. Meleger, J. C. Schneider, J. Snyder, A. S. Dorvlo and S. Al-Adawi, "Eye-movements and Ongoing Task Processing," *Perceptual and Motor Skills, 96,* (2003): 1330-1338.

28 W. B. Gumm, M. K. Walker and H. D. Day, "Neurolinguistic Programming: Method or Myth?" *Journal of Counseling Psychology, 29* (1982).

29 W. C. Coe and J. A. Scharcoff, "An Empirical Evaluation of the Neurolinguistic Programming Model," *International Journal of Clinical and Experimental Hypnosis, 33,* (1985): 310-318.

30 D. K. Fromme and J. Daniell, "Neurolinguistic Programming Examined: Imagery, Sensory Mode, and Communication," *Journal of Counseling Psychology, 31,* (1984): 387–390.

31 M. Elich., R. W. Thompson and L. Miller, "Mental Imagery as Revealed by Eye Movements and Spoken Predicates: A Test of Neurolinguistic Programming," *Journal of Counseling Psychology, 32,* (1985): 622-625.

32 B. Graunke and T. K. Roberts, "Neurolinguistic Programming: The Impact of Imagery Tasks on Sensory Predicate Usage," *Journal of Counseling Psychology, 32,* (1985): 525-530.

33 C. F. Sharpley, "Predicate Matching in NLP: A Review of Research on the Preferred Representational System," *Journal of Counseling Psychology, 31,* (1984): 238-248.

34 Ibid, 247.

35 ———, "Research Findings on Neurolinguistic Programming: Nonsupportive Data or Untestable Theory?" *Journal of Counseling Psychology, 34,* (1987): 103-107.

36 E. L. Einspruch and B. D. Forman, "Observations Concerning Research Literature on Neurolinguistic Programming," *Journal of Counseling Psychology, 32,* (1985): 589-596.

37 Sharpley, "Research Findings," 106.

38 T. E. Dowd and J. Pety, "Effect of Counselor Predicate Matching on Perceived Social Influence and Client Satisfaction," *Journal of Counseling Psychology, 29,* (1982): 206-209; T. E. Dowd and A. G. Hingst, "Matching Therapists' Predicates: An In Vivo Test of Effectiveness," *Perceptual and Motor Skills, 57,* (1983): 207-210; J. L. Ellickson, "Representational Systems and Eye Movements in an Interview," *Journal of Counseling Psychology, 30,* (1983): 339-345.

39 M. Krugman, I. Kirsch and C. Wickless, "NLP Treatment for Anxiety: Magic or Myth?," *Journal of Consulting and Clinical Psychology, 53,* (1985): 526-530.

40 W. J. Matthews, I. Kirsch and D. Mosher, "Double Hypnotic Induction: An Initial Empirical Test," *Journal of Abnormal Psychology, 94,* (1985): 92-95.

41 P. N. Dixon, G. D. Parr, D. Yarbrough and M. Rathael, "Neurolinguistic Programming as a Persuasive Communication Technique," *Journal of Social Psychology, 126,* (1986): 545-550.

42 J. A. Swets and R. A. Bjork, "Enhancing Human Performance: An Evaluation of "New Age" Techniques Considered by the U.S. Army," *Psychological Science, 1,* 85-86, (1990): 90.

43 M. A. Mercier and M. Johnson, "Representational System Predicate Use and Convergence in Counseling: Gloria Revisited," *Journal of Counseling Psychology, 31,* (1984): 161-169.

[44] A. L. Hammer, "Matching Perceptual Predicates: Effect on Perceived Empathy in a Counseling Analogue," *Journal of Counseling Psychology, 30,* (1983): 172-179.

[45] M. Buckner and N. M. Mera, "Eye Movement as an Indicator of Sensory Components in Thought," *Journal of Counseling Psychology, 34,* (1987): 283-287.

[46] E. H. Wertheim, C. Habib and G. Cumming, "Test of the Neurolinguistic Programming Hypothesis That Eye Movements Relate to Processing Imagery," *Perceptual and Motor Skills, 62,* (1986): 523-529.

[47] D. Durand, J. Wetzel and A. Hansen, "Computer Analysis of Sensory Predicate Use in Written and Oral Communication," *Psychological Reports, 65,* (1989): 675-684.

[48] M. P. Wilbur and J. Roberts-Wilbur, "Categorizing Sensory Reception in Four Modes: Support for Representational Systems," *Perceptual and Motor Skills, 64,* (1987): 875-886.

[49] M. D. Yapko, "Implications of the Ericksonian and Neurolinguistic Programming Approaches for Responsibility of Therapeutic Outcomes," *American Journal of Clinical Hypnosis, 27,* (1984): 137-143.

[50] E. L. Einspruch and B. D. Forman, "Observations Concerning Research Literature on Neurolinguistic Programming," *Journal of Counseling Psychology, 32,* (1985): 589-596.

[51] R. Rosenthal, "The 'File Drawer Problem' and Tolerance for Null Results," *Psychologicall Bulletin, 86,* (1979): 638-641.

[52] M. Heap, "Neurolinguistic Programming: An Interim Verdict," In *Hypnosis: Current Clinical, Experimental and Forensic Practices,* ed. M. Heap, (London: Croom Helm,1988), 268-280.

[53] F. J. Dorn, B. I. Brunson, I. Bradfor and M. Atwater, "Assessment of Primary Representational Systems with Neurolinguistic Programming: Examination of Preliminary Literature," *American Mental Health Counselors Association Journal, 5,* (1983): 161-168.

[54] Sharpley, "Predicate Matching."

[55] The results of this analysis was originally published as: T. Witkowski, "A Review of Research Findings on Neuro-linguistic Programming," *The Scientific Review of Mental Health Practice, 9(1),* T. (2012): 29-40.

[56] Ibid.

[57] Heap, "Neurolinguistic Programming."

[58] Einspruch & Forman, "Observations Concerning."

[59] W. O'Donohue and K. E. Ferguson, "Evidence-based Practice in Psychology and Behaviour Analysis," *The Behaviour Analyst Today, 7,* (2006): 335-349.

[60] J. C. Nocross, G. P. Koocher and A. Garofalo, "Discredited Psychological Treatments and Tests: A Delphi Poll," *Professional Psychology: Research and Practice Copyright, 37,* (2006): 515–522.

[61] Heap, "Neurolinguistic Programming," 276.

[62] http://www.dailymotion.com/video/xdwl8h_characteristics-of-pseudoscience_tech

[63] M. T. Singer and J. Lalich, *"Crazy" Therapies. What Are They? Do They Work?* (San Francisco: Jossey-Bass Publishers, 1996): 168.

[64] P. Tosey and J. Mathison, "Neuro-linguistic Programming and Learning Theory: A Response," *Curriculum Journal, 14,* 371-388, (2003): 371.

[65] Also known as neurolinguistic seduction (NLS).

[66] J. O'Connor and J. Seymour, *Introducing Neuro-linguistic Programming: Psychological Skills for Understanding and Influencing People*, (London, UK: Thorsons, 1993).

[67] C. H. Yorkshire, "The Bandler Method," *Mother Jones Magazine*, (February/March, 1989): 13.

[68] Ibid, 63

[69] Ibid, 64.

[70] Ibid, 17-18.

[71] R. Gołaś, "Współcześni szarlatani," *Ozon*, (December 2005): 36.

[72] Ibid., 37.

[73] L. M. Hall, "The Lawsuit That Almost Killed NLP," *Neuro-Semantics*, (September 20, 2010): http://www.neurosemantics.com/nlp/the-history-of-nlp/the-lawsuit-that-almost-killed-nlp

CHAPTER 18

HOWS AND WHYS OF INVENTING A NEW THERAPY: A PSYCHOLOGICAL SOKAL HOAX [1]

It was a beautiful June evening in 2007, echoing the intense sounds of awakened nature. After a long and hot day, the evening finally brought a well-deserved cool breeze. One of us sat back in his chair on the porch of his home in the beautiful countryside with a glass of cold white wine and plunged into contemplation. At that time, I didn't even realize how fraught that evening, together with accompanying reflections, would become.

I was reflecting on the old times, right after I got my degree in psychology and started my first work. I had experienced a series of very unpleasant professional surprises. Finally, after many years of self-obtained knowledge, I concluded that for all that time, I was diagnosing people with worthless diagnoses based on information acquired during my studies at the university, using for example, projective methods. I was working with real people using tools that are completely worthless. I was teaching pseudoscientific hocuspocus to psychology students and nobody had ever questioned it. Today, I know it was wrong; I know those diagnostic tools are worthless. Why did I do it then? As it often happens in similar cases, my ignorance and blind trust in authorities was to blame. I would never suspect that the curriculum of a state university could promote, teach and propagate complete nonsense, or that university teachers could spread pseudoscientific claims. It took years of experience, tons of scientific curiosity and a pinch of skepticism to dig into, research and understand many of those "truths". On this memorable June evening on my porch, I realized that I had done things that today not only am I ashamed of, but that could have harmed others as well.

We both have thought a lot about our university teachers and professional mentors. Their role was (and still should be) to open their student's eyes and teach them to be critical. Many academics know very well the pseudoscientific provenances of concepts taught by their colleagues during official curricular courses, but we never heard any critique of even the most bizarre charlatan claims that we came across during our studies. At most, we could see the ambiguous contempt from experimental scientists aimed at their colleagues with different orientations, and as it is easy to imagine, this contempt was usually reciprocated. However, it was usually very hard to find any rationale or substantive arguments in those relationships.

When we discovered that we were betrayed and cheated in the name of unjustified tolerance for any, even the most absurd ideas (whether it was homeopathy or psychoanalysis), and sucked into the whirl of lies and fraud, it only deepened our bitterness. Independently, we both made a promise to ourselves that we would talk about pseudoscience, regardless of how many others will be offended by hearing the truth. We felt that stupidity and deception do not deserve any delicacy, gracefulness or any respect. We were often accused of exaggerations, or given rebuttals that even if pseudoscience is widespread and common, then perhaps it is not as dangerous as we claim. Some criticize us of demonizing when we talk about tolerance because nobody spreads ridiculous "novel" ideas. Inevitably, the thought occurs that perhaps it might actually be worth convincing all those disbelievers to show them that ignorance towards new and ridiculous claims is dangerous, and that introducing a completely made-up, fake and ridiculous pseudo-concept into the popular scientific circulation requires nothing but a bit of cunning. In fact, it requires no resources, no experience and no knowledge at all and it does not have to be backed up by any research, authority or literature.

As the bottle of wine became definitely more empty than full, my brain started to shape a grotesque, pseudoscientific construct. I knew it was completely void of any value, but attributing it with superficial attractiveness gave me a lot of satisfaction. The next day, busy with everyday tasks, I forgot about my little mental exercise, but a few days later, the thoughts came back and started to shape my cunning idea. A few weeks and a few bottles of wine later, I opened up my laptop and transferred everything from my head onto the digital paper.

I would never believe that somebody could take my wine-fed delusions seriously, but I was tempted to give it try. After all, we all remember Sokal's hoax and many others who followed him. Would anyone treat me like the creators of the Center for Feeling Therapy? Would I end up with Bandler, Grinder and many other pseudoscientists? Could I use a popular science media, backed up by top authorities in Polish psychology, to actually introduce a worthless, scam therapy into nationwide circulation? Popular science magazines are known to generate interest in certain ideas in both the scientific community and the general public. Perhaps I could get interest from potential patients (clients)? This would further strengthen the position of my made up therapy on the market...

I chose to announce the "discovery" of a new therapy in the popular science magazine *Charaktery*.[2] Why did I choose *Charaktery*? Primarily because this monthly magazine, dedicated entirely to psychology, boasts a professional advisory team, which at the time the article was under review, was composed of eight experts (with professorial titles) and one Ph.D. Its editorial board included four doctors, one of whom had another advanced graduate degree. These facts, strengthened by the editors' declarations emphasizing that the monthly is a popular science magazine, may have led its readers to conclude that the content presented to them in the journal is valid, verified, and supported by solid scientific research.

Nevertheless, after a thorough analysis of the journal's content, I drew opposite conclusions. In recent years – in particular, those few that preceded my hoax – apart from a range of credible articles, the monthly published highly questionable content that sells just as easily as science. *Charaktery* has popularized neurolinguistic programming (NLP)[3], Bert Hellinger's family constellations therapy[4], Carl Simonton's cancer healing[5] and many more treatments of questionable (at best) scientific credibility.

Another argument that supported my choice of *Charaktery* was the fact that the journal regularly reaches a very specific and quite wide audience (over 50,000). It includes psychologists (for many of whom the journal is the key source of information), students of psychology, and therapists with little or no educational background in psychology. This target group is also composed of numerous former, present, and prospective patients—individuals who frequently discontinue their pharmacological treatments once they start psychotherapy. It is therefore difficult to imagine a better carrier for therapeutic news in Poland than *Charaktery*. *American Scientific Mind*[6] could be taken as its rough American equivalent, in particular taking into account that since April 2010, the editors of *Charaktery* are also the editors of the Polish version of the German magazine *Gehirn & Geist* and *American Scientific Mind*[7].

I will quote below select passages from the fabricated text composed. For the sake of clarity, I will pair them with narrative comments to highlight the absurdity of my claims, as well as their completely random and pseudo-scientific character.

Knowledge Straight from the Field

How is it possible that termites, blind by nature, know how to build intricate and well equipped nests, and do so anonymously and with the engagement of the entire community? How can large flocks of birds or schools of fish suddenly change direction without individual animals bumping into each other? How does it happen that successive generations of lab rats learn to navigate the maze much quicker than their ancestors? These and many other amazing puzzles have already been solved by the morphic resonance hypothesis. Attempts are made now to apply this knowledge in the domain of psychotherapy.

The patient lies on a 21st-century version of a couch. But instead of the psychoanalyst sitting at the headrest, we see a person in a white lab-coat who gazes intently at a computer screen connected to a huge CT scanner. The patient's head rests inside the machine and is currently being scanned using the fMRI method – functional magnetic resonance imaging – which means that the patient's brain activity is being mapped on a real-time basis. This is the first step of a new therapy – therapy of the future. Its goal is to diagnose disturbances of the morphogenetic field and thus determine what therapeutic measures should be used to induce the desired morphic resonance. For instance, the therapist will be able to prescribe the correct music treatment and to recommend how much music, what kind and in what social circumstances (e.g., inside a theater or at a full football stadium) should the patient be exposed to.

To bring therapy to a close, the measurement procedure will be repeated. If the therapist decides that the disturbances of the morphogenetic field have been eliminated, therapy is deemed to have been successfully completed. If, however, despite these efforts, the patient's field lacks stability, the therapist may resort to additional diagnostic methods, such as computer positron emission scanning, and suggest further treatment based on its results.

I deliberately included this ridiculous statement that if the therapy yields no results, the therapist should consider additional diagnostic tools. Presumably, the expression "computer positron emission scanning" sounded scientific enough for the editors to justify its usage in the text.

Therapy based on the assumption of existence of the morphogenetic field, paired with state of the art technical achievements in brain scanning, does away with lengthy analyses of one's childhood problems, embarrassing disclosures of sexual problems, psychodynamic resistance, and every other obstacle that hinders the patient's progress on the path to regaining mental balance.

Strasbourg Experiment

The above scenes should in no way be treated as clips from a science fiction movie. They depict actual events that comprise an experimental (for the time being) research program directed by Professor Daniel Gounot from Laboratoire de Neuroimagerie in Vivo affiliated with the Strasbourg Medical School in France.

The Laboratory to which I am referring is in fact located in Strasbourg, and Daniel Gounot conducts his research there. His research has involved brain mapping, but has nothing in common with the proposed therapeutic method, and by no means with morphogenetic field therapy! Some of Gounot's articles are publicly available online in English. A couple minutes of net surfing are more than sufficient to obtain a general idea of who he is and what he does. He had been notified by me about the hoax before the article went to press.

To a large extent, the program is based on the brilliant concept of the morphogenetic field announced in 1981 in *Science* magazine by biologist Rupert Sheldrake, and inspired by reflections of Henri Bergson, a French philosopher. As a matter of fact, wishing to mention all sources of inspiration, one should also pay tribute to Carl Jung, who was the first to indicate the existence of collective memory and who emphasized the existence of phenomena that he named acausal synchronicity.

The factual data related to the concept are true,[8] except for the statement that the concept itself is brilliant. Typing the expression "morphogenetic fields" into any Internet browser yields several hundred search results, with an overwhelming majority of links leading to websites related to paranormal events or bizarre therapies, such as Bert Hellinger's family constellations therapy.[9] Had the editorial team bothered to check relevant information, even on Wikipedia, they would have quickly obtained enough to form an opinion of the concept. They could have found out, for instance, that neither biologists nor physicists treat the concept seriously, rejecting it as at odds with scientific evidence. They could have also had an opportunity to learn that the concept was being developed mainly by science-fiction writers.

In the subsequent part of the article, I stated:

> Jacques Lacan, an outstanding French psychoanalyst, also made a valuable contribution to the concept's application in psychotherapy. He was the first to come up with the idea of using mathematical topology in the structure analysis of mental illnesses. It would not be an exaggeration to claim that Lacan, knowing nothing about present methods of brain mapping as we use them today, laid the mathematical foundations of mental illness analysis based on those methods.

Apart from the fact that Jacques Lacan was French, worked as a psychoanalyst, and wrote about the link between mental illness and mathematical topology, all the rest is rubbish. Lacan did not lay any mathematical foundation for the analysis of mental conditions, let alone those based on brain mapping. I invoked his name with the assumption that there are certain names in social sciences, including his, which should immediately ring a bell in the mind of a well-read intellectual. The psychoanalyst, together with such figures of postmodernism as Julia Kristeva, Luce Irigaray, and Jean Boudrillard, was largely ridiculed as a result of the Sokal hoax.[10] Surprisingly, this bait left deliberately in the text apparently did not catch the attention of its editorial team.

> According to Sheldrake's theory, the morphogenetic field is a field of unspecified physical nature filling the space, which – together with the genetic factor of DNA – gives form to living organisms. It also bears heavily on the behavior of living organisms and on their interactions with other organisms. What's more, the morphogenetic field is related to the notion of "formative causation". Sheldrake refers to it as an ability of each organism to convey a memory of frequently recurring events by recording them in the morphogenetic field. Subsequently, such information is passed down to ancestors and other living organisms through active contacts with their own fields of the same type, through morphic resonance. This happens when a critical number of representatives of a given species has learned a certain type of behavior or has acquired specific features of a given organism, which are then automatically – due to morphic resonance – much more quickly acquired by other members of the species. It is hard to explain the brisk pace of acquisition by natural learning processes alone. Intriguingly, research shows that the higher the concentration of a given population is, and – what necessarily follows – of morphogenetic fields, the more intense morphic resonance becomes.

Contrary to the claims of this passage, no researchers have ever confirmed either the existence of morphogenetic fields or morphic resonance. At the time I was engineering my hoax, searching the EBSCO scientific database for such keyword as *formative causation, morphic fields, morphogenetic fields* resulted in only ten publications related to the topic, including the book by Sheldrake himself, a review of it, a highly critical write-up of his concept, and no more than two studies that obtained disappointing results with reference to the quicker pace of the learning process when in the company of others.

> Occurrences of morphic resonance are studied by, among others, biologists who study odd animal behaviors, including animals gathering in enormous numbers in one place without any identifiable reason, for example so-called "kitty parliaments" that consist of a large group of cats living in one city who congregate in certain places. Birds hold similar "parliaments" that occur for other than migratory purposes. Animals gathering in one place do not fight or make any noises. They just spend some time together, after which they disperse. Particularly shocking are huge gatherings of snakes, with the animals crawling to one specific location year after year. According to some ethologists, the intended goal of such "get-togethers" is the inducement of morphic resonance.
>
> As a matter of fact, it is not difficult to notice a similar tendency for congregation in human behavior. This constant need to crowd in certain places, which we try to explain by such excuses as "star performance" or "sport competitions", is as interesting as it is puzzling. How else to explain this drive to form large gatherings if it is much cheaper and more comfortable to participate in those events virtually?

The concept's proponents do refer to such phenomena; nonetheless, the arguments related to human behavior were invented by me, and I am not aware of anyone who mentions them elsewhere.

Memory in the Field

The idea of morphic resonance means in practice that all living creatures owe their build and behavior to inherited memory. Due to long established patterns, morphogenesis and behavior of instinctive nature have already become habitual and can therefore be altered only in minor ways. New habit formation can be only observed in the case of new patterns of development and behavior.

The fragment marked in italics plagiarizes an article by Anna Opala that was published online.[11] This fragment was added to my article by the editorial team of *Charaktery* as they reviewed my submission. All subsequent citations in italics indicate similar plagiarisms as well.

For example, before WWII in England, blue tits (*Cyanistes caeruleus*) learned to open milk bottles with their beaks and to steal cream that was left on the doorstep. After the war, milk was no longer delivered in this way for many years. Nevertheless, when in 1952 milk bottles reappeared in front of people's houses, the birds mastered the skill with lightning speed, despite the fact that many generations of blue tits had passed since the war. But that is not all – two years later, as late as 1955, all species of tits in Europe were able to steal cream from milk bottles. As ethologists point out, it was not possible for such a skill to spread through imitation over such a huge geographic area. According to Sheldrake, this must mean that the memory of bottle opening has survived in the morphogenetic field of the species through which it has been conveyed.

The above fragment seemed enigmatic enough to the editors that they asked me for additional information – what tit species were, when (date) and where (the country) the described events had taken place. Their diligence was only perfunctory; it was easy to invent everything, and the mere act of filling in the details was enough to make the text credible. The fragment I quoted above was in fact something I remembered from literature whose origin and purpose I can no longer recall.

Interesting observations have been made by American cattle breeders. They would traditionally put electric fences that prevented cattle from straying and from causing damages. *Farmers from the Western states have discovered that they could save a lot of money by using fake fences – by painting stripes across the road. Fake fences worked just as well as the real ones – they made cows stop dead in their tracks at the sight of the painted obstacle... Is it possible that the young calves could have learned from older animals that is was better not to try risking confrontation with a device that might inflict severe pain? This seems rather unlikely, because even those herds which had never seen real fences before now avoided the painted ones like the plague. Ted Friend from the University of Texas experimented with several hundred head of cattle and concluded that painted fences were avoided by the same percentage of animals that had never seen the fences before as by those that had faced the real, steel fences. Similar results were noted with sheep and horses. This clearly indicates the existence of morphic resonance passed down by the preceding generations, which had learned to avoid such fences themselves.*

One can multiply such examples. Laboratory experiments with rats also prove that the phenomenon is a fact. The most known case is breeding several generations of rats that have mastered a skill of escaping from a water maze. With time, rats tested in laboratories all over the world, without any experiments or training, learned the trick faster and faster.

In Place of a Grammar Gene

Keeping in mind the phenomenon of morphic resonance, it is easier to understand the complexity of learning mechanisms, particularly foreign language acquisition. Given the reserves of collective memory from which each individual draws and to which each individual contributes, it becomes simpler to learn what our ancestors have already acquired.

This conclusion coincides with observations of linguists, such as Noam Chomsky. Chomsky pointed out, that small children learning foreign languages make rapid advances, which could not be attributed to simple imitation. It seemed as if children imbibed language structures with their mothers' milk. Steven Pinker, a famous evolutionist, describes many similar examples in his book The Language Instinct.

This is particularly apparent while creating new languages or local dialects, which in many cases happens very fast. When people of different nationalities who have no common language must communicate, they spontaneously create an improvised pidgin language that consists of single words and ungrammatical word clusters taken from different languages. Such dialects appeared frequently in colonies and among slaves, but often quickly transformed into legitimate languages. It was enough for children to be exposed to the pidgin during natural language acquisition period. Clearly, repeating illogically sequenced words was not sufficient for them any longer, therefore children systematized them into grammatical rules that had never been used before.

The evolution of sign languages was even more telling. For instance, in Nicaragua, they were not used until recently because deaf people had been isolated. The first schools for the hearing-impaired were established by the Sandinistas when they came to power in 1979. According to Pinker, however, students in those schools were mainly taught lip-reading and normal speech which did not produce any satisfactory results. Most importantly, however, children saw each other on school buses and out playing, and communicated by means of signs that they also used at home to communicate with their parents. Out of these signs, they created their own communication system, which soon became the official sign language presently known as the LSN – Lenguaje de Signos Nicaraguense (Nicaraguan Sign Language). It is still used by the hearing-impaired people who started learning at the age of ten or later. In contrast, deaf children who have been receiving language instruction since the age of four have developed the improved version of that language, with a richer vocabulary and a more systematized grammar. To differentiate between the two versions, the new variant was called Idioma de Signos Nicaraguenses (ISN), which, as Pinker stated, emerged "literally before our very eyes".

Both Chomsky and Pinker assume that language skills are passed down as information coded in the gene material which pertains to all languages. This should explain why small children from any given ethnic group are able to learn any given language. The morphic resonance theory provides an even simpler interpretation of that phenomenon. According to this theory, small children attune their speech not only to the people in their direct environment, but also to the millions of past users of that particular language, which means that morphic resonance enables children to acquire the language, as well as to learn anything else. In the same manner, a deaf person learns sign language from the inherited memory of other deaf people from the past. There are no genes determining the ability to learn specific languages, neither spoken nor signed.

Obviously, interpretation of language acquisition in terms of formative causation is controversial, as is the theory of the genetic origin of universal information related to all languages. After all, as Pinker points out, "no one has yet located the grammar gene".

These paragraphs were plagiarized as well. I wonder, how would a linguist comment on this fragment?

Brain Tuning

Years passed before Sheldrake's theory, at first highly criticized, actually gained acceptance. The green light was given by research discoveries made by Professor Louis Cozolino from Pepperdine University in Malibu. He demonstrated links between this therapeutic method and brain structure in a clear and accessible manner, and proved that all forms of psychotherapy, from psychoanalysis to behavioral interventions, are effective as long as they strengthen changes in the essential neural pathways.

I have randomly combined Louis Cozolino, Sheldrake's theory, and the entire story invented by myself. I have not even touched Cozolino's book[12] to remain in conformity with the principle I followed while writing the text – to use the most superficial knowledge possible. What I wrote in the article was taken from a review of the book available online. As it turned out again, it was possible to make total nonsense credible by supporting it with a name, a university, and references to research and brain structure.

However, a major turning point in the studies of morphic resonance came with the discovery that it was possible to measure morphogenetic fields with the use of modern methods of brain mapping. How was that achieved? As it is often the case with great discoveries, it all happened by chance. While conducting research on brain mapping with his team, aforementioned Professor Daniel Gounot grew suspicious that the presence of the person running the test had a certain negative impact on the variance of obtained results. He decided to check his conjectures on an experimental basis. As it soon turned out, even seemingly insignificant parameters are important - for example, the distance between the person running the test and the subject. Isolating the patient in a separate room always produced results of much lower variance, which directly suggested some kind of influence of one brain on the other. This was not an entirely new observation - research on synchronization of brains of two persons having a conversation had long been known. Brain synchronization was measured by the means of EEG, but apart from the phenomenon being confirmed, the discovery was not used in any way. In addition, in this case the communication process also took place. In Gounot's test, the mere presence of other people resulted in changes in the subjects' brains, which is why he had to move a few steps further. He started to test two and more subjects at the same time. What he was observing were systems (maps) of brain activity that influenced the other subjects, creating a catalogue of specific systems of brain activity which caused the "tuning" of the other brain. The search for the mechanism behind these phenomena led to the concept of morphogenetic fields and morphic resonance, which represented an ideal explanation of the examined regularities. The condition of the brain that is capable of tuning in to other brains, recorded in the form of an activity map, is an indicator of the morphogenetic field at work. This is how – from of a vague and strongly criticized philosophical notion – morphic resonance has transformed into a hard, by all means useful reality visible on a computer screen.

This entire quote is a figment of my imagination. The editors never asked me for the source.

With Wagner against Phobia

How can this approach be applied to psychotherapy? First of all, diagnostic analysis of morphogenetic fields aimed at a therapeutic intervention assumes that some of them, particularly those needing therapy, came into existence in isolation from the influence of the fields of experience of those individuals who had adjusted to the environment more efficiently. Take a very simple example: when they see a therapist, many people raise the issue of their inability to adapt to a competitive environment, failure to demonstrate assertive behaviors, their shyness in establishing relationships, etc. According to the morphic paradigm, one may hypothesize that those people have not had any opportunity in their lives to experience morphic resonance with individuals for whom such skills and experiences are part of everyday life. This is where problems in developing competences in the aforementioned areas stem from. The field is diagnosed by comparing the activity map of a person who has strongly established patterns of such behaviors, one whose habits allow us to expect that morphic resonance can be induced in a person who faces problems in this area. The analysis of differences conducted by the Strasbourg research team suggests numerous aspects of problematic behavior that may be solved by resonance-based therapy. Instead of lengthy psychoanalytic sessions, or ineffective assertiveness training sessions, the morphic approach proposes simply being in the environment that ensures dense morphogenetic fields of people who possess such desired qualities – one needs look no farther than the fan-packed stadium mentioned earlier, though it may seem like an oversimplified example. If you add sessions of music therapy in which works by Wagner will play a fundamental role, chances are that after a number of sessions, the morphogenetic field will change. The only thing left to do is to take hard measurements to check it.

Obviously, at this stage of research we are only able to diagnose and handle simple maladjustments and phobias. Besides, it would be unethical to use morphic resonance in patients suffering from psychoses or other serious disorders when we do not know yet the entire potential of the method and the consequences of inducing morphic resonance. Moreover, diagnosis applies only to a small part of the fields. The entire image will be eventually composed of tens or even hundreds of specific systems of brain activity. Undoubtedly, for the purpose of comprehensive analysis, we will need complex computer software to search those hundreds of images at high speed for subtle differences. Today we are not yet able to precisely define the entire area of the morphogenetic field, since its characteristics bear all the hallmarks of a quantum field and, likewise, escape any attempts to examine and describe it thoroughly. Nevertheless, as it appears from our present knowledge, the resonance method offers excellent opportunities for psychotherapy.

As before – utter fantasy with no scientific basis.

TEXT IN A FRAME NEXT TO THE MAIN TEXT
Experiments with Morphic Fields
The simplest way to directly prove the existence of morphic fields is to work with a group of organisms. Individual organisms may be separated in such a way that they cannot reach one another through the senses which are known to us. If, despite the separation, information is transmitted among individuals, this presents evidence for the existence of connections related to morphic field.

It is known, for instance, that blind termites start to build the nest at different sides to meet in the middle with amazing accuracy. They can do so even when a piece of odor-proof glass is placed in the middle of the nest.

Migration is an equally mysterious phenomenon. As one of tropical entomologist says - neither hunger, neither thirst nor invasion by natural enemies can explain why clouds of locusts unexpectedly soar up into the air and move to another place. As Professor Remy Chauvin comments on the issue of sudden migrations of huge flocks, migrations occur clearly against the instinct of species preservation, and they frequently lead to mass extinction of the animals. It seems as though the animals were thrown into a frenzy, one so contagious that it makes representatives of other species follow the migrating animals.

As a rule, scientists studying phenomena of that kind are unable to find explanations for them. Why do herds of African gazelles all of a sudden and for no apparent reason leave magnificent grazing lands and head for a desert to starve to death there? Are these "telepathic" behaviors also responsible for the phenomenon of "the collective mind" of some insects?

The phenomena I was writing about, which are also quoted by the followers of resonance field therapy, have been already sufficiently explained within the currently accepted paradigm in ethology. Even if some aspects remain an unsolved mystery, they may not be treated as evidence for the existence of morphogenetic fields.

Information on the link between the quantum field theory and the morphogenetic field was included in the text during early stages of editing. Again, it was plagiarized as well. This time, however, the passage was enigmatic and unclear enough to make the editor, who was working with me on revising the article, ask me to either present the analogy in a more accessible manner or delete it altogether. Because the quantum field analogy is likely to be the most serious theoretical abuse of Sheldrake's theory, I wondered whether I would be successful in attempting to publish that nonsense as well. I was.

TEXT IN A FRAME NEXT TO THE MAIN TEXT
Sheldrake's Fields Versus Quantum Physics
Critics of Sheldrake's theory ask about carriers of this field. But it seems to be a risky question, since it may undermine the existence of many other physical notions, accepted today without reservations. Let's focus for a moment on, say, gravitational fields. Nobody has ever found gravitons that would act as carriers of the field, and still we do not throw doubt on the existence of the gravitational field. Gravitational or electromagnetic fields are detected only by their effects, so in order to explain these effects, notions of the fields have been created. There are many more problems with the quantum field. When a quantum field embraces a particle, it influences it in such a way that behaviors at the quantum level are very subtle and have nothing to do with mechanics. The wave acts here as information. As we know, instead of any combination of components, spins of two different, not interrelated particles are always counter-directed. To speak more vividly: since every particle must have a property in opposition to the other, if we see that one is "black", the other must be "white". All this takes place on a concurrent basis, without any signals being exchanged. Measurement of polarization of one of the particles immediately provides us with information concerning the other, its twin. It is identical, except for the opposite sign. This odd and inexplicable phenomenon was called "quantum entanglement" and Einstein even pointed it out as "spooky action at a distance." According to de Broglie, what we call an atom is organized by a higher, or quantum field of information. This field gives the atom its characteristics. The quantum field contains information on the entire environment and the past, all information that governs the electron's present activity. The organizational field is everywhere. Quantum me-

chanics knows fields of information in the wave function and, quite likely, also super-quantum fields that govern the quanta fields themselves.

The morphogenetic fields act similarly by organizing the behavior of biological units. On the one hand, they are equipped with genetic action programs, but the shape of organisms, pace of acquisition of habits and skills, as well as communication are influenced by information transmitted by that field.

The analogies to quantum physics included in the text were meant to be my own deliberate tribute to Sokal.

While still waiting for the text to be published, I decided to write another article as a continuation of the first one. It found favor with the editorial team, who sent comments on how to correct the text, which I did accordingly. The new article explained the phenomena of empathy, altruism, de-individuation, Werther effect, mysterious animal suicides, missing elements in the concept of memetics, as well as the phenomenon of *psi*. Only the latter met with the editors' determined resistance, although in that case, I invoked a rather widely known article[13] published in *Psychological Bulletin* – one of the most prestigious psychological journals. Still in the process of writing, I came to the conclusion that I could explain anew, in a convincing manner, numerous more or less obvious psychological phenomena referring to the notions of morphic resonance and morphogenetic fields. All the same, since in my view I had sufficiently confirmed my hypotheses, I decided not to delay revealing my hoax until the publication of the second article. Extending the hoax would not contribute anything more relevant except for increasing the quantity of data that corroborated my hypothesis concerning the lax scientific standards of the magazine. I also concluded that I could not allow for the further deception of readers.

Having revealed my hoax, I offered assistance to the editorial team in preparation of the content-related disclaimer that, in accordance with local laws, should have been published. The editors refused to cooperate and they have not published such a disclaimer up to date. Only a letter from the editor-in-chief was published in the subsequent issue, in which he explained that the editors had been deceived by me and, at the same time, he passed moral judgment. The plagiarisms allowed by the editorial team in the text were called "an obvious technical mistake."

The day that I revealed my hoax, it was immediately of interest to the academic community and was compared with the Sokal hoax. That is why I deem it sensible to explain the differences in the goals that I had set for myself while preparing the hoax. I did not see Sokal's parody as an example to emulate. It became only a source of inspiration for my plans. I do not have sufficient knowledge to expose serious abuses of quantum physics or higher mathematics. Nor had I intended to take a stance on postmodern philosophy. Furthermore, I did not – and do not – see any point in repeating what Sokal had done. In my view, he demonstrated unequivocally what he had planned to show. My reference to quantum physics and mathematical topology was only of a symbolic nature, as the mark of recognition of Sokal.

Many criticized me for choosing a popular science journal – contrary to Sokal. They claimed that misleading the editors of a scientific journal would have been much more challenging – and convincing. Again, this was a deliberate choice; my aim was to show how pseudoscience found its way to the public. In psychology, both pseudoscience and parascience are rarely spread through the formal scientific press, nor are pseudotherapies or charlatanry. They come into existence largely in books, how-to guides, and popular psychology or popular science magazines. Most frequently, they imitate science or function at the edges of the scientific mainstream.

Of course, revealing this scam as described above resulted in an eruption of various emotions among the journal's readers and among those who heard about the hoax from other sources. I have since learnt whole lot about… my own motives for this provocation. I wasn't even conscious of any of them, but you can always count on anonymous commenters filled with good will and deep insight. I have learned, for example, that I published the hoax in order to promote my book or myself. Others claimed that the entire provocation was nothing but personal revenge or even an open war against the editorial team members, even though I have never, ever met any of them personally before the hoax was published. As it turned out, my entire perception of the world is severely impaired and skewed and I perceive the reality around me through the prisms of manipulation, deceit and conflict. Some even suggested that… I should undergo a therapy. I was called a hooligan and a squealer. Others concluded that the only reason I pulled off this hoax was because the activities of NLP practitioners (and others of their kind) prevent me from earning money. Yet other interpretations were that my actions were driven by bitterness and lack of academic career progression, or that I was lazy, driven by cowardice, or that I had problems in my personal life. Personally, I enjoyed the suggestion of my poor mental condition and suggestions to undergo a therapy…

The most surprising comment however was published in the online guest book on my personal web pages:

> Hmm, dear sir, I must say that your article is not a hoax. You might have made it all up, but how are you going to explain that I was thinking about the very same concepts and I have reached conclusions identical to yours??? I have reached them in a somewhat different way, but I am currently working on more detailed proofs. I must say that many of the things you have named will greatly speed up my works. Thank you and I would kindly like to ask you for your e-mail. I would like to talk to you about it in more detail.

As you can see, my wine-induced fantasies turned out to be very much in line with someone's "research." Perhaps unmasking my mystification prevented a new pseudoscience from emerging on the market.

In order not to rely purely on subjective opinions and to analyze the potential impact of my article published in *Charaktery*, I conducted brief tests among 172 persons who had never heard about the hoax.[14] After reading the paper, they were asked to answer a range of survey questions. Most interesting was the analysis of readers' responses to the paper. Nearly 77% of the readers would recommend the article to friends. Moreover, 72.5 % expressed their interest in broadening their knowledge regarding the therapy in question. Almost half the readers would enroll in a course on that type of therapy (42.9%). Nearly two out of three persons (61%) would recommend the therapy to a person who is struggling with psychological problems. Based on these outcomes, we can conclude that the text itself was characterized by high persuasion potential.

The publication of one article and the acceptance of a second article by a widely read journal with a respectable scientific title suggested that I could easily circulate this knowledge with impunity. What is more, I could even advertise my therapy in the journal, and, consequently, also conduct therapy. The latter, however, would go beyond intellectual provocation. Presumably, there is no scientific journal that would offer such opportunities for promoting new therapies. Reports on the empirically supported therapies regularly published by scientific journals[15] have not necessarily resulted in a rapid growth in the number of people conducting such therapies. Misleading the editorial team of a scientific journal could therefore be a means of achieving goals as established by Sokal or others.

While preparing the hoax, I placed an emphasis on exposing problems related to propagating pseudoscientific content in circles beyond the scientists themselves, that is, students as well as present and prospective clients for psychological services available on the free market. I believe that my hoax and its revelation were simple and clear enough to reach this group. The scientific press does not enable such opportunities. To explain fully (and only to well-educated readers) the meaning of what he had done, Alan Sokal wrote a whole book on his hoax.

Hoaxes, mystifications, and provocations have always raised doubts of an ethical nature. Critics of my undertaking have repeatedly put forward such accusations. Among them were some who stated that sending the article to the editors was nothing more than wasting their time, abusing their trust, misleading the readers, and so on. These accusations are worth analyzing in terms of the specific status achieved by the popular science press. It enjoys the privileges and recognition that are due to scientific journals, at the same time taking little responsibility for the published content. In scientific journals, many people (editors, reviewers and readers) monitor each publication. Even if a worthless or dishonest article has gone to press, sooner or later it will be detected by the scientific community, as exemplified by some of the publications of Cyril Burt.[16] Journals that do maintain the quality of their content quickly lose their impact factor, and their position and reputation.

The situation looks substantially different in the case of popular science journals. Their position, in the majority of cases, is closely related to their market potential, and the latter is determined by readers and advertisers. Editors select texts in such a way so as to increase the number of readers, and by doing so, to maximize their advertising income. Publishing unverified content may at worst lead to losing the best educated and most demanding readers, but those unfortunately belong to a minority. There is no specific scope of responsibility to be taken by members of academic boards or editorial teams of popular science journals. There is likewise no mechanism intended to verify their scientific value. In my view, this poses a considerable threat to readers of such magazines, in particular those devoted to health issues. In the case of the popular science press, this threat is even more serious because it reaches readers who are not sufficiently knowledgeable to verify the content offered to them. A journal about mental health practices, in which "specialists" in neuro-linguistic programming or other similar pseudo-therapies advertise is unlikely to print articles criticizing such services.

Accusations of an ethical nature aimed at my hoax would have been applicable had I perpetrated it on the editors of a scientific journal. Readers of scientific journals are generally skeptical about papers published therein. In addition, they typically possess knowledge and competence sufficient for recognizing scientific inaccuracy or, even more important, blatant falsehoods. Editors of scientific journals are primarily in charge of the scientific and methodological quality of printed articles. Editors of a popular science journal, in particular on mental health practices, are responsible for the welfare of their readers. Hence, as a reader of this type of press, I believe I have every right to expect the editors to protect me from harmful recommendations, false statements, and misinformation.

I would not like the conclusion drawn from my hoax to be limited to a simple statement that "anyone can be cheated." As I have demonstrated above, the text was far from sophisticated, it offered clear hints that allowed its demystification, and I used superficial knowledge. It would therefore be more accurate to say that "it is unusually easy to cheat the editors of at least one popular science journal." Yet, this conclusion does not contain everything that my hoax has exposed. Instead of protecting the credibility of published articles, the editorial team of *Charaktery* took an active role in deceiving themselves and readers.

It would also be an erroneous generalization to claim that the editors of *Charaktery* deceive their readers by publishing nonsensical texts and plagiarisms. My hoax provides no grounds for such reasoning as this magazine publishes a number of valuable articles. What should be concluded instead is that it may sometimes be extremely easy to sneak a pseudoscientific concept into popular science, and that editors may be ready to collaborate if this happens to coincide with the market goals of the magazine. If, instead of the author who had planned to reveal the hoax from the very beginning, there was a fraudulent scientist or pseudoscientist with a strong belief in his own arguments, access to readers would be wide open to him.

My hoax has revealed a range of questions of a more general nature related to science, especially concerning how it is popularized and the responsibility of psychologists for their activities within the field of mental health practices. The quest for answers in the wake of the revelation of my hoax was conducted in a number of public discussions, primarily by scientists. The most intense discussion was published in a separate issue of one of the most influential scientific psychological journals, *Social Psychology*, devoted entirely to threats revealed by the hoax.[17]

A complete list of issues raised by the hoax is too long to discuss separately. Nevertheless, the most important are undoubtedly worth discussing. Are representatives of scientific psychology responsible for how it is popularized and what type of knowledge is popularized? What is the difference between the tolerance of editors, therapists, and scientists for various forms of practicing science and conducting therapy, and the indifference to abuses in this sphere? Should we, as psychologists, abandon the market of psychological services, often referred to as psycho-business, to the rules of the free market? How much nonsense published in popular science and in scientific magazines remain forever hidden, begin to live a life of its own, exerting influence on people struggling with mental or health problems? We hope to answer at least some of these important and troubling questions in the next chapter of this book.

---

[1] All events described in this chapter were carried out by Tomasz Witkowski, therefore all narratives are first person singular.

[2] http://www.charaktery.eu

[3] E.g., B. Peczko, "Różne oblicza NLP," *Charaktery*, 5, (2006): 39-41.

[4] M. Fajkowska-Stanik, "Kto chce umrzeć, niech umiera," *Charaktery*, 3, (2005): 16-18; M. Gumowski, "Zimne serce odchodzi," *Charaktery*, 11, (2004): 27-28; M. Parzuchowski and M. Wilkirski, "W sieci rodzinnych uwikłań," *Charaktery*, 5, (2004): 24-25.

[5] M. Gumowski and J. Koźmińska-Kiniorska, "Uwierzyć by zwyciężyć," *Charaktery*, 4, (2006): 66-68.

[6] http://www.scientificamerican.com/sciammind/http://www.scientificamerican.com/sciammind/

[7] http://psychologiadzis.eu

[8] R. Sheldrake, "A New Science of Life," *New Science*, 90, (1981): 766-768.

[9] B. Ulsamer, *The Art and Practice of Family Constellations. Leading Family Constellations as Developed by Bert Hellinger,"* (Heidelberg: Carl-Auer-Systeme Verlag, 2001).

[10] A. Sokal, "Transgressing the Boundaries: Toward a Transformative Hermeneutics of Quantum Gravity," *Social Text*, 46-47, (1996): 217-252; A. Sokal, A and J. Bricmont, *Intellectual Impostures*, (London: Profile Books, 1998).

[11] A. Opala, *Pola morficzne według Ruperta Sheldrake'a*, (2007): http://www.ustawieniarodzin.pl/Bert-Hellinger-Psychoterapia-Warsztaty-Anna-Winnicka/pola-morficzne-wedlug-ruperta-sheldrake/

[12] L. Cozolino, *The Neuroscience of Psychotherapy: Building and Rebuilding the Human Brain*, (New York: W.W. Norton & Company, 2002).

[13] D. J. Bem and C. Honorton, "Does Psi Exist? Replicable Evidence for an Anomalous Process of Information Transfer," *Psychological Bulletin, 115*, (1994): 4-18.

[14] T. Witkowski and P. Fortuna, "O psychobiznesie, tolerancji odpowiedzialności czyli strategie czystych uczonych," ["On Psycho-business, Tolerance and Responsibility Or Strategies Employed by Pure Scientists], *Psychologia Społeczna, 4,* (2008): 295-308.

[15] Task Force on Promotion and Dissemination of Psychological Interventions. Division of Clinical Psychology, American Psychological Association, "Training in and Dissemination of Empirically Validated Psychological Treatments: Report and Recomendations," *The Clinical Psychologist, 48,* (1995): 3-23; D. L. Chambless, M. J. Baker, D. H. Baucom, L. E. Beutler, K. S. Calhoun, A. Daiuto, R. DeRubeis, et al., "Update on Empirically Validated Therapies, II," *Clinical Psychologist, 51,* (1998): 3–6; A. D. Reisner, "The Common Factors, Empirically Validated Treatments, and Recovery Models of Therapeutic Change," *Psychological Record, 55,* (2005): 377-399; W. C. Sanderson and S. Woody, *Manuals for Empirically Validated Treatments: A Project of the Task Force on Psychological Interventions,* (Oklahoma City: American Psychological Association, Division of Clinical Psychology, 1995).

[16] T. F. Gieryn and A. Figert, "Scientists Protect Their Cognitive Authority: The Status Degradation Ceremony of Sir Cyril Burt," in *The Knowledge Society, Sociology of the Sciences Yearbook, 10,* eds. G. Bohme and N. Stehr, (1986): 67-86.

[17] M. Lewicka and P. Śpiewak, eds., *Psychologia Społeczna, 4,* (2008).

CHAPTER 19

FROM KEEPING SILENT TO ARROGANT HOSTILITY: STRATEGIES EMPLOYED BY SCIENTISTS WITH REGARD TO PSEUDOSCIENCE

Doctoribus atque poetis omnia licent

France, 1784. Anton Mesmer, the charismatic German physician, enjoys success as the discoverer of animal magnetism. The method and its author are becoming future legends. Using magnets or just his own hands, Mesmer causes his patients to go into convulsions accompanied by moans or screams. Sometimes it even results in the loss of consciousness. The paroxysms heal the patients. According to Mesmer, all this is thanks to an invisible natural force called animal magnetism.

The success of the method prompts its author to take steps to have it formally approved by the French authorities. In reply to his demands, the King of France, Louis XVI, appoints a commission to investigate the method. The commission's main task is to answer the question of whether animal magnetism really exists. Members of the commission design and then carry out a series of experiments in order to, among others, eliminate psychological factors and self-healing properties of the body as variables responsible for the alleged healings. In addition, they refuse Mesmer's demands that they focus only on analyzing the cases of people who have been healed. As a result, they come to the conclusion that imagination without magnetism may provoke convulsions, but magnetism without imagination does not provoke anything, therefore a fluid, which does not exist, is useless[1].

The way the investigation was designed and conducted is regarded today as a model example.

> The logic of argument has a universality that transcends culture, and late eighteenth century debunking differs in no substantial way from modern efforts. Indeed, I write this essay because the most celebrated analysis of mesmerism, the report of the Royal Commission of 1784, is a masterpiece of the genre, an enduring testimony to the power and beauty of reason.[2]

Some authors even believe that we still have a lot to do to catch up with this 18th-century approach to pseudoscientific statements.[3] We couldn't agree more. With a methodology and critical thinking at their disposal, scientists should have a decisive say in judging pseudoscience, especially when it claims to be able to heal people. It is worth once again citing Albert Einstein at this point: "All our science, measured against reality, is primitive and childlike — and yet it is the most precious thing we have."

Unfortunately, using "the most precious thing we have" scientists all too often remain indifferent or adopt other strategies conducive to the spread of ridiculous concepts. This deserves condemnation, all the more so given the fact that often scientists achieve their position having been educated and having conducted research financed by public funds. Therefore, they should serve the public using their knowledge and defend people against pseudoscientific disinformation. Instead, they unfortunately often become involved in activities supporting pseudoscience.

In her book *Manufacturing Victims*, Tana Dineen demonstrates emphatically and convincingly that most psychologists have committed all their energy to creating something she calls the Psychology Industry. Her argument is so extensive that we will not repeat it, which is why our description will not include the strategy of active participation in the development of psychobusiness. Yet, as Dineen writes, among psychologists we can find honest people interested in the ethical pursuit of their profession, which is why from time to time we do hear 'voices in the wilderness' among the scientists. It is worth taking a closer look at them and reflecting on why they are so rare. In Part One, we tried to demonstrate a scholars' attitude to abuses in academic psychology, while in Part Two, to abuses in psychotherapy. Similarly, here we would like to see how scholars approach psychobusiness – a pseudoscientific activity with a strong economic drive. We will begin by describing their strategies from those least harmful to psychobusiness, and we will finish by analyzing those that can limit its omnipresence.

The task we have set for ourselves is methodologically quite difficult. As far as we know, no one has conducted such studies before; there is no official data in this respect and so we are forced to analyze primarily anecdotal materials. We will, however, take a look at scientists' reactions to the more spectacular psychotherapeutic scandals as well as to their statements accompanying the release of some critical books and to the exposure of the provocation we have described earlier.

The most common strategy adopted by scientists in the face of psychobusiness may be the ***strategy of remaining indifferent and keeping silent***. It is very difficult to define its reach, just as it is difficult to estimate the frequency of all situations when people fail to do something. This is how Tana Dineen describes it:

> Not all psychologists are allowing themselves to be swept along by seductive theorizing and popular beliefs. However, while continuing to distinguish fact from opinion and resisting becoming victim-makers themselves, most are hesitating to express their dissenting views. Fearful of jeopardizing their own financial security and reputations, afraid of personal attacks, or concluding that there

is nothing they can do, they are choosing to remain silent. There are those who have continued to take research seriously, acknowledging the limits of their knowledge and showing respect for people. But this is not because they have a Ph.D. or a license; it is because of who they are. Among them, there are a few individuals who are attempting to save what remains of the science and of the ethical practice of psychology, but they are very much "voices in the wilderness."[4]

The strategy can be seen very clearly when it is necessary to take a decisive stance. This was the case of the controversies surrounding recovered memory therapy. As we have written in the chapter devoted to memory recovery, neither the APA nor other organizations such as the Canadian Psychological Association or the British Association for Counselling took an unequivocal position on memory recovery therapy.

Keeping silent is especially characteristic of scientists upon who, at least in some way, rests the burden or a shadow of responsibility for disclosing irregularities, for example an accident during therapy, money being wangled by means of false promises, and so forth. The strategy of keeping silent was employed by many Polish scholars in the face of the provocation involving the *Charaktery* magazine that we described in the previous chapter. Avoiding a subject as much as possible is to limit the consequences of events. This strategy is often combined with a strong conviction that "one should not foul one's own nest" and that professional solidarity is a value in itself, regardless of the consequences caused by this solidarity for those outside the group. Keeping silent is also a passive strategy that requires no effort.

One of its more spectacular examples is the way in which the American Psychological Association (APA) treated the recommendations issued in 1978 and in 1990 by Task Forces on Self-Help Therapies. They suggested the following actions:

1. Develop a set of guidelines for psychologists similar to the standards that guide developers of psychological test materials. Such guidelines could clarify methodological outcome evaluation issues pertinent to the adequate development of self-help therapies.
2. Provide to psychologists a list of informational points that should be included in a commercially available self-help program. For example, books would contain a front page that discussed the extent to which the program was evaluated, recommended uses of the program, reading level of written instructions.
3. Provide a set of guidelines to aid psychologists who negotiate with publishers. The publication of sample contract clauses could significantly improve the position of psychologists who wish to set limits on claims or other promotional efforts.
4. Develop a short pamphlet to educate the public in the use of self-help therapies. The public could be informed as to how self-help therapies are used as adjuncts to therapist-assisted treatment, or by themselves. The issue of devel-

     oping realistic expectancies in light of sensationalized claims could be addressed.

5. Consider working in concert with other professional or consumer-advocate groups in an effort to educate the consumer public and possibly develop a review process to review current evidence of self-help programs. In time, it was suggested, standards for establishing a formal "approval seal" might be possible.[5]

Despite the fact that these actions were recommended twice over a period of more than ten years, none of them was taken into account by the APA. Even more interestingly, at the same time, some APA members were actively involved in the development, marketing and promotion of untested self-help materials. As Gerald Rosen writes:

This came about through APA's 1983 purchase of *Psychology Today* and the companion Psychology Today Tape Series. By 1985, psychologists on the staff of *Psychology Today* were contracting for new audiotapes to be added to the series. The criteria used to determine which audiotapes should be used were the prominence or credibility of the author and the face validity of the instructions. No attempt was made actually to test the ability of consumers to use the audiotapes.

In the context of this history, the reader of this article can now consider what was offered to the public. First, consumers who purchased the audiotapes received a brochure with the name of the APA right on the front cover. On the back of the brochure, it stated "Backed by the expert resources of the 87,000 members of the American Psychological Association, *The Psychology Today Tape Program* provides a vital link between psychology and you."[6]

Thus, instead of some sort of order being introduced into the self-help therapy market, for at least three years, customers were offered untested audiotapes recommended by the 87,000 members of the APA.

    When analyzing the strategy of keeping silent, we may have the impression that academics—who so often explain the phenomenon of conformism using spectacular examples of the indifference of witnesses to crimes committed in public areas—fail to see that they behave in exactly the same way when a crime is committed before their very eyes in the name of financial gain and against the values that constitute the essence of science.[7] This is also what happened in the case of the Center for Feeling Therapy described earlier. According to Paul Morantz, the lawyer who was in charge of lawsuits against the Center's therapists, "At one time I asked a California Psych Board investigator why Center licenses were not removed when their books were published detailing unethical conduct. He answered: 'And who reads those?'"[8]

    Another indicator of the popularity of the strategy of keeping silent and indifferent is the scant number of critical studies concerning the methods used by psychobusiness. Its potential victims often have no chance of coming across any criticism and, as Dineen writes: "Meanwhile the psychological associations remain impotently ambiguous, perhaps because their constituents have a greater investment in the psychotherapies and more to lose."[9]

It is also worth pointing out here that *The Scientific Review of Mental Health Practice*, "the only peer-reviewed journal devoted exclusively to distinguishing scientifically-supported claims from scientifically-unsupported claims in clinical psychology, psychiatry, social work, and allied disciplines"[10] ceased to exist in 2012, ten years after the publication of its first issue. The reason was rather prosaic – lack of funding. Yet the scientific community made no attempt to save this journal, probably the only periodical of its kind in the world.

The *playing down strategy* is even worse than keeping silent because it supports psychobusiness. Similar to the first strategy, it achieves comparable results, but it is proactively applied. It enables people to rationalize passivity. The authors of statements quoted below are psychologists-scientists and the statements come from an Internet discussion about the harm of pseudoscience:

What evidence do you have to support your claims that psychoanalysis, NLP, Silva's method, "adjustments" and similar scams actually harm patients? Is it just the fact that they don't work nor help them? Do vitamins and cold pills help?

You know well, that I have a similar stance on psychoanalysis, NLP and similar trends, but I believe that you overestimate how harmful they are – just like you overestimate the influence of "real" psychology on improving people's conditions. NLP will not help those with problems – but it will not harm those without them.

As we can see, the playing down strategy is all about denying the importance of the negative consequences of pseudotherapies or even their very existence, which in itself is tantamount to a clear consent to these therapies being practiced. Such a strategy is also commonly used by entire psychobusiness when it is confronted with the allegation that its methods are ineffective. There may be two likely reasons behind statements of this kind issued by scientists. First, the statements are formulated without any reflection, with their authors drawing on the available heuristic; after all, similar arguments are heard all the time, whether the matter concerns psychotherapy or alternative medicine. Secondly, they may result from a reduction of cognitive dissonance: if I do nothing against psychobusiness and come across information about its dubious effects, then either I will start doing something or I will justify to myself why I don't do anything. This second way of reducing dissonance certainly requires less effort.

At this point it is worth mentioning a practice followed by scholars studying the effectiveness of therapy, a practice that tends to immunize pseudoscience against critics and enforces psychobusiness. It consists of studies indicating that demonstrate the ineffectiveness of many treatments, while simultaneously highlighting additional undesireable effects. Historical sources of this strategy can be derived from Frank's affection for non-specific effects.[11] Non-specific effects are the range of influences on treatment outcome that arise out of the process of treatment, rather than directly from the active treatment components. In psychotherapy, non-specific effects could be caused by non-specific factors such as expertness, attractiveness, trustworthiness but also empathic understanding or respect of warmth. According to some researchers, non-specific effects could be responsible for as much as one third of all outcomes of psychotherapy.[12]

And now, whenever there appears another publication trying to demonstrate that there are no differences between various treatments, scholars unanimously blame non-specific factors and suggest that the treatments do work. Whenever there emerges a publication demonstrating a lack of effectiveness of a treatment in comparison with a control group, its apologists point to non-specific effects. Conclusion? Regardless of how absurd the assumptions of a treatment are, regardless of the results of studies, we can always rely on non-specific effects; therefore, therapies do work. Yet, there is a fundamental mistake behind this reasoning, a hidden assumption that non-specific factors occur solely in the therapeutic process and are absent during a conversation with friends in the garden, a walk with the kids in the park or during a game of billiards. What is worse, the belief in non-specific effects acquiring a form of psychotherapeutic mysticism is sometimes even cited when we are dealing with the negative effects of therapy.

The strategy of playing down the problem could also be seen in opinions expressed during the discussion about the provocation in *Charaktery*. This is how one of the adversaries actively justified why there was no need for new therapies to be tested scientifically:

> From the intellectual perspective, the most important question for me seems to be one of whether this or that social influence technique (therapeutic, personal change technique, etc.) has any justification in science. Thus, does NLP have any scientific background? Are Hellinger's constellations or the morphogenetic field based on any scientific evidence and if so, what evidence? If they are, they can be used, if they are not – they must be opposed. I think that these are wrongly formulated questions and, above all, that they blur the boundary between knowledge and social practice. We could just as well ask whether a fruit salad recipe is scientifically proven. Does stroking a child's head to calm the child down have any psychological background? What background? We would have to note from such a perspective that the first boats built by the indigenous people living by the sea in all parts of the world constituted the beginning of boat business which had no scientific background (and was quite dangerous to boot). I think that's not what this is all about. If, as is generally known, the effectiveness of all treatments is more or less similar, then it is obvious that what is of crucial importance is the client's susceptibility and not the specificity of the therapy [or] its scientific background. Thus, the appeal for therapeutic procedures to

be verified from scientific standpoints – unfortunately, resembling somewhat a call for introducing censorship – seems to be misguided.[13] Questions concerning the scientific underpinnings of a fruit salad recipe or the background of boat making can have simple answers, if we take into account the classic division into science, pseudoscience, parascience and protoscience, which the author does not seem to have considered. Both the fruit salad recipe and the principles according to which the first boats were constructed are manifestations of protoscience; just as protoscientific were the inquires of pre-Socratic Greek philosophers into the essence of nature. Pseudoscientific would be the term used to describe a salad recipe which would include belladonna and which would claim to have a scientific basis. Such concepts, emerging both in the kitchen and in the office of a therapist, regardless of the 'fruitfulness' of argumentation, are always immoral and any balancing act to justify why they should be left alone is dangerous because it facilitates their development.

The last sentence from the opinion quoted above contains another element quite frequently present in statements by scholars who try to convince others that it does not make sense to have therapeutic methods verified by scientists. I mean here the threat of censorship. It is a wholly emotional argument, the objective of which is to stimulate aversion to control. It is a particularly effective way of persuading people in countries that have experienced totalitarian rule. This thread also emerges in another opinion:

> However, it is impossible for psychologists to interfere in the free market in order to protect patients against [the] potentially harmful actions of various charlatans by limiting the patients' freedom to choose their treatment. We all know what taking control over the free market smacks of. Poles, unfortunately, had an opportunity to experience this in the People's Republic of Poland. Is the introduction of a communist ideology by psychologists a good solution? We sincerely doubt it.[14]

The authors of the fragment go further than the previous adversary because they use an eristic stratagem, catalogued by Schopenhauer under no. 13, the essence of which is to fabricate an absurd counter-proposition to a statement we disagree with and to give both sides to the listener to choose from.[15] Thus, when it comes to the situation in pseudoscience, the authors present two options: the current state of affairs or a return to the communist ideology. If we were to follow this line of reasoning, we could say that in Germany, the only possible alternative is a return to the Nazi ideology. Yet it is the Germans who have developed one of the best systems for protecting patients against pseudoscientific rubbish and no one describes it as Nazi ideology. What also requires a comment is another remark that fits in well with the strategy in question:

Psychobusiness is based on a voluntary contract. Those who offer psychological help (defined as broadly as possible) are on the one side, and on the other we have those who expect this help (in the form of therapy, training, marketing advice etc.).[16]

This statement and the following explanation are probably to show that fighting something which people freely choose is like trying to make them happy whether they like it or not. However, the author of this statement (deliberately?) fails to mention a huge area of therapy and rehabilitation used with regard to children, people with mental disabilities and people who are mentally ill, that is, cases in which it is difficult to speak of free choice. This area is particularly well explored by representatives of psychobusiness.

Summing up our reflections on this strategy, we should note that playing down the significance of pseudoscience and pseudotherapy is a common phenomenon. When analyzing the behavior of therapists and scientists, O'Donohue and Ferguson, quoted here many times, write about a lack of any willingness to oppose these negative phenomena, even on the part of scientifically-oriented practitioners. They look for sources of such an attitude in, for example: "A relativistic ethic in which all perspectives are held to have equal value."[17]

When we hear statements similar to such as: "HIV virus does not cause AIDS", "humans do not contribute to climate change", "vaccines cause autism", the only word that immediately springs into our minds is denialism. We use this expression every time when we hear someone say things that are totally opposite to current scientific consensuses. All false claims cited above lead to irreversible consequences. Some can quickly result in the death of innocent people; the consequences of others will be delayed. Denialism is conserved and spread also by scientists, although luckily the majority of the scientific community understands that current reality is described by empirical research. Denialists are often eager to accept a view of the world typical of advocates of various conspiracy theories. The phenomenon is widespread on internet, especially across social media. Denialism is also kindled by tabloids, but also by poorly educated journalists in mainstream media. Scientists do not easily tolerate denialists... with one particular exception: psychologists working in academia. What are the objects of their neglect and negation? These objects are the facts that show the negative consequences of pseudoscience and research data that reveal the damage done by pseudotherapies. Why aren't we using the word denialism to describe people who use the above described strategies, despite the fact that not doing so directly gives the green light to all kinds of charlatanisms?

A more active strategy used by scholars and which is conducive to the growth of psychobusiness is *depreciation of any criticism of psychobusiness*. It is also a kind of denialism. The essence of this strategy consists of undermining the competence, mental abilities and even mental health of authors who expose problems caused by psychobusiness. Often, it is accompanied by an attack on the methods of exposing the problems. We were by no means the first to notice this strategy. It is described by Carol Tavris and Elliot Aronson, who explained the mechanisms of reducing cognitive dissonance by psychologists and psychiatrists engaged in all kinds of malpractice. However, they describe it far more bluntly as "killing the messenger."[18] To give an example, they cite an extraordinary opinion expressed by D. Corydon Hammond:

> "I think it's time somebody called for an open season on academicians and researchers. In the United States and Canada in particular, things have become so extreme with academics supporting extreme false memory positions, so I think it's time for clinicians to begin bringing ethics charges for scientific malpractice against researchers, and journal editors – most of whom, I would point out, don't have malpractice coverage." Some psychiatrists and clinical psychologists took Hammond's advice, sending harassing letters to researchers and journal editors, making spurious claims of ethics violations against scientists studying memory and children's testimony, and filing nuisance aimed at blocking publication of critical articles and books. None of these efforts have been successful at silencing the scientist.[19]

It is worth adding that the call for an open season was made at the 14th International Congress of Hypnosis and Psychosomatic Medicine in San Diego in June 1997, that is, it was addressed not only to practitioners, but also to scientists dealing with the topic of the congress.

Unfortunately, killing the messenger or at least disparaging him is a strategy that can be encountered more often than it might seem. The person who experienced it to the fullest is the Canadian psychologist Tana Dineen, frequently quoted in this book. Described as a "renegade psychologist" by the *National Post* and the *San Diego Union Tribune*, a "dissident psychologist" by the *Ottawa Citizen* newspaper and a "heretic" by the *LA Daily Journal* (the largest newspaper for lawyers in US), this is how she writes about the reception of her book.

> When *Manufacturing Victims* was first released in 1996, it drew volatile reactions from within the Psychology Industry. It was attacked as "a conspiracy book" and called "the Ripley's Believe It Or Not of Psychology." Colleagues who had neither met me nor read the book offered their opinions, diagnosing me as suffering from some treatable malady such as "burnout" or "depression." One psychologist, after watching an interview of me on national television, lodged a formal complaint with my licensing board that led to an investigation in the name of "protecting the television watching public." After eighteen months, the board finally acknowledged my charter right to speak and my role as "a social critic," and dismissed the complaint.[20]

Many attacks on Tana Dineen were and are initiated by academics. Unfortunately, her example is not unique.

In the 1990s, renowned memory researcher and APS Past President Elizabeth F. Loftus, at the University of California, Irvine, drew considerably hostile reactions when her studies challenged people's claims that they had uncovered — often with the help of therapists — repressed memories of abuse, molestation, and even alien abduction. Loftus even had to have armed guards accompany her to lectures after she received death threats.[21]

Elizabeth Loftus' contribution to stemming the flood of false charges in court, charges formulated on the basis of memories recovered during therapy, is inestimable. However, it is just as much a thorn in the side of those who support the recovered memory therapy.

Another person who came under similar attack was Susan A. Clancy after the publication in 2010 of her book *The Trauma Myth*, in which she presents her conclusions from interviews with the victims of sexual abuse in childhood. Contrary to the prevailing opinion that abuse caused shock, fear and inevitably led to trauma, Clancy discovered that the abuse had been confusing for the child but not traumatic in the usual sense of the word. Only when the child grew old enough to understand exactly what had happened — sometimes many years later — did the fear, shock and horror begin. And only at that point did the experience become traumatic and begin its well-known destructive process. Her discoveries were in complete opposition to conventional wisdom or to approved models of sexual abuse.

Dr. Clancy reports that she became a pariah in lay and academic circles. She was "crucified" in the press as a "friend of pedophiles," colleagues boycotted her talks, advisers suggested that continuing on her trajectory would rule out an academic career. All that fuss about one little word — "trauma" — and a change in its timing. Why should it matter one way or the other?
Dr. Clancy suggests several reasons her data aroused such passion. For one thing, a whole academic and therapeutic structure rides on the old model of sexual abuse; her findings had the potential to undermine a host of expensive treatment and prevention projects.[22]

Another similar case occurred in 2009 after James Heilman, an emergency room physician known as an advocate for the improvement of Wikipedia's health-related content, helped Wikipedia publish the ten inkblots of the Rorschach test alongside the most common responses given to each. He did it after becoming frustrated by a debate on the website as to whether a single Rorschach inkblot plate should be taken down.

The nearly 100-year-old test had been the subject of much controversy among psychologists for years. In the 1959 edition of Mental Measurement Yearbook, Lee Cronbach, former President of the Psychometric Society and

American Psychological Association, and the world's leading expert on psychological testing, is quoted in a review: "The test has repeatedly failed as a prediction of practical criteria. There is nothing in the literature to encourage reliance on Rorschach interpretations."[23] In addition, a major reviewer Raymond J. McCall writes: "Though tens of thousands of Rorschach tests have been administered by hundreds of trained professionals since that time [of a previous review], and while many relationships to personality dynamics and behavior have been hypothesized, the vast majority of these relationships have never been validated empirically, despite the appearance of more than 2,000 publications about the test."[24] A moratorium on its use was called for in 1999.[25]

Although plenty of similar statements by eminent psychologists-psychometrists can be found in the literature, the Rorschach Test is still used in clinical diagnosis and in many countries, also in forensic diagnosis. In addition, at the time of its publication in Wikipedia, the text was not covered by any copyright. Yet the psychologist community responded like angry bees after realizing that someone was stealing honey from their hive. Heilman was showered with allegations and accusations. Eventually, in August 2009, two Canadian psychologists filed complaints about Heilman to his local doctors' organization.

> One of them, Andrea Kowaz of the College of Psychologists of British Columbia, complained that by including the inkblots on Wikipedia, Dr. Heilman was violating the test's secrecy and that if he were a psychologist, his behavior would be "viewed as serious misconduct."
> The other letter, from Laurene J. Wilson, a psychologist at Royal University Hospital in Saskatoon, echoed the concern about the test's security, but added that Dr. Heilman "shows disrespect to his professional colleagues in psychology and disparages them in the eyes of the public."[26]

Heilman responded that he considered the complaints to be nothing but an attempt to punish him for trying to demystify the psychological profession. "These are intimidation tactics," Dr. Heilman said. "They are trying to close the doors to scientific discourse. They don't want anybody other than themselves involved in a discussion about what they do."[27]

To explain to the public that many projective tests have poor or no validity, as well as to raise concerns among professional psychologists who still use such tests alone in clinical diagnosis or in legal proceedings, a campaign called *Psychology is Science not Witchcraft* was organized and coordinated by the Polish Skeptic Club (KSP) in March 2012. Information about this campaign was published in major influential nationwide journals, newspapers, radio stations, and on the largest Polish Internet portals. Over 140 people from nine large non-government organizations took part in the protest. Scientists, lecturers and students wore T-shirts with Rorschach inkblot and the slogan of the campaign on it for four days at universities, in their workplaces, and on the streets. Many open lectures and other events were organized within the campaign. We also published Rorschach inkblots in the Polish language version of the Wikipedia.[28]

Just like in other previously described cases, there were many side-effects of our campaign. Some of the participating university students reported harassment by lecturers at their universities. Some of the researchers and lecturers involved in our actions suffered ostracism from their academic community. Many accusations of self-promotion leveled against the organizers of the campaign appeared as expected. So did the accusations of copyright violation related to the publication of the Rorschach's test as well as to the violation of psychological ethic. What do the authors of these accusations understand by "copyright violations"?

It is quite difficult to probe this way of thinking because it is marked by double standards. It is best illustrated by the release of a book by Michał Stasiakiewicz entitled *The Rorschach's Test*.[29] It contains all inkblots used in the formal test with their detailed descriptions, answers and their interpretations. The book is easily available and anyone with about $10 to spare can purchase it. Many advocates of the Rorschach's test regard it as an important scholarly publication. However, we have not been able to find even a single fragment accusing its author of infringing on copyright or violating ethical principles by making this "diagnostic" method public. It's much easier to find enthusiastic reviews of the book. We cannot help thinking that the work of advocates of projective tests is called "scientific" only as long as supports their claims. Any critical work immediately becomes an offense committed by someone with no ethical principles.

Before we began our campaign, we had very thoroughly examined the problem of copyright on the Rorschach inkblots, assisted in this by the staff of the Intellectual Property Law Institute, Jagiellonian University. All our actions were completely legal. Our accusers probably did not make the slightest effort to check the legitimacy of the public allegations made in print against us. If they did, it puts them in an even worse light, as it suggests that they deliberately, manipulatively lied.

The accusation of unethical behavior is a tried and tested instrument of demagogic discussion. Whenever it appears, it is worth looking more closely at those who rather freely bandy it about. In this case, the defenders of ethics should take into account at least three additional aspects of the whole matter. Firstly, using any tests that are freely available to the public is not indicated, as it does not allow for correct diagnosis. Any freely available test is useless, and formulating opinions on its basis is against psychologists' code of ethics in most countries. The Rorschach's inkblot test has been publicly available on Wikipedia since 2009 in at least 14 languages. Also, as we wrote earlier, thanks to Stasiakiewicz, it has been available in Poland since 2004. Moreover – and we had checked this before we launched our campaign – all inkblots with detailed descriptions (in Polish) could be found on at least a dozen websites.

Have these facts ever deterred at least one "diagnostician" from using the test? Probably not.

The fact that many psycho-diagnostic tests are widely available, not just those discredited, is a big problem. In just under an hour of googling, you could get hold of the more important psychological tests with answer keys and interpretation guidelines. However, it is virtually impossible to find at least one sentence by a Polish psychologist who would be concerned about such a state of affairs; we bet dollars to doughnuts that not one of them has ever taken any step to protest against it or to change it. So why do so many passionately accuse us of making the test publicly available and of violating the code of professional ethics?

When in 1616☐ Galileo Galilei was summoned to Rome to appear in court and accused of spreading heresy; the adjudicating panel was composed of scientists working in the same field as he did. These scientists formulated the charges. They were the reason Galileo was sentenced to house arrest for the rest of his life. It seems that today, most scientists are able to report their findings without worrying about drastic sentences from the state, but presented examples show that it is only a kind of illusion. They still face hostility from people who simply deny the empirical results or who have a vested interest in keeping the *status quo*. Interest groups, ideologues, industry lobbies, and even policymakers all freely lash out at researchers whose work threatens their belief systems or their economic interests. "These attacks are not new, but modern communications technologies have given science deniers far more potent tools to blast everything from climate science to vaccines. In addition to harassing phone calls and letters, they now can pummel researchers with hostile emails, or assail their integrity on blogs and other social media tools — all in relative anonymity. And in addition to questioning the validity of the science, these critics often resort to personal attacks on the scholars as a way to discredit the data."[30]

Cyber-bullying and public abuse, harassment by vexatious freedom-of-information requests, complaints, legal threats or actions, (and perhaps most troubling), the intimidation of journal editors who are acting on manuscripts that are considered inconvenient by deniers, are so common that several scientists who experienced them in a particular way wrote a paper about methods of preventing them. They sum up their reflections in the following manner:

> How should the scientific community respond to the events just reviewed? As in most cases of intimidation and bullying, we believe that daylight is the best disinfectant. This article is a first step in this effort towards transparency. Knowledge of the common techniques by which scientists are attacked, irrespective of their discipline and research area, is essential so that institutions can support their academics against attempts to thwart their academic freedom. This information is also essential to enable lawmakers to improve the balance between academic freedom and confidentiality of peer review on the one hand, and the public's

right to access information on the other. Finally, this knowledge is particularly important for journal editors and professional organizations to muster the required resilience against illegitimate insertions into the scientific process.[31]

The experiences of scientists attacked by their colleagues are, unfortunately, familiar to us. Protesting against pseudoscience, we have often encountered criticism and ostracism on the part of the academic community. In response to our exposure of the provocation published in *Charaktery*, one of the members of the journal's academic advisory board simply said: "If you refer to my opinion, then I'll frankly say that the method you have chosen is unworthy of the title of doctor which you are using."[32] This man continues to sit on the academic advisory board of the journal, which is actively promoting therapies for the adult children of alcoholics, and recently started publishing articles arguing that there is a "disease" called Adult Children of Corporations Syndrome. He probably believes that this is worthy of the title of professor, which he uses.

As we are writing these words, pseudoscientific claims are not only circulated by popular psychology journals. APA Continuing Education Credits can be earned by taking part in a course on *Energy Psychology: Progression or Retrogression in Understanding and Treating Psychological Disorders*.[33] This is by no means an isolated incident. As Lilienfeld and his associates say,

> [The] APA has been accepting advertisement for a plethora of invalidated treatments, including Thought Field Therapy… and Imago Relationship Therapy, two techniques for which essentially no published controlled research exists. Among the workshops for which the APA has recently provided CE credits to practicing clinicians are courses in calligraphy therapy, neurofeedback… Jungian sandplay therapy, and the use of psychological theater to "catalyze critical consciousness"…. The APA has also recently offered CE credits for critical incident stress debriefing, a technique that has been shown to be harmful in several controlled studies…. Some state psychological associations have not done much better. Very recently, the Minnesota Board of Psychology approved workshops in rock climbing, canoeing, sandplay therapy, and drumming meditation for CE credits.[34]

The situation is just as bad in higher education institutions; in 2009, seventy-four percent of students in the US believed that projective tests, including the Rorschach's inkblot test, were valuable in psychiatric diagnoses.[35] In Poland, the Polish Psychological Society (equivalent of the APA in the US) makes money from selling projective tests, and universities run programs promoting pseudoscience without any hindrance. Many of them are part of the psychology, pedagogy, rehabilitation or even medical curricula.

Patients diagnosed with cancer are led to believe (by radical forgiveness "therapists") that science has long established that the basic cause of cancer stems from problems with forgiving someone in the past and from nursing a grudge. The advocates of the Doman Method offer the parents of children with mental disabilities apparently "scientifically-proven and effective" methods of therapy. These include hanging the children upside down or administering carbon dioxide a dozen of times a day and so forth. In their opinion, these methods raise the children's intelligence to the levels of Einstein, Da Vinci or Mozart. In a huge number of nurseries, kindergartens, pre-schools, psychological-pedagogical clinics and other centers working with children around the world, children are instructed to do 'lazy 8s,' a basic exercise recommended by neuro-kinesiologists. This is often supported by local governments, educational authorities, or even financed by the European Union. People suspected of child sexual abuse are convicted and put behind bars solely on the basis of results of tests involving useless projective methods. In our belief – those actually guilty can easily get hold of those tests in advance and get away from justice…

Meanwhile, faced with the popularity of such practices, eminent psychologists wonder whether pseudoscience is a problem at all because there may be as many psychologies as there are psychologists.[36] They ask questions about whether it is possible to find a demarcation line between science and pseudoscience[37] or try to show that demonstrating a lack of scientific background for practices like hanging children upside down to make them more intelligent is illogical and intellectually worthless.[38]

It seems that in discussions similar to the one that followed the publication of the Rorschach inkblots by James Heilman or the exposure of the provocation in *Charaktery*, we often deal with another strategy so willingly employed by scientists confronted by pseudoscience: ***intellectualization***. We use this term fully aware of its psychoanalytical origins and significance. Intellectualization of a problem seems to be giving scholars the illusion of action. No matter how profound the reflection on the current state of affairs is, how dazzling the arguments are or how scathing the criticism, one thing remains certain: they will not change anything in the social reality apart from changing how the authors of these statements feel. If that was not enough, they will strengthen the belief that charlatans of all kinds are free to do as they please with impunity.

Whenever pseudoscience is criticized in the media and the criticism has a chance to reach a wider audience, immediately "scientists" will pop out and say that more scientific debate is needed and that the press is not the right place for such debate. Of course, the press is an excellent vehicle for popularizing pseudoscience and supporting psychobusiness, and these scientists hardly ever protest against this; yet they demand that the two should be criticized in the privacy of university walls. Although we whole-heartedly support freedom of speech and expressing one's views, we are strongly against organizing discussions with representatives of psychobusiness at universities. Why?

First of all, there is no way that will change the sad reality in any way. Conclusions from such discussions have no chance of reaching a wider audience—or the potential victims of psychobusiness. The basic function of such actions is to facilitate the implementation of the strategy we have called intellectualization. Proud of their intellectual skirmishes and brilliant arguments, scientists will leave the debate venue with their self-assessment intact and with a sense of having done their duty, while representatives of pseudoscience will be confirmed in their conviction that they are dealing with diehard, conservative professors.

However, these are not the worst effects of such debates. Much worse is the marketing impact pseudoscientists get from them. For example: the fact that Bert Hellinger was invited to a discussion at the University of Social Sciences and Humanities in Warsaw was perceived by Hellinger's potential customers as a recognition of his skills! This happened regardless of the nobility of the intentions of the organizers who simply wanted to show students how groundless his concepts were. Ordinary people get the following message: Bert Hellinger, addressed as professor throughout the meeting, was invited to a debate with the most distinguished representatives of academic psychology by one of the finest universities in Central and Eastern Europe. After few such events, regardless of their outcome, the advertising material of such psycho-stars begin to boast lists of renowned institutions to which the "star" was invited. We have seen many advertisements of this kind and have protested on many occasions against universities inviting clairvoyants,[39] healers,[40] NLP gurus and other sorcerers. Unfortunately, this seems to have become a permanent feature of the universities' mode of operation, all in the name of open discussion and freedom of intellectual information exchange.

Our opinions about discussions with pseudoscience can be summarized by the blog post of an experienced advocate of skeptical thinking and the author of the *Skeptoid* podcast, Brian Dunning. Explaining why it is not worth engaging in a debate with pseudoscientists, he writes:

> When you advertise a debate, maybe 1,000 people will attend. And let's say you do a smashing job and manage to convince that entire handful of convincible attendees that science is real. Great, you won over five people. But what you're forgetting is that for those 1,000 attendees, there are 5,000 people out there who heard about the debate (they saw the ads or flyers or whatever) who did not attend. What you unintentionally communicated to those 5,000 people is that your scientific discipline is academically comparable to the pseudoscientific version, and that both are equally valid. The fact that the debate exists at all struck a blow to the public's perception of the credibility of science that far outweighs any progress you may have made in the room.[41]

The American physicist and cosmologist, Lawrence M. Krauss, seconded him: "Merely having a debate inevitably suggests that each side has some credibility. As a result, opponents of the scientific method like creationists try very hard to appear in debates with scientists. Merely being on the same stage represents a victory!"[42]

Now think about this: it is very easy to legitimize pseudoscience by giving it the same attention and level of debate that is reserved for academic disputes. Many people actually demand such debates in the walls of academic institutions, whether it is a discussion between creationists and evolutionists, homeopaths and medics, chiropractors and orthopedic surgeons or between psychoanalysts and psychiatrists. Unfortunately, apart from legitimizing pseudoscientific concepts, not much else will result from such debates. Shall we also invite the advocates of "the stork theory" to debate with embryologists and obstetricians on how humans are conceived and born? And if we do, who and how will benefit from such debate? So if you have ever wondered why the advocates of pseudoscience demand discussion with academics, you should now have the answer to this question.

The intellectualization strategy was revealed in its entirety during the *Psychology Is Science, Not Witchcraft* campaign, which we described earlier. There was also a range of events organized independently as an effect of our campaign (such as local debates, invited lectures or conferences). Discussions were held at major universities for over three months. They involved not only psychologists, but also lawyers or people that were diagnosed (or even sentenced) by expert witnesses with invalid tests. As a result, a large national conference took place in November 2012 where the most renowned scientists met with advocates of projective tests to discuss *Conditions for the Use of Projection Tests in Psychological and Forensic Diagnosis*. The very title of the conference can be a model example for an analysis of sentences with hidden assumptions. The sentence unequivocally does away with any doubts over the validity of using projective methods. The organizers of the conference had already responded to the doubts and had decided to just specify in what conditions the methods could be used. It is a bit like organizing a debate about the conditions of using weapons of mass destruction after protests have been held against them. We know from people participating in the conference that it provided an excellent opportunity for strengthening the faith of the believers. Not one among the participants and organizers of the *Psychology Is Science, Not Witchcraft* campaign received information about the conference, not to mention an invitation to it. Many less aware observers got a simplified message: projective methods are diagnostic tools to which scientists devote a lot of attention.

Summing up this strategy, we cannot refrain from an observation which can best be conveyed by words of Stanislav Andreski, a critic of social sciences, quoted in this book several times:

The pioneers of rationalism inveighed against the traditional dogmas, ridiculed popular superstitions, campaigned against priests and sorcerers, and castigated them for fostering and preying upon the ignorance of the masses – hoping that a final victory of science would banish forever the evils of unreason and organized deception. Little did they suspect that a Trojan Horse would appear in the camp of enlightenment, full of streamlined sorcerers clad in the latest paraphernalia of science.[43]

Our overview of the strategies would be incomplete, if we failed to take into account positive, though not necessarily active, attitudes to the criticism of psychobusiness. In spite of everything, some academics do express their approval of actions aiming at restricting the impact of psychobusiness. This can be seen, for instance, in the feedback given verbally or in writing to the author of the provocation described in the book, as well as in some posts on Internet forums. Here are some examples:

> You did a great job showing how easy it is to bamboozle people with the so-called science, which in this era of various madmen, swindlers and other "savants" is quite important.
> The provocation highlights a very important problem – [the] integrity of scientists at various levels of practicing science: from preparing a research project and formulating hypotheses to collecting results, analyzing them and interpreting them.

These opinions came mainly from people dealing with empirical psychology and having an unequivocally negative opinion about pseudoscience and psychobusiness. However, we have called this strategy **passive acceptance of criticism**, because apart from expressing their endorsement, this part of the academic community rarely undertakes any action in the public sphere.

With all due respect to people who with their work and uncompromising attitude to the principles of science pursue its main goals, we need, nevertheless, to ask whether such an attitude is sufficient. Should representatives of science, from which pseudoscientific strands emerge almost constantly, be more active with regard to them? We are not talking only about activity in the case of events associated with the provocation, but about a general attitude towards abuses generated by psychologists. We fear that passivity will have the same consequences as the strategy of keeping silent that we described earlier. Unfortunately, such attitudes are quite common. As Carol Tavris writes in the *Foreword* to the book *Science and Pseudoscience in Clinical Psychology*, "Pseudoscientific therapies will always remain with us because there are so many economic and cultural interests promoting them. But their potential for harm to individuals and society is growing, which is why it is more important than ever for psychological scientists to expose their pretensions and dangers. As Richard McNally says, the best way to combat pseudoscience is to do good science. Indeed good psychological science has already helped slow, if not overturn the hysterical epidemics of our recent history that wrought so much harm."[44]

It would be hard not to agree with this; some studies, which we will write more about when examining the next strategy, hit pseudoscience very hard.

However, this happened only when scientists planned and carried out studies in response to problems generated by pseudoscience. All too often, scientists undertake methodologically correct research into completely irrelevant problems, producing junk science, which they themselves and their colleagues regard as "good science". Even if the results of their research could provide excellent ammunition in the fight against pseudoscience, we often don't know about it because unfortunately, popularizing science is something scientists despise, or at best, it is something to which they remain indifferent. In addition, academia does not have any rewards at its disposal that could prompt such actions.

Mario Buge brilliantly conveys the attitude of scientists to pseudoscience and psychobusiness: "Scientists and philosophers tend to treat superstition, pseudoscience, and anti-science as harmless rubbish, or even as proper for mass consumption; they are far too busy with their own research to bother about such nonsense. This attitude is most unfortunate ... superstition, pseudoscience and anti-science are not rubbish that can be recycled into something useful; they are very intellectual viruses that can attack anybody, layman or scientist, to the point of sickening an entire culture and turning it against scientific research."[45]

We couldn't agree more; therefore, it is worth describing the last (and most fruitful) strategy: *open, matter-of-fact criticism*. This is our term for behavior which involves looking carefully at the problems generated by psychobusiness and then trying to fight these negative phenomena. Unfortunately, if we try to estimate the number of scholars using this strategy, we would inevitably conclude that it is the one least popular among psychologists. An exemplary response of scientists to the popularization of pseudoscience comes from a discipline other than psychology.

Worlds in Collision, by Dr. Immanuel Velikovsky is the most recent of the four theories. When the book was published in 1959, after a shrewd publicity campaign, the first reaction of most professional astronomers was to regard it as a hoax. The second response – when it became clear that not only was Velikovsky in deadly earnest but publishers and editors seemed equally convinced he had something important to say – was one of rage. The flood of indignant letters to the publisher from scientists who threatened to boycott the firm's textbooks, led to dismissal of the associate editor who brought the manuscript to the company's attention.[46]

It is very likely that today, as scholars chase publication opportunities and increasingly compete for readers, the above response is just a story from a time when the fundamental values of science had a bigger influence on the behavior of its representatives.

Some of those few scholars who fight psychobusiness despite the indifference or even hostility of their peers have already been mentioned or quoted in the book. However, it is worth adding some more examples. Jean A.

Mercer is a psychologist whose whole life has been changed by her contact with pseudoscience and psychobusiness. In the late 1990s, Mercer encountered psychobusiness in the form of Attachment (Holding) Therapy and Rebirthing. The encounter was very brutal, because it took place in court, during a trial concerning the case of 10-year old Candace Elizabeth Newmaker, who was a victim of child abuse and was killed during a 70-minute attachment therapy session purported to treat reactive attachment disorder. The treatment used that day included a rebirthing script, during which Candace was suffocated.

Jean Mercer, who testified in court as one of the expert witnesses in the therapist-caused death of Candace Newmaker in 2000, began to speak out and publish critiques of "alternative" mental health interventions. Moreover, together with other psychologists who took part in the trial, she wrote a book on that case entitled *Attachment Therapy on Trial: The Torture and Death of Candace Newmaker*. Much of the text's material is drawn from court testimony in the therapists' trial, and from 11 hours of videotape made while several adults forcibly held Candace beneath a blanket during the therapy. This book also presents the history connecting attachment therapy to century-old fringe treatments, explaining why they may appeal to an unsophisticated public.

In addition, Jean Mercer has founded an organization called Advocates for Children in Therapy, which opposes a number of psychotherapeutic techniques that it considers potentially or actually harmful to children who undergo treatment. The group's mission is to provide advocacy by "raising general public awareness of the dangers and cruelty" of practices related to attachment therapy. Now Mercer is the foremost spokesman for science-based and humane psychotherapy for adopted and foster children, and her story can be the basis for many a film script showing how scientific knowledge can be used for the protection and good of other people.

Just as good material can be found in the biography of Elizabeth Loftus, quoted many times in this book. "She has been called a whore by a prosecutor in a courthouse hallway, assaulted by a passenger on an airplane shouting, 'You're that woman!', and has occasionally required surveillance by plain-clothes security guards at lectures. The war over memory is one of the great and perturbing stories of our time, and Elizabeth Loftus, an expert on memory's malleability, stands at the highly charged center of it."[47]

How did this happen? In the early 1990s, the focus of Loftus' work shifted to investigating whether it was possible to implant false memories for entire events that had never taken place. The impetus for this new line of research was a case for which Loftus had been asked to provide expert testimony in 1990. The unique point in this case was that George Franklin stood accused of murder, but the only evidence against him was provided by his daughter, Eileen Franklin-Lipsker, who claimed that she had initially repressed the memory of him raping and murdering her childhood friend, Susan Nason 20 years earlier, but had only recently recovered it while undergoing therapy. Loftus gave evidence about the malleability of memory, but had to concede that she did not know of any research about the particular kind of memory Franklin-Lipsker was claiming to have.

It is worth pointing out here that Loftus followed the advice of Richard McNally, who said that "the best way to combat pseudoscience is to do good science," but not in the way most scholars understand it – by simply by continuing to do their job. If that had been the case, she would still be conducting research devoted to the nature of eyewitness memory. Rather, her contact with the psychobusiness of recovered memory therapy encouraged her to carry out studies showing there was no rational evidence for such an approach. She has done this job admirably well despite the numerous attacks she has experienced and is still experiencing from her opponents.

Scientists pursuing the strategy of open, matter-of-fact criticisms of psychobusiness have published their papers in journals like the above-mentioned *The Scientific Review of Mental Health Practice*. They also write for the *Skeptical Inquirer* and *Skeptic Magazine*; they write blogs, are active in skeptics' societies, give lectures and some even help the victims of psychobusiness. Comparing their efforts with the aggressive marketing of psychobusiness is like comparing David with Goliath. The former had only cunning and intelligence at his disposal, the latter, size, force and weaponry. Unfortunately, a clash between a weak and a strong party rarely ends like that Biblical story.

Vehement opponents of psychobusiness include a group of scholars who in good faith act to the detriment of science and indirectly strengthen pseudoscience. These are scholars who, when confronted with pseudoscience, follow the strategy of **arrogant hostility**. Mumbo-jumbo, bullshit, voodoo science, psychobabble, neuro-babble, quackery, snake oil, charlatanism, witchcraft, sorcery, flimflam, fraud, trickery, con, magic, bogus science, sham, scam, swindle, flapdoodle. These are just some of the terms used by critics of pseudoscience. This is not surprising, for we ourselves use some of these terms, as it is sometimes difficult to maintain intellectual distance in the face of psychobusiness' insistent persuasion. Yet some of such responses, especially the consistently arrogant attitude of some scholars, cause irreparable damage to science. This is what may have happened, not with regard to psychology but to science in general, such as in the case described below.

Government Chief Scientific Adviser John Beddington is stepping up the war on pseudoscience with a call to his fellow government scientists to be "grossly intolerant" if religious or political groups are misusing science. In closing remarks to an annual conference of around 300 scientific civil servants on 3 February, in London, Beddington said that selective use of science ought to be treated in the same way as racism and homophobia. "We are grossly intolerant, and properly so, of racism. We are grossly intolerant, and properly so, of people who [are] anti-homosexuality...We are not—and I genuinely think we should think about how we do this—grossly intolerant of pseudo-science, the building up of what purports to be science by the cherry-picking of the facts and the failure to use scientific evidence and the failure to use scientific method," he said. ...

Beddington also had harsh words for journalists who treat the opinions of non-scientist commentators as being equivalent to the opinions of what he called "properly trained, properly assessed" scientists. "The media see the discussions about really important scientific events as if it's a bloody football match. It is ridiculous."[48]

As scientists involved for years in campaigns against pseudoscience, we whole-heartedly agree with Beddington's arguments. Yet the way and the place in which they were formulated may have brought more harm than benefit to science. Why? Because terms like "war against bad science", "intolerance" or "bloody football match" repeated hundreds of times in the media and combined with "racism", "homophobia" and with highlighting the ignorance of journalists have created an image of aggressive, arrogant and impertinent scientists in the mind of the public. It is very easy to generalize and extend such convictions to include science as a whole. As a result, the proponents of pseudoscience and irrationality use isolated cases of arrogance to present the image of science in the following manner:

The worship of science is the great superstition of our age. The scientific adviser speaks and we are all supposed to believe him, whether he is promoting crops genetically modified to withstand huge doses of poisonous weed killers and pesticides, or tampering with the origin of human life itself in so-called stem cell research. Those who dare question scientists are demonized for their irrationality. Global warming may or may not be a certainty, but anyone who queries it has his sanity questioned. Cast doubt on these gods of certainty and you are accused of wanting to suppress free expression - which is the argument now being used by Nutt and pals against the Home Secretary. In fact, it is the arrogant scientific establishment which questions free expression. Think of the hoo-ha which occurred when one hospital doctor dared to question the wisdom of using the MMR vaccine. The point here is not whether he was right or wrong - it was the way in which the scientific establishment closed ranks in order to assassinate him. There was a blanket denunciation of his heresy, just as there is if anyone dares to point out some of the mistakes made by that very fallible genius Charles Darwin. Science rules - and it does so with just as much energy as the old Spanish Inquisition that refused to allow any creed other than Catholicism, and with the Inquisition's need to distort arguments and control the brains of men and women who might otherwise think for themselves.[49]

Although this is not a direct response to the previous events, it does show how the other side formulates its opinions on the basis of one blistering statement, how it uses associations with emotion-tinged terms like "heresy," "Inquisition" and so forth.

Finally, who else if not the scientists, and psychologists in particular, know the principles of persuasion, influence and methods of changing people's attitudes? Why don't they use these methods to influence public opinion? After all, we know very well that a one-sided message containing strong emotional expressions wonderfully strengthens attitudes similar to those of the person delivering the message but fails with those who are not convinced, who are hesitant. Moreover, it can even be detrimental because aggression and arrogance usually put people off. In the case of people with contrary views, it causes only reactance and strengthens attitudes that the author of the message is trying to change. To put it simply, this strategy is nothing more than converting the converted.

These rules are not some arcane knowledge. They are taught to psychology students who learn about them from social psychology textbooks; unfortunately, scientists trying to change the attitude of people seduced by pseudoscience and psychobusiness often are unable to use them. Are we perhaps dealing with the proverbial shoemaker's children going barefoot? Open-mindedness and closed-mindedness are qualities of people, not science. As long as we are unable to convince people that this is indeed the case, they will be fed with stories of how Copernicus' concept was negated, how impertinent scholars rejected Semmelweis discoveries, how the authors of the hypothesis of cosmic origins of meteorites were laughed at and so forth. All such stories are used to illustrate the closed mindedness of science and thus to "prove" that pseudoscientific rubbish can make sense, which the impertinent scholars merely fail to see. While most of such rubbish indeed does not make sense, the way it is presented to the potential audience is very powerful. Failure to note this power is the greatest impertinence of scholars. That is why, as Tana Dineen writes, "Academic psychology and the profession of psychiatry lost their hold on the Psychology Industry as the market became flooded with techniques ranging from transactional analysis, hypnosis, existential analysis, Rolfing and primal therapy to re-evaluation therapy, Reichian sensitivity, bioenergetics and journeys into consciousness."[50]

We can bemoan this, but we should also be aware that it has been our own great failure. As scientists armed with the entire available knowledge of how to shape people's attitudes, charlatans have defeated us using only their intuition. The fact that today the same courts commonly accept DNA results and psychological opinions based on projective tests is a result of education and the popularization of knowledge of biology and psychology. The fact that today no one sells perpetual motion machines in shopping malls but that there exist those who sell water and sugar calling them "medicines" is also a result of how we have transmitted scientific knowledge to ordinary people. The fact that most people are able to calculate how much they should pay for the energy they consume, and at the same time, are willing to spend money on to make their brains more efficient is a result of education. Finally, the fact that patients trust the interpretation of what a radiologist saw on an X-ray and what a psychoanalyst "discovered" in their subconscious is a result of the strategies we adopted in our confrontations with pseudoscience and psychobusiness. There is no "third factor" we could blame. We, as representatives of science, are wholly responsible and should be blamed for this state of affairs

[1] F. A. Pattie, *Mesmer and Animal Magnetism: A Chapter in the History of Medicine.* (Hamilton, NY: Edmondston, 1994), 151.

[2] S. J. Gould, "The Chain of Reason vs. The Chain of Thumbs," *Natural History*, *14*, (1989): 12-21.

[3] T. S. Ball and D. D. Alexander, "Catching Up with Eighteenth Century Science in the Evaluation of Therapeutic Touch," *Skeptical Inquirer*, (July/August, 1998): 169-174.

[4] Dineen, *Manufacturing Victims*, 4.

[5] S. O. Lilienfeld, S. J. Lynn, and J. M. Lohr, *Science and Pseudoscience in Clinical Psychology*, (New York: Guilford, 2004), 411.

[6] G. M. Rosen, "Self-help or Hype? Comments on Psychology's Failure to Advance Self-care," *Professional Psychology: Research and Practice, 24(3)*, (1993): 340-345.

[7] See the case of Kitty Genovese described in many textbooks: E.g. S. Kassin, S. Fein and H. Markus, *Social Psychology.* (Belmont, CA: Wadsworth, 2014), 406-408.

[8] Morantz, "Escape from the"

[9] Dineen, *Manufacturing Victims*, 139.

[10] "About the Scientific Review of Mental Health Practice," (2007): from http://www.srmhp.org/about.html

[11] J. D. Frank, *Persuasion and Healing: A Comparative Study of Psychotherapy*, 3rd ed. (Baltimore and London: John Hopkins University Press, 1993).

[12] D. G. Kewman, L. Mercier, and M. Hovell, "The Power of Nonspecific Effects in Healing: Implications for Psychosocial and Biological Treatments," *Clinical Psychology Review, 13*, (1993): 375-391.

[13] Ibid.

[14] G. Gustaw and K. Brocławik, "Homeopatia Witkowskiego" [Yet another hoax]. *Psychologia Społeczna, 4*, K. (2008): 329-335.

[15] A. Schopenhauer, *The Art of Always Being Right*, A. (2013). [Kindle version]. Retrieved from Amazon.com

[16] W. Łukaszewski, "Kto kogo? Komentarz do artykułu Tomasza Witkowskiego i Pawła Fortuny 'O psychobiznesie, tolerancji i odpowiedzialności, czyli strategie czystych uczonych,'" ["By Who to Whom? Commentary on the Article by Tomasz Witkowski and Paweł Fortuna: 'On psycho-business, tolerance and responsibility or strategies employed by pure scientists,'"]. *Psychologia Społeczna, 4*, (2008): 349-352.

[17] O'Donohue, & Ferguson, "Evidence-based Practice," 344.

[18] C. Tavris, and E. Aronson, *Mistakes Were Made (But Not By Me): Why We Justify Foolish Beliefs, Bad Decisions, and Hurtful Acts.* (Orlando, FL: Harcourt, E. 2007), 124.

[19] Ibid. 124-125.

[20] Dineen, *Manufacturing Victims*, 132.

[21] S. Sleek, "Inconvenient Truth-tellers. What Happens When Research Yields Unpopular Findings," *Observer*, *26(9)*, (2013): http://www.psychologicalscience.org/index.php/publications/observer/2013/november-13/inconvenient-truth-tellers.html

[22] A. Zuger, "Abusing Not Only Children, but Also Science," *New Your Times*, (2010): http://www.nytimes.com/2010/01/26/health/26zuger.html?_r=2&

[23] L. J. Cronbach, "Assessment of individual differences," *Annual Review of Psychology*, 7, (1956): 173-196.

[24] R. M. Dawes, "Giving Up Cherished Ideas: The Rorschach Ink Blot Test," *Institute for Psychological Therapies Journal, 3(4,)* (1991): 154.

[25] H. N. Garb, "Call for a Moratorium on the Use of the Rorschach Inkblot Test in Clinical and Forensic Settings," *Assessment, 6(4),* (1999): 313–8.

[26] N. Cohen, "Complaint over Doctor Who Posted Inkblot Test," *The New York Times,* (August 23, 2009): http://www.nytimes.com/2009/08/24/business/24inkblot.html?_r=0

[27] Ibid.

[28] More about it: http://sceptycy.org/?page_id=555

[29] M. Stasiakiewicz, *Test Rorschacha,* (Warsaw: Scholar, 2004).

[30] Sleek, "Inconvenient Truth-tellers."

[31] S. Lewandowsky, M. A. Mann, L. Bauld, G. Hastings, and E.F. Loftus, "The Subterranean War on Science," *Observer, 26(9),* (2013): http://www.psychological science.org/index.php/publications/observer/2013/november-13/the-subterranean-war-on-science.html

[32]The post has been removed from the journal's Internet forum. The quote comes from the authors' archive.

[33] C. L. Fracasso and H. L. Friedman, "Energy Psychology: Progression or Retrogression in Understanding and Treating Psychological Disorders?" (n.d.): https://www.healthforumonline.com/Our-Courses/Courses/47/search__energy/productId__98/categoryId__4/secure__true/

[34] Lilienfeld et al., *"Science and Pseudoscience,"* 463.

[35] M.A. Lenz, K. Ek, and A.Mills, *Misconceptions in Psychology,* Presentation at 4th Midwest Conference on Professional Psychology, Owatonna, Minnesota. As cited in Lilienfeld, et al., *50 Great Myths.*

[36] K. Mudyń, "Miejsce nauki i psychologii w kulturze zdominowanej ideologią wolnego rynku," ["The Place of Science and of Psychology in a Culture Dominated by the Ideology of a Free Market,"]. *Psychologia Społeczna, 4,* (2008): 353-361.

[37] M. Dymkowski, "W sprawie teoretycznego zaplecza psychoterapii," ["On the Theoretical Basis of Psychotherapy,"] *Psychologia Społeczna, 4,* (2008): 318-320.

[38] Łukaszewski, "Kto kogo?"

[39] http://sceptycy.org/?p=1094

[40] http://sceptycy.org/?page_id=25

[41] B. Dunning, "Should Science Debate Pseudoscience?" [Blog post] (August 18, 2009): http://skeptoid.com/episodes/4167

[42] L. Krauss, "Odds Are Stacked When Science Tries to Debate Pseudoscience," *The New York Times,* (April 30, 2002): http://www.nytimes.com/2002/04/30/science/essay-odds-are-stacked-when-science-tries-to-debate-pseudoscience.html

[43] Andreski, *Social Sciences,* 237-238.

[44] C. Tavris, Foreword, in Lilienfeld et al., *Science and Pseudoscience,* (2004), xvi.

[45] M. Bunge, "What is Pseudoscience?" *The Skeptical Inquirer, 9,* (1984): 36-46.

[46] M. Gardner, *Fads and Fallacies in the Name of Science,* (Mineola, NJ: Dover, 1957), R. 3, 1.

[47] J. Neimark, "The Diva of Disclosure, Memory Researcher Elizabeth Loftus," *Psychology Today, 29(1),* (1996, January): 48.

[48] J. Dwyer and L. Hood, "Beddington Goes to War Against Bad Science," *Research*, (2011):
http://www.researchresearch.com/index.php?option=com_news&template=rr_2col &view=article&articleId=1032320

[49] A. N. Wilson, "Yes, Scientists Do Much Good. But a Country Run by These Arrogant Gods of Certainty Would Truly Be Hell on Earth," *Mail Online*. (November 4, 2009): http://www.dailymail.co.uk/debate/article-1224858/Yes-scientists-good-But-country-run-arrogant-gods-certainty-truly-hell-earth.html#ixzz2vBWXEAaJ

[50] Dineen, *Manufacturing Victims*, 122.

TO BE CONTINUED...

Every practicing psychologist who displays at least some curiosity must eventually come across therapeutic methods developed at someone else's desk that are used by others without a single thought of reflection, diagnostic tools sanctioned by nothing but tradition or concepts completely void of any empirical fundaments. The origins of this book date back to the year 2005 when one of us, tired of constantly stumbling upon vast areas of completely useless psychological "knowledge," decided to separate the wheat from the chaff and write a piece of work that would help him (and his future readers) to clean up his own front yard. The idea seemed simple; how hard could it be to identify myths, simplifications and misconceptions deeply rooted in clinical, popular and forensic psychology, to classify them and then publish them?

In order to achieve this, an outline for a book composed of three main sections, nine chapters and few dozen subchapters was drafted. It initially seemed that there would not be enough material to fill in the outlined book. Seriously, how many false ideas could have slipped into a 100-year-old, cherished field of psychology? In 2009, an entire book was published, representing only the first section of the tome. Yet, it contained less than a third of the full scope of what was initially planned. In 2013, the second book-length section was published, and there was still more than enough material to fill another one!

Preparations of the English version of the first section of the tome resulted in a substantial increase in the amount of facts and images that revealed the full picture of the very poor state of psychology at every possible level. It was terrifying to watch how every attempt to uncover a single problem revealed dozens of other pathologies that were swept under the carpet for years. The more we learned, the better we could see how all those problems were connected and how the oceans of harm, human misery and suffering have emerged today. Pandora's box is a barely sufficient metaphor.

The Polish version of our book entitled *Forbidden Psychology: Between Sorcery and Science* started to live its own life even before it was published. The simple concept of writing a short monograph has turned into long-term project. Readers are constantly asking for more, but it is primarily those who were victims of psychologists and therapists who motivate us to do more work. This book has already engaged some curious journalists, who started to write and question previous taboos. Today, more people than ever are ready to write about the negative consequences of psychotherapy, about useless courses of neuro-linguistic seduction, the involvement of psychologists in judicial systems and about their ethical conduct.

Those changes are subtle. We feel them rather as a breeze of fresh of air, something we can briefly appreciate when the humid air on a hot day moves momentarily. It will soon be gone and the heat will become unbearable again. A long time has to pass before the air will thicken enough for a real storm to form that would bring more permanent relief. History teaches us that meteorology is not necessarily useful in describing the quality or atmosphere of the social sciences, but it is always worth finding some kind of refreshing shadow of hope.

We invite all of our readers who were personally affected, witnessed or experienced abuse from psychologists, psychotherapists, expert witnesses (in psychology) or scientists, to contact us via our webpages at www.forbiddenpsychology.wordpress.com.

Psychology has many more skeletons hidden in its closets. Psychologists harm people with their diagnoses. Poorly trained, and usually completely free of consequences, expert witnesses are often the masters (literally) of life in many courtrooms around the world. The abyss of psychobusiness constantly creates progressively more repulsive monsters. Psychologists-scientists are responsible not only for all the sins covered in this tome, but also for all the misconduct, neglect and thousands of unanswered questions that result from cementing stereotypes that create completely false images of individuals in the eyes of general public. We will tell you more about the above (including what difference this book has made) next time. To be continued…

CPSIA information can be obtained at www.ICGtesting.com
Printed in the USA
BVOW06s1527260116

434293BV00008B/154/P